3

THE GEORGE FISHER BAKER
NON-RESIDENT LECTURESHIP
IN CHEMISTRY AT
CORNELL UNIVERSITY

Angular Momentum

Angular Momentum

Understanding Spatial Aspects in Chemistry and Physics

Richard N. Zare

Department of Chemistry
Stanford University
Stanford, California

A Wiley-Interscience Publication

John Wiley & Sons

New York • Chichester • Brisbane • Toronto • Singapore

Library of Congress Cataloging in Publication Data:

Zare, Richard N.
    Angular momentum.

    Based on lectures given to the Chemistry Dept., Cornell
University, in the autumn of 1980 as part of their Baker
lecture series.
    "A Wiley-Interscience Publication"
    Includes bibliographies and index.
    1. Angular momentum (Nuclear Physics)   I. Title
II. Title: Baker lecture series.

QC793.3.A5Z37   1987   530.1'2   87-16204
ISBN 0-471-85892-7

Printed in the United States of America

10 9 8 7 6 5 4 3 2 1

To
Dudley R. Herschbach

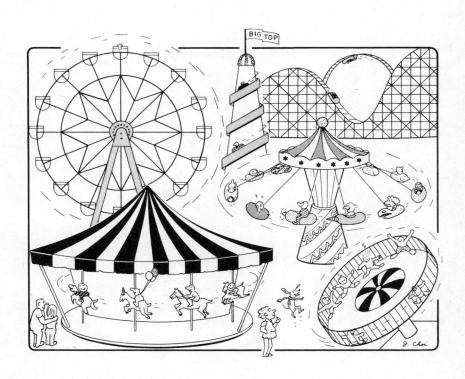

PREFACE

From merry-go-rounds to spinning tops, angular momentum has always fascinated me, confused me, and at times filled me with dizzying excitement. This is a book about angular momentum in quantum mechanics and its applications to chemistry and physics. It grew out of twenty lectures given to the Chemistry Department, Cornell University, in the autumn of 1980 as part of their Baker Lecture series. Thereafter for many years I taught a graduate course on "Advanced Quantum Mechanics" in the Chemistry Department, Stanford University, and I distributed this material to my class in lieu of a suitable textbook.

The material in this text is intended for a one-semester graduate course. The chapters are organized primarily as a way to learn about angular momentum theory, secondarily as a reference source. A working knowledge of basic quantum mechanics is presumed, the type of knowledge gained traditionally in every graduate introductory course on quantum mechanics. It has been my experience that angular momentum theory often acts as a major roadblock to understanding many contemporary research problems. Moreover, this topic seems to be difficult to learn only by reading about it. Indeed, mastery appears to come from more than one exposure to this topic and from solving real problems of interest. Consequently, I have placed at the end of each chapter extensive examples and applications of the text. I would urge members of my class to work on these so-called problem sets and applications individually, and then collectively. I regard this material as an integral part of the text, a way to make clear, alive, and concrete the more abstract explication of principles that come before in each chapter. For someone to read the text and skip the problem sets and applications is like someone reading a book on how to play the piano without ever sitting before a keyboard.

Many problems in chemistry and physics involve homogeneous media and isotropic space, for which the charms of angular momentum theory seem remote. However, as soon as we become concerned about direction in describing some process, it is necessary to speak of angles and momenta conjugate to them. Thus angular momentum theory arises naturally in discussing such phenomena as how a beam of radiation interacts with matter or how a beam of atoms and molecules scatter from other collision targets. Its mastery is essential for a detailed understanding of such microscopic phenomena. As proficiency increases in this topic, it is possible to disentangle purely geometrical factors from dynamical ones. This separation, embodied in the Wigner–Eckart theorem, may be regarded as the ultimate goal of angular momentum theory whereby we exploit the full symmetry inherent in some physical process to analyze it into its essential components.

This is not the first book on angular momentum theory, but it differs from others in the emphasis placed on making it a learning text for those with a minimum background in quantum mechanics. It also differs in the choice of examples that are drawn almost entirely from atomic and molecular phenomena. I believe that it is not possible to present this material too simply to anyone learning angular momentum theory for the first time. Consequently, many intermediate steps are left in the text, which, to the initiated, may appear inelegant, if not annoying. At the same time this text serves a secondary purpose of being a reference work; the vast majority of formulas needed to solve any problem in angular momentum theory are contained in this book.

Angular momentum theory is central to understanding and unifying photon and particle collision phenomena. It is hoped that what follows serves as a good preparation for experiencing the pleasures contained in the rich and growing literature on directionality in chemistry and physics.

RICHARD N. ZARE

Stanford, California
August 1986

CONTENTS

THE GEORGE FISHER BAKER
NON-RESIDENT LECTURESHIP
IN CHEMISTRY AT
CORNELL UNIVERSITY

Angular Momentum

Chapter 1

ANGULAR MOMENTUM OPERATORS AND WAVE FUNCTIONS

1.1 DEFINITION OF ANGULAR MOMENTUM OPERATORS

Let a particle of mass m and velocity $\mathbf{v}$ be located at a position $\mathbf{r}$ measured from some origin. Then according to classical mechanics[1], the particle has a linear momentum $\mathbf{p}$ given by

$$\mathbf{p} = m\mathbf{v} \tag{1.1}$$

and an angular momentum ℓ given by

$$\ell = \mathbf{r} \times \mathbf{p}, \tag{1.2}$$

The quantum transcription[2] of Eq. (1.2) is obtained by replacing $\mathbf{p}$ by $(\hbar/i)\nabla$, where $\hbar = h/2\pi$ is Planck's constant divided by 2π, i is the square root of minus one, and ∇ is the gradient operator whose Cartesian components are

$$\nabla = \frac{\partial}{\partial x}\hat{\mathbf{x}} + \frac{\partial}{\partial y}\hat{\mathbf{y}} + \frac{\partial}{\partial z}\hat{\mathbf{z}} \tag{1.3}$$

In Eq. (1.3) (and elsewhere) a superscript caret denotes a unit vector. For convenience, we drop the burden of carrying around $\hbar$ by introducing a system of units in which $\hbar = 1$. Thus the Cartesian components of $\mathbf{p}$ are

$$p_x = -i\frac{\partial}{\partial x}, \qquad p_y = -i\frac{\partial}{\partial y}, \qquad p_z = -i\frac{\partial}{\partial z} \tag{1.4}$$

1

and those of ℓ are

$$\ell_x = yp_z - zp_y = -i\left(y\frac{\partial}{\partial z} - z\frac{\partial}{\partial y}\right)$$

$$\ell_y = zp_x - xp_z = -i\left(z\frac{\partial}{\partial x} - x\frac{\partial}{\partial z}\right)$$

$$\ell_z = xp_y - yp_x = -i\left(x\frac{\partial}{\partial y} - y\frac{\partial}{\partial x}\right) \tag{1.5}$$

The commutator $[\mathbf{A}, \mathbf{B}] = \mathbf{AB} - \mathbf{BA}$ of two operators $\mathbf{A}$ and $\mathbf{B}$ plays a central role in quantum mechanics[2]; the necessary condition for the observables A and B to be simultaneously measurable is that the corresponding operators $\mathbf{A}$ and $\mathbf{B}$ commute[3], that is, $[\mathbf{A}, \mathbf{B}] = 0$. From Eq. (1.4) it is readily seen that the position vector of a particle and its momentum satisfy the basic commutation relations:

$$[x, p_x] = i, \qquad [x, p_y] = 0, \qquad [x, p_z] = 0 \tag{1.6}$$

with all cyclic permutations. Thus, for example, it is not possible to measure simultaneously along the same direction the position and linear momentum of a particle to arbitrary precision.

The commutation relations of the Cartesian components of ℓ are also readily derived:

$$[\ell_x, \ell_y] = i\ell_z, \qquad [\ell_y, \ell_z] = i\ell_x, \qquad [\ell_z, \ell_x] = i\ell_y \tag{1.7}$$

Equation (1.7) has the interpretation that quantum states cannot be specified by any more than one of the labels (eigenvalues) of the three components of angular momentum. The "good" quantum numbers corresponding to the largest set of mutually commuting operators represent the maximum information that can be known about a quantum mechanical system. The measurement of another variable corresponding to an operator not commuting with this set necessarily introduces uncertainty into one of the variables already measured. A sharper specification of the system is, therefore, not possible.

Because of the importance of the commutator, it is natural to define[4] a general angular momentum operator $\mathbf{j}$ as one whose Cartesian components obey the commutation rules

$$[j_x, j_y] = ij_z, \qquad [j_y, j_z] = ij_x, \qquad [j_z, j_x] = ij_y \tag{1.8}$$

in analogy to Eq. (1.7). This extended definition returns an unexpected dividend. As we shall see in the next section, it permits the existence of spin—a quantity that has no classical analogy. We will reserve ℓ for orbital angular momentum and use $\mathbf{j}$ for general angular momentum.

1.2 EIGENVALUES AND MATRIX ELEMENTS OF ANGULAR MOMENTUM OPERATORS

The square of the total angular momentum is defined as

$$\mathbf{j}^2 = j_x^2 + j_y^2 + j_z^2 \tag{1.9}$$

This operator has the commutation properties that

$$[\mathbf{j}^2, j_x] = [\mathbf{j}^2, j_y] = [\mathbf{j}^2, j_z] = 0 \tag{1.10}$$

Hence we can construct states $|jm\rangle$ that are simultaneously eigenfunctions of $\mathbf{j}^2$ and any one component of $\mathbf{j}$, say, j_z; that is,

$$\mathbf{j}^2 |jm\rangle = \lambda_j |jm\rangle$$

$$j_z |jm\rangle = m |jm\rangle \tag{1.11}$$

We proceed to determine the eigenvalues $\lambda_j = \langle jm| \mathbf{j}^2 |jm\rangle$ and $m = \langle jm| j_z |jm\rangle$.

The operator $j_x^2 + j_y^2 = \mathbf{j}^2 - j_z^2$ is diagonal in the $|jm\rangle$ representation. Moreover, it has positive definite (nonnegative) eigenvalues

$$(j_x^2 + j_y^2) |jm\rangle = (\lambda_j - m^2) |jm\rangle \tag{1.12}$$

because the expectation value of the square of a Hermitian operator, that is, the square of a real eigenvalue, is greater than or equal to zero. Hence we conclude that the value of m is bounded from both above and below in that m^2 cannot exceed λ_j. This implies that for a given $\mathbf{j}$ there exist minimum and maximum values of m, denoted by m_{min} and m_{max}, respectively.

Let us introduce the raising and lowering operators $j_\pm$, defined by

$$j_+ = j_x + ij_y, \qquad j_- = j_x - ij_y \tag{1.13}$$

From Eqs. (1.8) and (1.10) it is readily shown that these operators satisfy the commutation rules:

$$[\mathbf{j}^2, j_\pm] = 0$$

$$[j_z, j_\pm] = \pm j_\pm$$

$$[j_+, j_-] = 2j_z \tag{1.14}$$

Let us examine the behavior of the function $j_\pm |jm\rangle$. We find

$$\mathbf{j}^2 j_\pm |jm\rangle = j_\pm \mathbf{j}^2 |jm\rangle = \lambda_j j_\pm |jm\rangle \tag{1.15}$$

and

$$j_z j_\pm |jm\rangle = (j_\pm j_z \pm j_\pm) |jm\rangle = (m \pm 1) j_\pm |jm\rangle \qquad (1.16)$$

Thus $j_\pm |jm\rangle$ is an eigenfunction of $\mathbf{j}^2$ with the eigenvalue λ_j and an eigenfunction of j_z with the eigenvalue $m \pm 1$. It follows that $j_\pm |jm\rangle$ is proportional to the normalized eigenfunction $|jm \pm 1\rangle$; that is

$$j_\pm |jm\rangle = C_\pm |jm \pm 1\rangle \qquad (1.17)$$

where $C_\pm$ is a proportionality constant. The ability of the raising and lowering operators $j_\pm$ to alter m by ± 1 unit, respectively, while preserving λ_j gives them their names. Note that the $j_\pm$ operators are also referred to in the literature as *step-up* and *step-down operators*, *ladder operators*, or *shift operators*.

Since the values of m are bounded between $m_{\min}$ and $m_{\max}$, it follows that

$$j_+ |jm_{\max}\rangle = 0 \qquad (1.18)$$

and

$$j_- |jm_{\min}\rangle = 0 \qquad (1.19)$$

By applying j_- to Eq. (1.18) and j_+ to Eq. (1.19) and by using the identity

$$j_\mp j_\pm = \mathbf{j}^2 - j_z(j_z \pm 1) \qquad (1.20)$$

we obtain the two equations

$$\lambda_j - m_{\max}(m_{\max} + 1) = 0$$

$$\lambda_j - m_{\min}(m_{\min} - 1) = 0 \qquad (1.21)$$

Elimination of λ_j yields

$$m_{\max}(m_{\max} + 1) = m_{\min}(m_{\min} - 1) \qquad (1.22)$$

or

$$(m_{\max} + m_{\min})(m_{\max} - m_{\min} + 1) = 0 \qquad (1.23)$$

One of these two factors must vanish. Because $m_{\max} \geq m_{\min}$, the only solution to Eq. (1.23) is

$$m_{\max} = -m_{\min} \qquad (1.24)$$

Successive values of m differ by unity [see Eq. (1.16)]. Therefore, $m_{max} - m_{min}$ is a positive definite integer, which we may denote by $2j$, where j is an integer or half-integer. Then from $m_{max} - m_{min} = 2j$ and $m_{max} + m_{min} = 0$ we conclude that

$$m_{max} = j, \qquad m_{min} = -j \qquad (1.25)$$

and there are $2j + 1$ possible values of m, $m = j, j - 1, \ldots, -j + 1, -j$ for each value of j. Substitution of Eq. (1.25) into Eq. (1.21) yields the additional result

$$\lambda_j = j(j + 1) \qquad (1.26)$$

We are now in a position to evaluate the proportionality constant $C_{\pm}$ appearing in Eq. (1.17). We find

$$|C_{\pm}|^2 = \langle jm| j_{\mp} j_{\pm} |jm\rangle = \langle jm| \mathbf{j}^2 - j_z(j_z \pm 1) |jm\rangle$$

$$= j(j + 1) - m(m \pm 1) \qquad (1.27)$$

where we have used Eq. (1.20) and the property that the adjoint (complex conjugate transpose) of $j_{\pm}$ is $j_{\mp}$. (Note that the $j_{\pm}$ operators are not Hermitian, i.e., not self-adjoint, although j_x, j_y, and j_z are.) From Eq. (1.27) it is seen that the absolute value of $C_{\pm}$ is determined but its phase is arbitrary. We choose $C_{\pm}$ to be real, that is

$$C_{\pm} = [j(j + 1) - m(m \pm 1)]^{\frac{1}{2}} \qquad (1.28)$$

This agrees with the standard phase convention[5], namely, that the matrix elements of j_x are real while those of j_y are purely imaginary.

In summary, we write down all the matrix elements of the angular momentum operators in which $\mathbf{j}^2$ and j_z are diagonal:

$$\langle jm| \mathbf{j}^2 |j'm'\rangle = j(j + 1)\delta_{jj'}\delta_{mm'} \qquad (1.29)$$

$$\langle jm| j_z |j'm'\rangle = m'\delta_{jj'}\delta_{mm'} \qquad (1.30)$$

$$\langle jm| j_{\pm} |j'm'\rangle = [j(j + 1) - m'(m' \pm 1)]^{\frac{1}{2}}\delta_{jj'}\delta_{m,m'\pm 1} \qquad (1.31)$$

$$\langle jm| j_x |j'm'\rangle = \tfrac{1}{2}[j(j + 1) - m'(m' \pm 1)]^{\frac{1}{2}}\delta_{jj'}\delta_{m,m'\pm 1} \qquad (1.32)$$

$$\langle jm| j_y |j'm'\rangle = \mp\tfrac{1}{2}i[j(j + 1) - m'(m' \pm 1)]^{\frac{1}{2}}\delta_{jj'}\delta_{m,m'\pm 1} \qquad (1.33)$$

In Eqs. (1.29)–(1.33) $\delta_{ii'}$, called the *Kronecker delta*, is defined as having the properties that $\delta_{ii'} = 0$ for $i \neq i'$ and $\delta_{ii'} = 1$ for $i = i'$. We introduce the practice of placing some of the most frequently used results in a box.

The angular momentum quantum number j can have any of the values 0, $\frac{1}{2}$, 1, $\frac{3}{2}$, 2, ... in units of $\hbar$. Integral values of j correspond to orbital angular momenta ℓ, while half-integral values of j are referred to as *spin angular momenta*. The use of the commutation rules as the definition of the angular momentum operators puts orbital and spin angular momenta on the same footing.

1.3   ANGULAR MOMENTUM WAVE FUNCTIONS

So far we have considered the angular momentum eigenfunctions $|jm\rangle$ in an abstract vector space of $2j + 1$ dimensions. For integral j, denoted by ℓ, it is often useful to work in an explicit coordinate representation for ℓ and its states $|\ell m\rangle$ by introducing the spherical polar coordinates

$$x = r \sin \theta \cos \phi$$

$$y = r \sin \theta \sin \phi$$

$$z = r \cos \theta \qquad (1.34)$$

which have the inverse relations

$$r^2 = x^2 + y^2 + z^2$$

$$\cos \theta = z/r$$

$$\tan \phi = y/x \qquad (1.35)$$

Then Eq. (1.5) becomes

$$\ell_x = i \left(\sin \phi \frac{\partial}{\partial \theta} + \cot \theta \cos \phi \frac{\partial}{\partial \phi} \right)$$

$$\ell_y = i \left(-\cos \phi \frac{\partial}{\partial \theta} + \cot \theta \sin \phi \frac{\partial}{\partial \phi} \right)$$

$$\ell_z = -i \frac{\partial}{\partial \phi} \qquad (1.36)$$

This result is readily obtained using the nine partial derivatives in which the appropriate *Cartesian* coordinates are held constant

$$\frac{\partial r}{\partial x} = \sin \theta \cos \phi, \qquad \frac{\partial r}{\partial y} = \sin \theta \sin \phi, \qquad \frac{\partial r}{\partial z} = \cos \theta$$

$$\frac{\partial \theta}{\partial x} = \frac{\cos \theta \cos \phi}{r}, \qquad \frac{\partial \theta}{\partial y} = \frac{\cos \theta \sin \phi}{r}, \qquad \frac{\partial \theta}{\partial z} = -\frac{\sin \theta}{r}$$

$$\frac{\partial \phi}{\partial x} = -\frac{\sin \phi}{r \sin \theta}, \qquad \frac{\partial \phi}{\partial y} = \frac{\cos \phi}{r \sin \theta}, \qquad \frac{\partial \phi}{\partial z} = 0 \qquad (1.37)$$

and applying the chain rule for differentiation. For example,

$$\ell_x = -i \left(y \frac{\partial}{\partial z} - z \frac{\partial}{\partial y} \right)$$

$$= -i \left[r \sin \theta \sin \phi \left(\frac{\partial r}{\partial z} \frac{\partial}{\partial r} + \frac{\partial \theta}{\partial z} \frac{\partial}{\partial \theta} + \frac{\partial \phi}{\partial z} \frac{\partial}{\partial \phi} \right) \right.$$

$$\left. -r \cos \theta \left(\frac{\partial r}{\partial y} \frac{\partial}{\partial r} + \frac{\partial \theta}{\partial y} \frac{\partial}{\partial \theta} + \frac{\partial \phi}{\partial y} \frac{\partial}{\partial \phi} \right) \right]$$

$$= -i \left[r \sin \theta \sin \phi \left(\cos \theta \frac{\partial}{\partial r} - \frac{\sin \theta}{r} \frac{\partial}{\partial \theta} \right) \right.$$

$$\left. -r \cos \theta \left(\sin \theta \sin \phi \frac{\partial}{\partial r} + \frac{\cos \theta \sin \phi}{r} \frac{\partial}{\partial \theta} + \frac{\cos \phi}{r \sin \theta} \frac{\partial}{\partial \phi} \right) \right]$$

$$= i \left(\sin \phi \frac{\partial}{\partial \theta} + \cot \theta \cos \phi \frac{\partial}{\partial \phi} \right) \qquad (1.38)$$

as advertised, and so on. From Eq. (1.36) it follows that

$$\ell_\pm = e^{\pm i\phi} \left(\pm \frac{\partial}{\partial \theta} + i \cot \theta \frac{\partial}{\partial \phi} \right) \qquad (1.39)$$

and

$$\ell^2 = - \left[\frac{1}{\sin^2 \theta} \frac{\partial^2}{\partial \phi^2} + \frac{1}{\sin \theta} \frac{\partial}{\partial \theta} \left(\sin \theta \frac{\partial}{\partial \theta} \right) \right] \qquad (1.40)$$

In this representation the eigenvalue problem

$$\ell^2 \left| \ell m \right\rangle = \ell(\ell + 1) \left| \ell m \right\rangle$$

$$\ell_z \left| \ell m \right\rangle = m \left| \ell m \right\rangle \qquad (1.41)$$

yields two partial differential equations whose solutions are the spherical harmonic functions[2]

$$Y_{\ell m}(\theta, \phi) \equiv |\ell m\rangle \tag{1.42}$$

The spherical harmonics may be written as the product of two functions, one that depends only on the polar angle θ and the other on the azimuthal angle ϕ

$$Y_{\ell m}(\theta, \phi) = \Theta_{\ell m}(\theta)\Phi_m(\phi) \tag{1.43}$$

The $\Phi_m(\phi)$ satisfy the differential equation

$$\left[\frac{d^2}{d\phi^2} + m^2\right]\Phi_m(\phi) = 0 \tag{1.44}$$

and have the explicit form

$$\Phi_m(\phi) = (2\pi)^{-\frac{1}{2}}\exp(im\phi) \tag{1.45}$$

These functions are normalized so that

$$\int_0^{2\pi}\Phi_m^*(\phi)\Phi_{m'}(\phi)\,d\phi = \delta_{mm'} \tag{1.46}$$

The $\Theta_{\ell m}(\theta)$ satisfy the differential equation

$$\left[\frac{1}{\sin\theta}\frac{d}{d\theta}\left(\sin\theta\frac{d}{d\theta}\right) + \ell(\ell+1) - \frac{m^2}{\sin^2\theta}\right]\Theta_{\ell m}(\theta) = 0 \tag{1.47}$$

and have the explicit form for $m \geq 0$

$$\Theta_{\ell m}(\theta) = \frac{(-1)^m}{2^\ell \ell!}\left[\frac{2\ell+1}{2}\frac{(\ell-m)!}{(\ell+m)!}\right]^{\frac{1}{2}}(\sin\theta)^m$$

$$\times \left[\frac{d}{d(\cos\theta)}\right]^{\ell+m}(\cos^2\theta - 1)^\ell \tag{1.48}$$

These functions are normalized so that

$$\int_0^\pi \Theta_{\ell m}^*(\theta)\Theta_{\ell'm}(\theta)\,\sin\theta\,d\theta = \delta_{\ell\ell'} \tag{1.49}$$

For many purposes it is useful to relate the $\Theta_{\ell m}(\theta)$ to the so-called associated Legendre functions

$$P_\ell^m(\cos\theta) = \sin^m\theta\left[\frac{d}{d(\cos\theta)}\right]^m P_\ell(\cos\theta) \tag{1.50}$$

where

$$P_\ell(\cos\theta) = \frac{1}{2^\ell \ell!} \left[\frac{d}{d(\cos\theta)} \right]^\ell (\cos^2\theta - 1)^\ell \tag{1.51}$$

are the ordinary Legendre polynomials. Comparison of Eq. (1.48) with Eqs. (1.50) and (1.51) shows that we may make the identification

$$\Theta_{\ell m}(\theta) = (-1)^m \left[\frac{2\ell+1}{2} \frac{(\ell-m)!}{(\ell+m)!} \right]^{\frac{1}{2}} P_\ell^m(\cos\theta) \tag{1.52}$$

for $m \geq 0$. For negative m we choose to define

$$\Theta_{\ell,-|m|} = (-1)^m \Theta_{\ell|m|} \tag{1.53}$$

so that $\Theta_{\ell m}(\theta)$ for $m < 0$ is still given by Eq. (1.48) or (1.52) but with $|m|$ in place of m and the factor $(-1)^m$ omitted. This choice of phase is consistent with the previously made choice of sign concerning the matrix elements of ℓ_x and ℓ_y[5].

For the special case of $m = 0$

$$Y_{\ell 0}(\theta,\phi) = (2\pi)^{-\frac{1}{2}} \Theta_{\ell 0}(\theta)$$

$$= \left(\frac{2\ell+1}{4\pi} \right)^{\frac{1}{2}} P_\ell(\cos\theta) \tag{1.54}$$

Because $P_\ell^m(1)$ vanishes except for $m = 0$, it follows that

$$Y_{\ell m}(0,\phi) = \left(\frac{2\ell+1}{4\pi} \right)^{\frac{1}{2}} \delta_{m0} \tag{1.55}$$

Finally, we also note that

$$Y_{\ell m}^*(\theta,\phi) = (-1)^m Y_{\ell,-m}(\theta,\phi) \tag{1.56}$$

For reference purposes we list the first few spherical harmonics:

$$Y_{00}(\theta,\phi) = \frac{1}{\sqrt{4\pi}}$$

$$Y_{10}(\theta,\phi) = \sqrt{\frac{3}{4\pi}} \cos\theta = \sqrt{\frac{3}{4\pi}} \frac{z}{r}$$

$$Y_{1,\pm1}(\theta,\phi) = \mp\sqrt{\frac{3}{8\pi}} e^{\pm i\phi} \sin\theta = \mp\sqrt{\frac{3}{8\pi}} \frac{(x \pm iy)}{r}$$

$$Y_{20}(\theta, \phi) = \sqrt{\frac{5}{16\pi}}(3\cos^2\theta - 1) = \sqrt{\frac{5}{16\pi}}\frac{2z^2 - x^2 - y^2}{r^2}$$

$$Y_{2,\pm 1}(\theta, \phi) = \mp\sqrt{\frac{15}{8\pi}}e^{\pm i\phi}\cos\theta\sin\theta = \mp\sqrt{\frac{15}{16\pi}}\frac{(x\pm iy)z}{r^2}$$

$$Y_{2,\pm 2}(\theta, \phi) = \sqrt{\frac{15}{32\pi}}e^{\pm 2i\phi}\sin^2\theta = \sqrt{\frac{15}{32\pi}}\frac{(x\pm iy)^2}{r^2} \qquad (1.57)$$

The spherical harmonics constitute an orthonormal set over the unit sphere

$$\langle \ell m|\ell'm'\rangle = \int_0^{2\pi}d\phi\int_0^{\pi}\sin\theta\,d\theta\,Y_{\ell m}^*(\theta, \phi)Y_{\ell'm'}(\theta, \phi)$$

$$= \delta_{\ell\ell'}\delta_{mm'} \qquad (1.58)$$

In particular for the special case $m = m' = 0$, we obtain the orthonormality relations for the Legendre polynomials

$$\int_0^{\pi}\sin\theta\,d\theta\,P_\ell(\cos\theta)P_{\ell'}(\cos\theta) = \frac{2}{2\ell+1}\delta_{\ell\ell'} \qquad (1.59)$$

by substituting Eq. (1.54) into Eq. (1.58) and carrying out the integration over $d\phi$. Equation (1.58) allows us to identify $Y_{\ell m}^*Y_{\ell m}d\Omega$ with the probability that the position vector **r** points into the solid angle element $d\Omega = \sin\theta d\theta d\phi$ when the expectation value (average value) of ℓ_z is m. Note that $Y_{\ell m}^*Y_{\ell m}$ is independent of ϕ. This leads naturally to a geometric interpretation of ℓ as a vector making a constant projection m on the z axis when the system is in the state $|\ell m\rangle$. We discuss the implications of this statement more fully in the next section.

It is not possible to express the basis vectors $|jm\rangle$ for half-integral values of j and m in terms of single-valued continuous functions on a unit sphere, as we just have done for $|\ell m\rangle$. However, it is possible to build up the $|jm\rangle$ from the spin eigenfunctions for $j = \frac{1}{2}$, denoted by

$$\alpha = \left|\tfrac{1}{2}\tfrac{1}{2}\right\rangle = \begin{pmatrix}1\\0\end{pmatrix} \qquad (1.60)$$

and

$$\beta = \left|\tfrac{1}{2}-\tfrac{1}{2}\right\rangle = \begin{pmatrix}0\\1\end{pmatrix} \qquad (1.61)$$

Using Eqs. (1.29)–(1.33) we can easily show that the angular momentum operators j_x, j_y, j_z have the 2×2 matrix representation

$$j_x = \frac{1}{2}\sigma_x = \frac{1}{2}\begin{pmatrix}0 & 1\\1 & 0\end{pmatrix}$$

$$j_y = \frac{1}{2}\sigma_y = \frac{1}{2}\begin{pmatrix} 0 & -i \\ i & 0 \end{pmatrix}$$

$$j_z = \frac{1}{2}\sigma_z = \frac{1}{2}\begin{pmatrix} 1 & 0 \\ 0 & -1 \end{pmatrix} \tag{1.62}$$

where the σ_x, σ_y, σ_z are called *Pauli spin matrices*. It follows that

$$j_+ = j_x + ij_y = \begin{pmatrix} 0 & 1 \\ 0 & 0 \end{pmatrix}$$

$$j_- = j_x - ij_y = \begin{pmatrix} 0 & 0 \\ 1 & 0 \end{pmatrix}$$

$$\mathbf{j}^2 = j_x^2 + j_y^2 + j_z^2 = j_z^2 - j_z + j_+j_- = \frac{3}{4}\begin{pmatrix} 1 & 0 \\ 0 & 1 \end{pmatrix} \tag{1.63}$$

Here α is often called *spin up* and β *spin down* for $j_z\alpha = (+\frac{1}{2})\alpha$ and $j_z\beta = (-\frac{1}{2})\beta$. The eigenfunctions α and β are an example of what are called *spinors* because of their special transformation properties under rotation, which are discussed in Section 3.5.

Equations (1.62) and (1.63) are one particular representation of the angular momentum operators. Symbolically, the operators j_+, j_-, and j_z may also be written as differential operators in this spin space

$$j_+ = \alpha\frac{\partial}{\partial\beta}, \qquad j_- = \beta\frac{\partial}{\partial\alpha}$$

$$j_z = \frac{1}{2}\left(\alpha\frac{\partial}{\partial\alpha} - \beta\frac{\partial}{\partial\beta}\right) \tag{1.64}$$

from which it is readily verified in terms of these spinor differential operators

$$j_+\alpha = 0, \qquad\qquad j_+\beta = \alpha$$

$$j_-\alpha = \beta, \qquad\qquad j_-\beta = 0$$

$$j_z\alpha = \frac{1}{2}\alpha, \qquad\qquad j_z\beta = -\frac{1}{2}\beta \tag{1.65}$$

Combining Eqs. (1.63) and (1.64), we may express $\mathbf{j}^2$ in the notation of spinor differential operators as

$$\mathbf{j}^2 = \frac{1}{4}\left(\alpha\frac{\partial}{\partial\alpha} + \beta\frac{\partial}{\partial\beta}\right)\left(\alpha\frac{\partial}{\partial\alpha} + \beta\frac{\partial}{\partial\beta} + 2\right) \tag{1.66}$$

from which it is easily checked that $\mathbf{j}^2\alpha = \frac{3}{4}\alpha$ and $\mathbf{j}^2\beta = \frac{3}{4}\beta$.

Using this notation, we show that an arbitrary eigenfunction $|jm\rangle$ of $\mathbf{j}^2$ and j_z may be constructed in terms of α and β by

$$|jm\rangle = \frac{(\alpha)^{j+m}(\beta)^{j-m}}{[(j+m)!(j-m)!]^{\frac{1}{2}}} \tag{1.67}$$

This is proved at once by applying Eqs. (1.64) and (1.66) to (1.67), from which it is found that

$$j_+ |jm\rangle = [j(j+1) - m(m+1)]^{\frac{1}{2}} |jm+1\rangle$$

$$j_- |jm\rangle = [j(j+1) - m(m-1)]^{\frac{1}{2}} |jm-1\rangle$$

$$j_z |jm\rangle = m |jm\rangle$$

$$\mathbf{j}^2 |jm\rangle = j(j+1) |jm\rangle \tag{1.68}$$

It follows that Eq. (1.67) provides a representation of the angular momentum eigenfunctions $|jm\rangle$ in terms of the spinor differential operators. Moreover, this result is valid for both integral and half-integral j.

1.4   THE VECTOR MODEL

The angular momentum vector $\mathbf{j}$ can never point exactly along the z axis. The maximum value of $\langle j_z \rangle = m$ is when m takes on the value j while the length of the vector $\mathbf{j}$ is given by $(\mathbf{j} \cdot \mathbf{j})^{\frac{1}{2}} = \langle \mathbf{j}^2 \rangle^{\frac{1}{2}} = [j(j+1)]^{\frac{1}{2}}$. This result is consistent with the fact that there must be an uncertainty in the values of j_x and j_y. However, as $\mathbf{j}$ becomes large, the eigenvalue of $\mathbf{j}^2$, which may be written as $j^2(1+1/j)$, approaches j^2, allowing us to make the correspondence between the quantum number j and the classical angular momentum for integral j. Indeed, the presence of the $1/j$ term is a quantum mechanical effect that reflects our inability to measure simultaneously all three components of $\mathbf{j}$ and hence specify precisely the direction of $\mathbf{j}$.

Let us attempt to quantify this matter. The spread in the measurements of an observable A corresponding to the Hermitian operator $\mathbf{A}$ is conveniently described in terms of the variance of A, defined as

$$(\Delta A)^2 = \langle (\mathbf{A} - \langle \mathbf{A} \rangle)^2 \rangle = \langle \mathbf{A}^2 \rangle - \langle \mathbf{A} \rangle^2 \tag{1.69}$$

It is customary to call the positive square root of the variance of A, ΔA, the *uncertainty in A*. In a representation that diagonalizes $\mathbf{A}$, $(\Delta A)^2$ vanishes and A can be determined with no uncertainty, that is, to arbitrary precision in principle.

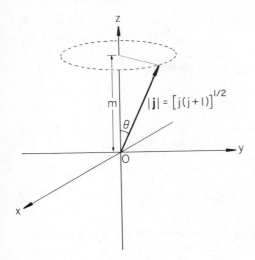

FIGURE 1.1   Vector model for the state $|jm\rangle$, which is represented by a vector **j** precessing about the axis of quantization (the z axis) making a constant projection m on it.

Let us consider the sum of the variances of j_x and j_y in the $|jm\rangle$ representation. According to Eq. (1.69), we have $(\Delta j_x)^2 = \langle j_x^2\rangle$ and $(\Delta j_y)^2 = \langle j_y^2\rangle$ since $\langle j_x\rangle$ and $\langle j_y\rangle$ both vanish [see Eqs. (1.32) and (1.33)]. But $\langle \mathbf{j}^2\rangle = \langle j_x^2\rangle + \langle j_y^2\rangle + \langle j_z^2\rangle$; that is,

$$(\Delta j_x)^2 + (\Delta j_y)^2 = j(j+1) - m^2 \tag{1.70}$$

Thus the sum of the variances $(\Delta j_x)^2 + (\Delta j_y)^2$ is constant for a given value of m. Moreover, the value of this sum reaches a minimum for $|m| = j$, that is, when the angular momentum vector points as nearly as possible along $+z$ or $-z$.

We are thus led to a picture, called the *vector model*[6], in which the eigenstate $|jm\rangle$ is represented by an angular momentum vector **j** of length $[j(j+1)]^{\frac{1}{2}}$ that precesses about the z axis (the axis of quantization) with a constant projection m (see Figure 1.1). Thus **j** moves in some uniform but unobservable manner on the surface of a cone whose apex half-angle θ satisfies the relation

$$\cos\theta = m/[j(j+1)]^{\frac{1}{2}} \tag{1.71}$$

In this picture the motion of **j** must be uniform so that **j** spends as much time pointing along $+x$ as along $-x$ or $+y$ as along $-y$, causing both $\langle j_x\rangle$ and $\langle j_y\rangle$ to vanish; in other words, the projection of **j** along the x and y axes averages to zero. Hence a classical orbit is ascribed to the state $|jm\rangle$ for integral j, that is, to the state $|\ell m\rangle$. Of course, the uncertainty principle prevents us from taking this picture too literally. However, the vector model does suggest how to regard the state $|\ell m\rangle$ in the correspondence limitindexcorrespondence[6]. Because large values of the angular momentum are commonplace in molecular problems, the vector model can often give insight into the interpretation of angular momentum theory.

To explore further the implications of the vector model and the interpretation of $|\ell m\rangle$ when the orbital angular momentum ℓ becomes large, we develop an asymptotic approximation[7] to the Legendre polynomials. Equation (1.47) may be rewritten for $m = 0$ as

$$\left[\frac{d^2}{d\theta^2} + \cot \theta \frac{d}{d\theta} + \ell(\ell + 1) \right] P_\ell(\cos \theta) = 0 \qquad (1.72)$$

The substitution

$$P_\ell(\cos \theta) = \frac{\chi_\ell(\theta)}{(\sin \theta)^{\frac{1}{2}}} \qquad (1.73)$$

leads to the differential equation for χ_ℓ

$$\left[\frac{d^2}{d\theta^2} + (\ell + \tfrac{1}{2})^2 + \tfrac{1}{4} \csc^2 \theta \right] \chi_\ell(\theta) = 0 \qquad (1.74)$$

which does not contain the first derivative of χ_ℓ with respect to θ and is similar to a one-dimensional Schrödinger equation. For large ℓ the term $(\ell + \tfrac{1}{2})^2$ is much larger than $\tfrac{1}{4} \csc^2 \theta$ everywhere except for angles very close to $\theta = 0$ or $\theta = \pi$. This suggests that we approximate Eq. (1.74) by the differential equation

$$\left[\frac{d^2}{d\theta^2} + (\ell + \tfrac{1}{2})^2 \right] \chi_\ell(\theta) = 0 \qquad (1.75)$$

whose solution is

$$\chi_\ell(\theta) = A_\ell \sin \left[(\ell + \tfrac{1}{2})\theta + \alpha \right] \qquad (1.76)$$

where A_ℓ and α are constants[8]. Thus for large ℓ and for $\theta \gg \ell^{-1}$ and $(\pi - \theta) \gg \ell^{-1}$, we obtain

$$P_\ell(\cos \theta) \longrightarrow A_\ell \frac{\sin \left[(\ell + \tfrac{1}{2})\theta + \alpha \right]}{(\sin \theta)^{\frac{1}{2}}} \qquad (1.77)$$

and we may also write under the same conditions[9]

$$Y_{\ell 0}(\theta, \phi) \longrightarrow \left(\frac{\ell + \tfrac{1}{2}}{2\pi} \right)^{\frac{1}{2}} A_\ell \frac{\sin \left[(\ell + \tfrac{1}{2})\theta + \alpha \right]}{(\sin \theta)^{\frac{1}{2}}} \qquad (1.78)$$

Thus

$$|Y_{\ell 0}(\theta, \phi)|^2 \longrightarrow \frac{(\ell + \tfrac{1}{2})}{2\pi} A_\ell^2 \frac{\sin^2 \left[(\ell + \tfrac{1}{2})\theta + \alpha \right]}{\sin \theta} \qquad (1.79)$$

For large ℓ the factor $\sin^2 \left[(\ell + \tfrac{1}{2})\theta + \alpha \right]$ oscillates very rapidly and can be replaced by its average value of $\tfrac{1}{2}$. Then if we insist that the integral of $|Y_{\ell 0}(\theta, \phi)|^2$ over $d\Omega$ be unity, we find that

$$|Y_{\ell 0}(\theta, \phi)|^2 = \frac{1}{2\pi^2 \sin \theta} \qquad (1.80)$$

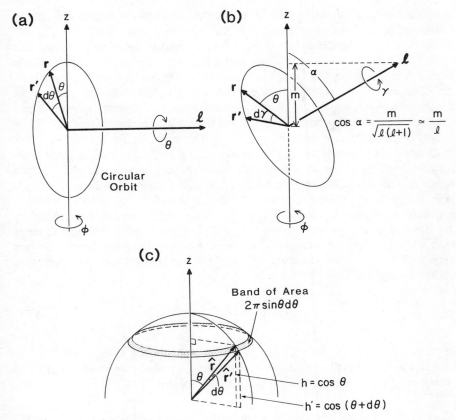

FIGURE 1.2   Classical orbits corresponding to (a) $|Y_{\ell 0}(\theta, \phi)|^2$ and (b) $|Y_{\ell m}(\theta, \phi)|^2$ for large ℓ. Note that $\boldsymbol{\ell}$ and $\mathbf{r}$ are perpendicular to one another. The region between θ and $\theta + d\theta$ defines a band shown in (c).

This result holds for large ℓ and for all θ except very close to $\theta = 0$ and $\theta = \pi$.

We may arrive at the same result by use of the vector model. For large ℓ the particle performs a circular orbit about $\boldsymbol{\ell}$ (see Figure 1.1); for $m = 0$, $\boldsymbol{\ell}$ is perpendicular to the z axis. A typical orbit is shown in Figure 1.2a. Here θ is the angle measured about $\boldsymbol{\ell}$; ϕ is the angle measured about z (in the xy plane). The probability of finding the particle between θ and $\theta + d\theta$ is uniform. But $\boldsymbol{\ell}$, itself, may have any azimuthal orientation about z, because once we specify the values of $\boldsymbol{\ell}^2$ and ℓ_z, we cannot locate the position of $\boldsymbol{\ell}$ in the xy plane. Hence the chance of finding $\boldsymbol{\ell}$ between ϕ and $\phi + d\phi$ is also uniform, and we must take this into account. The probability of finding the particle between θ and $\theta + d\theta$ is simply $d\theta/\pi$ since we choose θ to range from 0 to π. Because ϕ is not specified, the region between θ and $\theta + d\theta$ defines a band on the unit sphere as shown in Figure 1.2c. Then the probability $d\theta/\pi$ must be spread over this band, which is bounded by two spherical segments, one of height $h = \cos\theta$ and the other of height $h = \cos(\theta + d\theta)$ on the unit sphere. Recall that the area of a spherical segment of radius r and height h is $2\pi rh$. The area of the band is the difference between the areas of the two spherical segments, that is, $2\pi[\cos\theta - \cos(\theta + d\theta)] = 2\pi\sin\theta d\theta$. Thus the probability density function is given by the probability $(d\theta/\pi)$ per unit area $(2\pi\sin\theta d\theta)$, that is, by $(2\pi^2\sin\theta)^{-1}$, in agreement with Eq. (1.80). Another way to visualize this result is to consider the density of intersections of the longitudes and latitudes on a globe (sphere). These intersections crowd together at the poles and spread apart at the equator; that is, the density of intersections is inversely proportional to $\sin\theta$.

Consider the situation for $m \neq 0$, as shown in Figure 1.2b. Here α is the angle between $\boldsymbol{\ell}$ and the z axis, θ the angle between the z axis and the particle's position vector $\mathbf{r}$, ϕ the azimuthal angle about the z axis and γ the angle measured about $\boldsymbol{\ell}$. Once again, the probability of finding the particle is uniform in the angle γ while $\boldsymbol{\ell}$ may have any azimuthal orientation about z as long as α is constant. The probability that the particle is between γ and $\gamma + d\gamma$ is $d\gamma/\pi$ since again γ is chosen to range from 0 to π; moreover, this probability is spread over the zone $2\pi\sin\theta\, d\theta$ (see Figure 1.2c). Hence the probability density of finding the particle at some point with angle θ is $(d\gamma/\pi)/2\pi\sin\theta\, d\theta$, that is

$$|Y_{\ell m}(\theta, \phi)|^2 = \left(\frac{d\gamma}{d\theta}\right)\frac{1}{2\pi^2\sin\theta} \tag{1.81}$$

In the $m = 0$ case, $\gamma = \theta$ and Eq. (1.81) reduces to Eq. (1.80), as expected.

Actually, γ is the dihedral angle[10] between the planes containing z and $\boldsymbol{\ell}$ and containing $\boldsymbol{\ell}$ and $\mathbf{r}$. We can relate γ to the angles θ and α by

$$\cos\theta = \cos\alpha\cos\frac{\pi}{2} + \sin\alpha\sin\frac{\pi}{2}\cos\gamma = \sin\alpha\cos\gamma \tag{1.82}$$

Thus

$$\sin\theta d\theta = \sin\alpha\sin\gamma d\gamma \tag{1.83}$$

from which it follows that

$$|Y_{\ell m}(\theta, \phi)|^2 = \frac{1}{2\pi^2\sin\alpha\sin\gamma} = \frac{1}{2\pi^2\sin\alpha(1 - \cos^2\theta/\sin^2\alpha)^{\frac{1}{2}}}$$

$$= \frac{1}{2\pi^2(\sin^2\alpha - \cos^2\theta)^{\frac{1}{2}}} \tag{1.84}$$

This expression holds for $\sin^2\alpha > \cos^2\theta$. The region where $\sin^2\alpha < \cos^2\theta$ corresponds to either $\theta < \frac{\pi}{2} - \alpha$ or $\theta > \frac{\pi}{2} + \alpha$ and is classically forbidden. In Figure 1.3 we present a pictorial summary of this behavior. As compared with the semiclassical limit, the quantal result for large ℓ has rapid oscillations in the classically allowed region and dies exponentially in the classically forbidden region[6]. Moreover, the

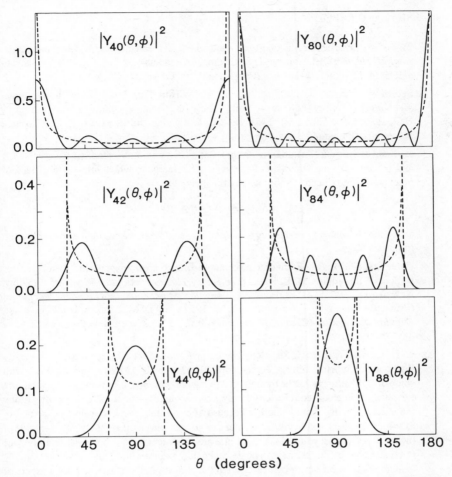

FIGURE 1.3 Probability density of $|Y_{\ell m}(\theta,\phi)|^2$ as a function of θ for selected values of ℓ and m. In each plot the dashed curve is the semiclassical probability density given by $1/[2\pi^2(\sin^2\alpha - \cos^2\theta)^{1/2}]$, where $\cos^2\alpha = m^2/[\ell(\ell+1)]$. Each curve is normalized so that $\int_0^{2\pi}\int_0^{\pi}|Y_{\ell m}(\theta,\phi)|^2 \sin\theta\, d\theta\, d\phi = 1$.

quantal result rounds off the sharp singularities predicted by this semiclassical treatment. Figure 1.3 suggests the value and deficiencies of the vector model. Apart from rapid oscillations, the probability density $|Y_{\ell m}(\theta,\phi)|^2$ behaves like that of a classical particle uniformly moving in a circular orbit for large ℓ. Moreover, as shown in Figure 1.3, the vector model limit is rapidly approached as ℓ increases in magnitude.

NOTES[11] AND REFERENCES

1. Two of my favorite introductory texts about classical mechanics are H. Goldstein, *Classical Mechanics* (Addison-Wesley, Reading, Massachusetts, 1950) and L. D. Landau and E. M. Lifschitz, *Mechanics* (Pergamon Press, Oxford, 1960).

2. See, for example, L. Pauling and E. B. Wilson, Jr., *Introduction to Quantum Mechanics* (McGraw-Hill, New York, 1935); D. Bohm, *Quantum Theory* (Prentice Hall, Englewood Cliffs, NJ, 1951); and L. D. Landau and E. M. Lifschitz, *Quantum Mechanics* (Pergamon Press, New York, 1974).

3. If **A** and **B** are two Hermitian operators that do not commute, then the observables A and B cannot be measured simultaneously. The degree to which an inevitable lack of precision is introduced is expressed by the inequality

$$(\Delta A)(\Delta B) \geq |[\mathbf{A}, \mathbf{B}]/2\,i|$$

called the Heisenberg *uncertainty principle*[2]. Once again the commutator has a special importance.

4. Three standard texts on angular momentum theory are indispensable to the serious student of this topic: M. E. Rose, *Elementary Theory of Angular Momentum* (Wiley, New York, 1957); A. R. Edmonds, *Angular Momentum in Quantum Mechanics* (Princeton University Press, Princeton, NJ, 1957); and D. M. Brink and G. R. Satchler, *Angular Momentum* (Clarendon Press, Oxford, 1962). The preceeding represents a blend of the above.

5. E. U. Condon and G. H. Shortley, *Theory of Atomic Spectra* (Cambridge University Press, 1935; reprinted in paperback form, 1963). E. U. Condon once told me with some disappointment that he thought that this work was most often cited in the literature for setting this phase convention! In my first encounter with phase conventions I took the attitude, "Why should I bother because after all, the phase is arbitrary?" Another example of a phase convention is driving a motor vehicle on the right- or left-hand side of the road. As long as you and everyone else are consistent in this choice, it certainly does not matter, but inconsistency can be detrimental to your health! This is the first of many instances where care must be exercised in the choice of phase, a most vexing aspect of angular momentum theory. Finally, a warning should be made to the uninitiated that phase conventions must be checked in going from one literature reference to another just as one checks driving habits in going from one country to another.

6. The classical correspondence of angular momentum operators and related quantities is explored in detail by P. J. Brussaard and H. A. Tolhoek, *Physica*, **23**, 955 (1957). The vector model appears to have been first introduced by A. Sommerfeld, *Ann. Physik*, **51**, 1 (1916).

7. We follow closely here L. D. Landau and E. M. Lifschitz, *Quantum Mechanics* (Pergamon Press, London, 1958), pp. 166–168. Please note that $\cos\theta$ in Eq. (1.72) and what follows is not the same $\cos\theta$ as in Eq. (1.71) and Figure 1.1 but refers instead to the polar angle between the z axis and the particle vector **r**.

8. By replacing $\cot\theta$ by $1/\theta$ for $\theta \ll 1$ and $\ell(\ell+1)$ by $(\ell+\frac{1}{2})^2$, Eq. (1.72) becomes

$$\left[\frac{d^2}{d\theta^2} + \frac{1}{\theta}\frac{d}{d\theta} + \left(\ell+\tfrac{1}{2}\right)^2\right] P_\ell(\cos\theta) = 0$$

This is the differential equation satisfied by the zero-order Bessel function J_0 with argument $(\ell + \frac{1}{2})\theta$, that is,

$$P_\ell(\cos \theta) = J_0 \left[\left(\ell + \tfrac{1}{2} \right) \theta \right]$$

for $\theta \ll 1$. It is well known that the asymptotic expansion of the zero-order Bessel function for $x \gg 1$ is

$$J_0(x) \to \left(\frac{2}{\pi x} \right)^{\frac{1}{2}} \sin \left(x + \frac{\pi}{4} \right)$$

Hence we have

$$P_\ell(\cos \theta) \to \left[\frac{2}{\pi(\ell + \frac{1}{2})\theta} \right]^{\frac{1}{2}} \sin \left[\left(\ell + \tfrac{1}{2} \right) \theta + \frac{\pi}{4} \right]$$

for $\theta(\ell + \frac{1}{2}) \gg 1$. Comparison with Eq. (1.77) permits the identification

$$A_\ell = \left[\frac{2}{\pi(\ell + \frac{1}{2})} \right]^{\frac{1}{2}}$$

and $\alpha = \pi/4$. Hence for $\theta \gg \ell^{-1}$ or $\pi - \theta \gg \ell^{-1}$

$$P_\ell(\cos \theta) = \left[\frac{2}{\pi(\ell + \frac{1}{2})} \right]^{\frac{1}{2}} \frac{\sin \left[(\ell + \frac{1}{2})\theta + \pi/4 \right]}{(\sin \theta)^{\frac{1}{2}}}$$

9. The condition $(\ell + \frac{1}{2})^2 \gg \frac{1}{4} \csc^2 \theta$ implies that $2(\ell + \frac{1}{2})|\sin \theta| \gg 1$, which for small θ may be restated as $\theta\ell \gg 1$ or $(\pi - \theta)\ell \gg 1$.

10. Let the three line segments AO, BO, and CO intersect at the common point O. Denote the included angle $\angle AOB$ by ϕ_{AOB}, $\angle AOC$ by ϕ_{AOC}, and $\angle BOC$ by ϕ_{BOC}. Then according to spherical trigonometry, the dihedral angle ψ_{ABC}, defined as the angle between the AOB and BOC planes, is related to the ϕ values by

$$\cos \psi_{ABC} = \frac{\cos \phi_{AOC} - \cos \phi_{AOB} \cos \phi_{BOC}}{\sin \phi_{AOB} \sin \phi_{BOC}}$$

11. Notes are intended to offer further information, often of a peripheral nature, in a manner that does not interrupt the flow of the main text; they are not meant to be skipped. See, for example, A. Held and P. Yodzis, *General Relativity and Gravitation*, **13**, 873 (1981).

PROBLEM SET 1

1. Derive Eq. (1.20).

2. Prove Eq. (1.56).

3. In the $|jm\rangle$ representation in which $\mathbf{j}^2$ and j_z are diagonal, the matrix elements of an arbitrary operator $\mathcal{O}$ diagonal in $\mathbf{j}^2$ are given by

$$\mathcal{O}_{mm'} = \langle jm|\,\mathcal{O}\,|jm'\rangle$$

Thus each such operator may be represented by a $(2j+1) \times (2j+1)$ matrix where the rows are labeled by m and the columns by m'. Such matrices are called *representations*.

   A. For the $j = 2$ case, write down explicitly the 5×5 matrices for the operators

$$j_x, j_y, j_z, j_+, j_-, \mathbf{j}^2$$

   B. By carrying out the indicated matrix operations, show that

$$j_x j_y - j_y j_x = i j_z$$

   C. Find the specific matrix element

$$\langle j = 2\,, m = 0\,|\,j_x j_y j_x\,|\,j = 2\,, m = 1\rangle$$

4. Given that $\langle j_z \rangle = \langle jm|j_z|jm\rangle = m$, and that m ranges from $-j$ to $+j$ in unit steps, we explore here an alternative procedure for identifying $\langle \mathbf{j}^2 \rangle = \langle jm|\mathbf{j}^2\,|jm\rangle$ with $j(j+1)$. This is accomplished by equating the expectation value of an operator $\mathcal{O}$ with its spatial average, defined by

$$(\mathcal{O})_{\text{sp}} = \frac{\sum_m \langle jm|\,\mathcal{O}\,|jm\rangle}{\sum_m \langle jm|jm\rangle} \tag{1.85}$$

Because the choice of coordinates is arbitrary and space is isotropic,

$$\langle \mathbf{j}^2 \rangle = \langle j_x^2 \rangle + \langle j_y^2 \rangle + \langle j_z^2 \rangle = 3(j_z^2)_{\text{sp}} \tag{1.86}$$

Perform the spatial average and show that $\langle \mathbf{j}^2 \rangle = j(j+1)$.
Sums of the form

$$S_{2k}(j) = \sum_{m=-j}^{j} m^{2k} \tag{1.87}$$

are frequently encountered where k is an integer but j may be integral or half-integral. We discuss here the general evaluation of Eq. (1.87). Of course we need not bother with $S_{2k+1}(j)$ since this sum vanishes when m ranges from $-j$ to $+j$ in integral steps. We begin by examining a related problem, namely, the sums

$$S_k(n) = \sum_{i=0}^{n} i^k \tag{1.88}$$

The $k = 0$ case is particularly simple, involving just counting:

$$S_0(n) = \sum_{i=0}^{n} i^0 = \sum_{i=0}^{n} 1 = n + 1 \tag{1.89}$$

However, we may recast Eq. (1.89) into a particularly provocative form as

$$S_0(n) = \sum_{i=0}^{n} [(i + 1) - i] \tag{1.90}$$

where enumeration of this series, that is, $[1 - 0] + [2 - 1] + \cdots + [(n+1) - n]$, shows that the only surviving term is $n + 1$. This suggests a general scheme for the evaluation of $S_{k+1}(n)$ in terms of the known $S_k(n)$ for k less than $k + 1$ by considering the sum $\sum_{i=0}^{n} [(i+1)^{k+1} - i^{k+1}]$, which must equal $(n+1)^{k+1}$. For example, we illustrate this procedure by deriving $S_1(n)$ with our knowledge of $S_0(n)$:

$$\sum_{i=0}^{n} [(i + 1)^2 - i^2] = (n+1)^2 = \sum_{i=0}^{n} [(i^2 + 2i + 1) - i^2]$$

$$= 2S_1(n) + S_0(n) \tag{1.91}$$

from which it follows that

$$S_1(n) = \frac{(n+1)^2 - S_0(n)}{2} = \frac{(n+1)^2 - (n+1)}{2}$$

$$= \frac{n(n+1)}{2} \tag{1.92}$$

In a similar manner it is shown that

$$S_2(n) = \frac{n(n+1)(2n+1)}{6} \tag{1.93}$$

$$S_3(n) = \frac{n^2(n+1)^2}{4} \tag{1.94}$$

$$S_4(n) = \frac{n(n+1)(2n+1)(3n^2 + 3n - 1)}{30} \tag{1.95}$$

For integral j we rewrite Eq. (1.87) as

$$S_{2k}(j) = 2 \sum_{m=0}^{j} m^{2k} - 0^{2k} \tag{1.96}$$

and we obtain with the help of Eqs. (1.89) and (1.92)–(1.95) the following:

$$S_0(j) = 2(j + 1) - 1 = 2j + 1 \tag{1.97}$$

$$S_2(j) = \frac{j(j + 1)(2j + 1)}{3} \tag{1.98}$$

$$S_4(j) = \frac{j(j + 1)(2j + 1)\,[3j(j + 1) - 1]}{15} \tag{1.99}$$

For half-integral j, Eq. (1.87) may be reexpressed as

$$S_{2k}(j) = 2 \sum_{i=0}^{j-\frac{1}{2}} \left[(2i + 1)/2\right]^{2k} \tag{1.100}$$

and explicit evaluation of Eq. (1.100) with the help of Eqs. (1.89) and (1.92)–(1.95) yields the same expressions as shown in Eqs. (1.97)–(1.99); that is, Eqs. (1.97)–(1.99) are valid for both integral and half-integral j. An alternative proof of the validity of Eqs. (1.97)–(1.99) for half-integral j is to recognize that

$$S_{2k}(j)/2^{2k} = S_{2k}(j/2) + S_{2k}(j/2 - 1/2) \tag{1.101}$$

APPLICATION 1
SCATTERING THEORY

Angular momentum plays an important role in describing and understanding collision processes in addition to the dynamical behavior of bound systems. The following is intended as a cursory introduction to the beautiful topic of elastic scattering as well as a good review of many aspects of Chapter 1. First a classical treatment is presented, then a quantum one; finally, their connection is briefly outlined.

Classical Treatment

The quantum corrections to the bulk behavior of gases, such as transport or equilibrium properties, are normally rather unimportant except for light gases at low temperatures[1, 2]. In contrast, scattering experiments are a much more delicate tool for observing two-body interactions. As we shall see later, quantum mechanics does play an essential role in many scattering effects. Nevertheless, classical mechanics is capable of illuminating the outline of this subject, is more intuitive, and thus provides a useful conceptual framework.

In the center-of-mass frame, a collision may be pictured as the interaction of a mass point μ with a central force potential $V(r)$, chosen to be located at the origin. The position vector $\mathbf{r}$ describes the location of the mass point with respect to the origin. Then the angular momentum about the origin is given by

$$\mathbf{L} = \mathbf{r} \times \mathbf{p} = \mu(\mathbf{r} \times \dot{\mathbf{r}}) \tag{1}$$

and the time rate of change of $\mathbf{L}$ is

$$\dot{\mathbf{L}} = \mu \frac{d}{dt}(\mathbf{r} \times \dot{\mathbf{r}}) = \mu[(\dot{\mathbf{r}} \times \dot{\mathbf{r}}) + (\mathbf{r} \times \ddot{\mathbf{r}})] = 0 \tag{2}$$

for the force $\mu\ddot{\mathbf{r}}$ is always directed along $\mathbf{r}$ if the potential $V(r)$ is only a function of the magnitude of $\mathbf{r}$. Hence $\mathbf{L}$ is a vector of constant length, that is, a conserved quantity during the course of the collision. Since $\mathbf{r}$ and $\dot{\mathbf{r}}$ must both be perpendicular to the fixed direction of $\mathbf{L}$ in space, the trajectory of the mass point is confined to a plane perpendicular to $\mathbf{L}$. This argument does not appear to hold if $\mathbf{L} = 0$, but then the motion must be along a straight line through the origin as $\mathbf{L} = 0$ requires $\mathbf{r}$ and $\dot{\mathbf{r}}$ to be parallel. Thus central force motion may always be regarded as motion in a plane. We arbitrarily choose the z axis along $\mathbf{L}$ so that the trajectory is confined to the xy plane.

A typical trajectory is pictured in Figure 1. For elastic scattering, the initial and final asymptotic motions differ only in the direction of the velocity $\mathbf{v}$, which has been rotated through the angle χ, called *the deflection angle*. The asymptotic speed v and

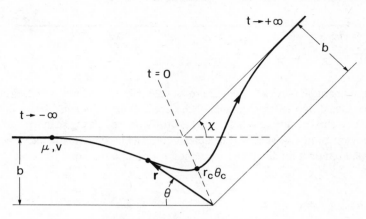

FIGURE 1   Elastic scattering trajectory.

impact parameter b (which is the distance of closest approach if $V(r) = 0$) are related to the total energy E and the magnitude of the angular momentum L by

$$E = \tfrac{1}{2}\mu v^2 \tag{3}$$

and

$$L = \mu|\mathbf{r} \times \dot{\mathbf{r}}| = \mu v b \tag{4}$$

Note that the trajectory is symmetric about the point of closest approach ($r = r_c$, $\theta = \theta_c$). We choose $t = 0$ at this point.

Conservation of angular momentum provides three independent constants of the motion, one for each Cartesian component of $\mathbf{L}$. Two of these suffice to express the direction of $\mathbf{L}$, which is constant during the trajectory, and the third determines the magnitude. An additional constant of the motion, the total energy E, is provided by conservation of energy $E = T + V$, where $T = \tfrac{1}{2}\mu\dot{\mathbf{r}} \cdot \dot{\mathbf{r}}$ is the kinetic energy and $V = V(r)$ is the potential energy.

A. Show that

$$L = \mu r^2 \dot{\theta} \tag{5}$$

and

$$E = \tfrac{1}{2}\mu(\dot{r}^2 + r^2\dot{\theta}^2) + V(r) \tag{6}$$

Hint: Introduce polar coordinates

$$x = r \cos \theta$$

$$y = r \sin \theta$$

to describe the motion in the scattering plane and evaluate

$$L = \mu(x\dot{y} - y\dot{x})$$

$$E = \frac{1}{2}\mu(\dot{x}^2 + \dot{y}^2) + V(r)$$

By eliminating $\dot{\theta}$ in Eqs. (5) and (6) we obtain an equation for the radial motion

$$E = \frac{1}{2}\mu\dot{r}^2 + V(r) + \frac{L^2}{2\mu r^2} \tag{7}$$

Equation (7) describes the one-dimensional motion of a particle of mass μ with total energy E in an effective potential

$$U_L(r) = V(r) + \frac{L^2}{2\mu r^2} \tag{8}$$

composed of the "true" potential $V(r)$ and the centrifugal potential $L^2/2\mu r^2$. In molecular scattering, the influence of the centrifugal potential is often dominant. The reason is that large values of L, corresponding to large impact parameters, often contribute most to the scattering process. The effective radial force is given by

$$\mu\ddot{r} = -\frac{\partial U_L}{\partial r} = -\frac{\partial V}{\partial r} + \frac{L^2}{\mu r^3} \tag{9}$$

We observe that the centrifugal contribution is always repulsive, whereas that from $V(r)$ is usually attractive for large r but repulsive for small r.

The complete solution for the scattering motion can be obtained from Eqs. (5) and (7) by integrating the differential equations

$$dr = \pm\left[\frac{2}{\mu}\left(E - V(r) - \frac{L^2}{2\mu r^2}\right)\right]^{\frac{1}{2}} dt \tag{10}$$

$$d\theta = \frac{L}{\mu r^2} dt \tag{11}$$

and taking into account the initial starting conditions of r and θ. Hence in the center-of-mass frame a classical trajectory is specified by six parameters, the three Cartesian position vectors and three Cartesian velocity vectors of the representative mass point μ, or alternatively, the direction and magnitude of $\mathbf{L}$, the value of the total energy E and the initial values of r and θ. However, we are not interested in the complete time-dependent solution but instead in the one observable of elastic scattering, namely, the deflection angle χ. From Eqs. (10) and (11) we find

$$\theta(r) = \int_0^\theta d\theta = -\int_\infty^r \frac{(L/\mu r^2)\,dr}{[(2/\mu)(E - V - L^2/2\mu r^2)]^{\frac{1}{2}}}$$

$$= -b \int_\infty^r \frac{dr}{r^2 \left(1 - b^2/r^2 - V/E\right)^{\frac{1}{2}}} \tag{12}$$

where the collision starts at $t = -\infty$ where $\theta = 0$ and $r = \infty$. Note that in Eq. (12) the square root in Eq. (10) is taken with a negative sign since $dr/d\theta$ is negative.

With respect to the angle of closest approach, θ_c,

$$2\,\theta_c + \chi = \pi \tag{13}$$

Hence the angle of deflection is given by

$$\chi = \pi - 2 \int_{r_c}^\infty \frac{(L/\mu r^2)\,dr}{[(2/\mu)(E - V - L^2/2\mu r^2)]^{\frac{1}{2}}}$$

$$= \pi - 2b \int_{r_c}^\infty \frac{dr}{r^2 \left(1 - b^2/r^2 - V/E\right)^{\frac{1}{2}}} \tag{14}$$

where the radial distance of closest approach is determined from the condition that $dr/d\theta = 0$ or equivalently $\dot{r} = 0$. From Eq. (7) we have

$$E = \frac{L^2}{2\,\mu r_c^2} + V(r_c) \tag{15}$$

which may be restated as

$$b^2 = r_c^2 \left[1 - \frac{V(r_c)}{E}\right] \tag{16}$$

B. Derive Eq. (16) from Eq. (15).

Once the potential function $V(r)$ is specified, the deflection angle χ may be calculated by using Eq. (14). Moreover, if $V(r_c) < 0$ (attractive potential at the turning point of the radial motion), then $r_c < b$, whereas if $V(r_c) > 0$ (repulsive potential at the turning point), $r_c > b$.

Consider a beam of particles incident on some scattering center. Collisions occur with all possible impact parameters (angular momenta), giving rise to a corresponding distribution in angles of scattering, which is described by a *differential cross section*. Let the intensity of the incident beam be characterized in terms of the flux I_0, where I_0 is the number of particles crossing a unit area normal to the beam direction per unit time. The differential cross section $I(\chi)$ is defined so that $I(\chi)\,d\Omega$ is the number of particles per unit time scattered into a solid angle element $d\Omega$ divided by the incident flux. Note that $I(\chi)$ has the dimensions of area per steradian.

Because of the spherical symmetry of the force field, the deflection pattern is axially symmetric about the incident beam direction. Hence $I(\chi)$ depends only on χ. Accordingly, the solid angle element $d\Omega$ may be taken between the cones defined by χ and $\chi + d\chi$, that is, $d\Omega = 2\pi \sin \chi d\chi$. Even if the force field were nonspherical, the averages over all possible impact parameters and all orientations of the target would cause any azimuthal dependence to vanish unless one of the reactants were initially polarized.

For a given initial velocity v, the fraction of the incident flux with impact parameter between b and $b + db$ is $2\pi b\,db$; these particles undergo deflections between χ and $\chi + d\chi$ if $d\chi/db > 0$ or between χ and $\chi - d\chi$ if $d\chi/db < 0$. This gives the relation $I(\chi)\,d\Omega = 2\pi b\,db$, from which it follows that

$$I(\chi) = \frac{b}{\sin \chi \, |d\chi/db|} \tag{17}$$

where χ is related to b through Eq. (14). Hence a knowledge of $\chi(b)$ for a given v serves to determine the differential cross section, $I(\chi)$. This is related to the *total cross section* σ by

$$\sigma = 2\pi \int_0^\pi I(\chi) \sin \chi d\chi \tag{18}$$

where we disregard the possibility of divergence. Here σ is a measure of the probability for removal by scattering from the incident beam (i.e., attenuation) and σ has the dimensions of area.

C. Consider scattering from an impenetrable sphere of radius a so that

$$V(r) = \infty, \qquad r < a$$
$$V(r) = 0, \qquad r > a$$

Show that the classical differential cross section is independent of the deflection angle and that the classical total cross section is just the area of a circle (bull's eye) of radius a. *Hint*: Begin by proving that $\chi = 2 \arccos (b/a)$ for $b \leq a$, $\chi = 0$ for $b > a$.

The deflection angle often resembles the behavior shown in Figure 2 as a function of impact parameter. For large impact parameters χ is small and negative (net

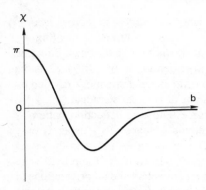

FIGURE 2   Deflection angle as a function of the impact parameter (generic form).

attraction), and as b decreases, the deflection becomes more negative (increasing attraction). The deflection then either approaches negative infinity, corresponding to *classical orbiting*, or reaches a finite minimum. Classical orbiting occurs when the impact parameter equals the maximum in the centrifugal barrier and the initial velocity is such that the (radial) velocity vanishes at the maximum in the centrifugal barrier. For still smaller impact parameters, the effect of the short range repulsive part of the potential causes the deflection angle to increase, passing through zero (where the net attraction and repulsion cancels) and finally to approach $180°$ as $b \to 0$. As Eq. (17) shows, the differential cross section becomes singular when either $\chi = 0°$, called *forward glory scattering*, or $d\chi/db = 0$ at the minimum, called *rainbow scattering*. The rainbow deflection angle χ_r will depend, of course, on the velocity v.

The calculation of the differential cross section is considerably simplified if we consider only those collisions at large impact parameters since they contribute only to scattering at small deflection angles, that is, in the forward direction. Let the initial momentum of the particles be along the x axis. Denote by $\mathbf{p}'$ the momentum of the particle after scattering. Then, $p' \sin \chi = p'_y$, and for small deflection angles, we take $\sin \chi \simeq \chi$ and write

$$\chi \simeq \frac{p'_y}{p'} = \frac{\text{momentum transfer}}{\text{momentum}} \tag{19}$$

Since the time derivative of the momentum is the force, the momentum transferred perpendicular to the initial direction is given by integrating the force component F'_y over the trajectory. But

$$F'_y = -\left(\frac{\partial V}{\partial y}\right) = -\left(\frac{\partial V}{\partial r}\right)\left(\frac{\partial r}{\partial y}\right) = -\left(\frac{\partial V}{\partial r}\right)\frac{b}{r} \tag{20}$$

where we use the facts that $r^2 = x^2 + y^2$ and $y \simeq b$. Hence

$$\chi = \frac{p'_y}{\mu(2E/\mu)^{\frac{1}{2}}}$$

$$= -b(2\mu E)^{-\frac{1}{2}} \int_{-\infty}^{\infty} \left(\frac{\partial V}{\partial r}\right) \frac{dt}{r}$$

$$= -b(2\mu E)^{-\frac{1}{2}} \left(\frac{2E}{\mu}\right)^{-\frac{1}{2}} \int_{-\infty}^{\infty} \left(\frac{\partial V}{\partial r}\right) \frac{dx}{r}$$

$$= -\frac{b}{E} \int_{b}^{\infty} \left(\frac{\partial V}{\partial r}\right) (r^2 - b^2)^{-\frac{1}{2}} dr \tag{21}$$

where $x = (2E/\mu)^{1/2} t$ and x varies from $-\infty$ to ∞ as r varies from ∞ to b and back.

D. Use Eq. (21) to examine the small-angle elastic scattering from the potential $V(r) = C_s r^{-s}$ for $s > 0$. Show that

$$\chi = \frac{sC_s \pi^{\frac{1}{2}}}{2b^s E} \frac{\Gamma\left[(s+1)/2\right]}{\Gamma\left[s/2+1\right]} \tag{22}$$

from which it is concluded that χE is a function of b alone.

Complete this study by evaluating $I(\chi)$, using Eq. (17) to prove that

$$I(\chi) = \frac{1}{s} \left\{ \frac{sC_s \pi^{\frac{1}{2}}}{2E} \frac{\Gamma\left[(s+1)/2\right]}{\Gamma\left[s/2+1\right]} \right\}^{2/s} \chi^{-(2+2/s)} \tag{23}$$

for small values of χ. Hence a log–log plot of the center-of-mass differential scattering cross section $I(\chi)$ as a function of the scattering angle χ at fixed energy should give a straight line with slope $-(2 + 2/s)$ from which the value of the s can be determined[2–5]. For the usual case of neutral closed-shell systems, $s = 6$ (van der Waals long-range attractive potential) and $I(\chi)$ is proportional to $E^{-1/3} \chi^{-7/3}$.

Quantum Treatment

The quantum formulation of scattering in a central force field may be found in many texts[5–11] and is not developed in detail here. Briefly, the scattering is determined by the asymptotic form of the wave function

$$\psi(r, \chi) \xrightarrow[r \to \infty]{} A \left[e^{ikz} + \frac{f(\chi)}{r} e^{ikr} \right] \tag{24}$$

where A is a normalization constant and

$$k = 1/\lambda = \mu v/\hbar \tag{25}$$

is the magnitude of the initial propagation vector **k** that is directed along $\chi = 0$. In Eq. (24) the term $A \exp(ikz)$ represents a plane wave incident on the scattering center and the term $Af(\chi) \exp(ikr)/r$ represents an outgoing spherical wave. Note that the amplitude of the scattered wave is inversely proportional to r since the radial flux must decrease as the inverse square of the distance. If the incoming and scattered fluxes of particles are separated from each other, by collimating slits, for example, the differential cross section is related to the scattering amplitude $f(\chi)$ by

$$I(\chi) = |f(\chi)|^2 \tag{26}$$

Thus the asymptotic form of the wave function determines the differential scattering cross section but cannot be found without solving the wave equation throughout all space. This may be carried out by the method of partial waves in which the general solution is represented as an infinite sum of Legendre polynomials

$$\psi(r,\chi) = \sum_{\ell=0}^{\infty} R_\ell(r) P_\ell(\cos\chi) \tag{27}$$

The boundary condition that $R_\ell(r)$ remains finite at $r = 0$ determines the asymptotic form of the solution, up to the normalization constant.

In the absence of a potential $V(r) = 0$, the wave function is simply $A \exp(ikz)$, which can be cast into the form of Eq. (27) for large r as

$$e^{ikz} \xrightarrow[r\to\infty]{} \sum_{\ell=0}^{\infty}(2\ell+1)\exp\left(\frac{i\ell\pi}{2}\right)\frac{\sin(kr - \ell\pi/2)}{kr}P_\ell(\cos\chi)$$

$$= \frac{1}{2i}\sum_{\ell=0}^{\infty} i^\ell(2\ell+1)P_\ell(\cos\chi)\left\{\frac{e^{i(kr-\ell\pi/2)}}{kr} - \frac{e^{-i(kr-\ell\pi/2)}}{kr}\right\} \tag{28}$$

Equation (28) has an appealing physical interpretation, namely, the incident plane wave is equivalent to a superposition of an infinite number of incoming and outgoing spherical waves in which each term in Eq. (28) corresponds to an orbital angular momentum of magnitude

$$L = \sqrt{\ell(\ell+1)}\,\hbar \simeq \left(\ell + \tfrac{1}{2}\right)\hbar \tag{29}$$

about the scattering center. Classically, this angular momentum would correspond to an impact parameter

$$b = \frac{L}{\mu v} \simeq \frac{\left(\ell + \tfrac{1}{2}\right)}{k} = \left(\ell + \tfrac{1}{2}\right)\lambda \tag{30}$$

Thus we can picture the incident beam as divided into cylindrical zones such that the ℓth zone contains particles with impact parameters between $\ell\lambda$ and $(\ell + 1)\lambda$.

E. In quantum mechanics only integral values of ℓ are allowed, yet we view b as being continuous. How will this affect our results when treating chemical systems?

Provided $V(r)$ falls off more rapidly than $1/r$ for large r, the general solution [Eq. (27)] can be shown to have the asymptotic form

$$\psi(r, \chi) \xrightarrow[r \to \infty]{} \sum_{\ell=0}^{\infty} (2\ell + 1) \exp \left[i \left(\frac{\ell\pi}{2} + \eta_\ell \right) \right] \frac{\sin(kr - \ell\pi/2 + \eta_\ell)}{kr} P_\ell(\cos \chi)$$

$$= \frac{1}{2i} \sum_{\ell=0}^{\infty} i^\ell (2\ell + 1) P_\ell(\cos \chi) \left\{ e^{2i\eta_\ell} \frac{e^{i(kr - \ell\pi/2)}}{kr} - \frac{e^{-i(kr - \ell\pi/2)}}{kr} \right\}$$

$$(31)$$

where η_ℓ is called the *phase shift* and must be a real number. Comparing Eq. (31) with the solution for $V(r) = 0$, namely, Eq. (28), we conclude that the effect of the scattering potential is to introduce a change of phase in the asymptotic form of the radial wave functions. Moreover, the phase shift occurs only in the outgoing part of the asymptotic wave function. This is indicated schematically in Figure 3. A repulsive potential causes a decrease in the relative velocity of the particles for small r so that the wavelength is increased. Thus the scattered wave is "pushed out" relative to that for $V = 0$, and consequently the phase shift η_ℓ is negative. An attractive potential "pulls in" the radial wave function and produces a positive phase shift. Comparing $\ell = 0$ to $\ell \neq 0$, we also see that the repulsive centrifugal potential contributes a negative phase shift of $-\ell\pi/2$.

Further analysis shows that the scattering amplitude is given by

$$f(\chi) = \frac{1}{2ik} \sum_{\ell=0}^{\infty} (2\ell + 1) \left(e^{2i\eta_\ell} - 1 \right) P_\ell(\cos \chi) \qquad (32)$$

which may be derived by comparing Eqs. (24), (28), and (31). Then the differential cross section has the form

$$I(\chi) = \lambda^2 \left| \sum_{\ell=0}^{\infty} (2\ell + 1) e^{i\eta_\ell} \sin \eta_\ell P_\ell(\cos \chi) \right|^2 \qquad (33)$$

Equation (33) shows that interference between the terms with different values of ℓ plays an important role in determining the differential cross section. In terms of

(a) V > 0, Repulsive Potential

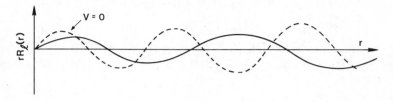

(b) V < 0, Attractive Potential

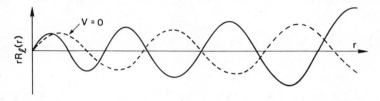

FIGURE 3   Form of the radial wave function (solid curve) versus separation distance for (a) a repulsive potential ($V(r) > 0$) and (b) an attractive potential ($V(r) < 0$). For comparison, the radial wave function (dashed curve) for no potential ($V(r) = 0$) is also shown.

our previous picture, the outgoing partial waves scattered from the various zones of incident impact parameters are superimposed with weighting factors that depend on the phase shift associated with each zone. Whenever the phase shift is zero or an integral multiple of π, the corresponding partial wave does not contribute to the scattering, since it has been "pushed out" or "pulled in" an integral number of half wavelengths so that the nodes and extrema in its asymptotic form match those of the incident partial wave, that is, the $V(r) = 0$ solution.

The total cross section is obtained by integrating Eq. (33) over all directions:

$$\sigma = 2\pi \int_0^\pi I(\chi) \sin \chi \, d\chi$$

$$= 4\pi \lambda^2 \sum_{\ell=0}^\infty (2\ell + 1) \sin^2 \eta_\ell \tag{34}$$

Cross terms in the expansion of Eq. (33), which represent the interference of different partial waves, cancel on integration because of the orthogonality of the Legendre polynomials. Equation (34) shows that each partial wave contributes separately to σ in proportion to $\sin^2 \eta_\ell$ and its statistical weight ($2\ell + 1$). The coefficient 4π represents the integral over the solid angle of scattering while $\lambda^2 = (\lambda/2\pi)^2 = 1/k^2$ is associated with the squared wavelength of the incident particles. From another viewpoint, however, the physical interpretation of Eq. (34) is somewhat strange. The flux of particles incident with impact parameters within the ℓth zone is proportional to the cross sectional area of the zone, namely, $\pi \lambda^2 [(\ell+1)^2 - \ell^2] = \pi \lambda^2 (2\ell+1)$. This disagrees with Eq. (33) by a factor of 4. Actually, this factor is the result of quantum mechanical diffraction effects that are not properly included in our simple physical picture. Nevertheless, we did find the correct form in which a $\sin^2 \eta_\ell$ weighting factor appears to express the effectiveness of the potential in scattering the ℓth partial wave.

The uncertainty principle provides a clarification concerning the convergence of Eq. (18) for the total cross section. In a classical treatment, σ diverges unless the potential vanishes outside some finite radius. If the potential extends to infinity, then even very large values of the impact parameter b produce a small deflection and thus contribute to the scattering cross section. However, if the potential is sufficiently weak at large r, the collisions with large impact parameters produce such slight deflections that the scattering angle χ is smaller than the directional uncertainty required by the uncertainty principle. For such collisions, the particle cannot be regarded as scattered, and thus the uncertainty principle, in effect, causes a cutoff in the impact parameter b that can contribute to the scattering process, making the total cross section finite. This is yet another example where quantum mechanics rounds off a classical singularity.

Let us make these considerations more quantitative. There is a largest value of b, denoted by b_c, for which χ is just large enough to be observable. Let $\Delta p'_y$ be the uncertainty in the transverse component of the momentum. Then according to the uncertainty principle, $b_c \Delta p'_y = h$ or $\Delta p'_y = h/b_c$. Thus it follows from Eq. (19) that the deflection angle χ_c associated with b_c must be

$$\chi_c = \Delta p'_y / p = h/[b_c(2\mu E)^{\frac{1}{2}}] \tag{35}$$

However, we also found according to Eq. (22) that for a potential of the form $V(r) = C_s r^{-s}$, χ_c is related to the impact parameter b_c by

$$\chi_c = \frac{sC_s \pi^{\frac{1}{2}} \Gamma[(s+1)/2]}{2 b_c^s E \Gamma[s/2 + 1]} \tag{36}$$

Combining Eqs. (35) and (36), we have

$$b_c = \left[\frac{sC_s \pi^{\frac{1}{2}} \Gamma[(s+1)/2]}{hv \Gamma[s/2 + 1]} \right]^{1/(s-1)} \tag{37}$$

where v is the initial velocity. It follows that the cross section cannot exceed πb_c^2 and that its velocity dependence is $v^{-2/(s-1)}$. For $s = 6$, $\sigma(v)$ is proportional to $v^{-2/5}$.

Actually, an undulatory velocity dependence of $\sigma(v)$ is often observed superimposed on the general trend predicted above[2–5]. This is brought about by the quantum interference arising from trajectories at very large impact parameters ($b \geq b_c$) and glory trajectories $b \simeq b_g$ since both contribute to $\chi \leq \chi_c$. Nevertheless, a plot of $\ln \sigma(v)$ versus $\ln v$ can be used to reveal the power law behavior of the long- and short-range parts of the potential and even to obtain an estimate of C_s.

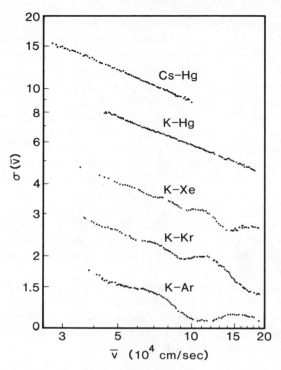

FIGURE 4　Reprinted with permission from D. Beck and H. J. Loesch, *Z. Phys.*, **195**, 444 (1966).

To illustrate this procedure, we present in Figure 4 the velocity dependence of the total cross section when a velocity-selected beam of K crosses a thermal beam of Cs, Hg, Xe, Kr, or Ar [see D. Beck and H. J. Loesch, *Z. Phys.*, **195**, 444 (1966)]. The data, which have been corrected for the velocity distribution of the target gas, show a slope close to $-2/5$, the value expected for an $s = 6$ long-range attractive potential. Glory scattering undulations in the total cross section are also apparent. The amplitudes of the glory undulations contain information on the width of the potential well. Total and differential cross section measurements have been widely used to obtain information about intramolecular potential energy functions[2–5].

Transition from Quantum to Classical Scattering

The phase shift, which so naturally arises in a quantum description of the scattering process, may also be defined semiclassically by comparing the number of wavelengths contained in two different paths, the actual trajectory, and the path that would be followed if the scattering potential were "switched off":

$$\eta_\ell^{SC} = \lim_{R \to \infty} \left[\int_{r_c}^R \frac{dr}{\lambda_r} - \int_b^R \frac{dr}{\lambda_r} \right] \tag{38}$$

Here R is the radius of a sphere whose center coincides with the scattering center, and

$$\lambda_r = \frac{\hbar}{\mu \dot{r}} = \frac{\hbar}{\mu v \left[1 - V(r)/E - b^2/r^2 \right]^{\frac{1}{2}}} \tag{39}$$

is the "local" de Broglie wavelength associated with the radial motion. Thus the semiclassical phase shift is given by

$$\eta_\ell^{SC} = k \left\{ \int_{r_c}^\infty \left[1 - \frac{V(r)}{E} - \frac{b^2}{r^2} \right]^{\frac{1}{2}} dr - \int_b^\infty \left[1 - \frac{b^2}{r^2} \right]^{\frac{1}{2}} dr \right\} \tag{40}$$

This formula can be put in a more familiar form by introducing $k_r = \mu \dot{r}/\hbar$ and rewriting Eq. (38) as

$$\eta_\ell^{SC} = \lim_{R \to \infty} \left[\int_{r_c}^R k_r \, dr - \int_{r_c}^R k \, dr + \int_{r_c}^R k \, dr - \int_b^R k \left(1 - \frac{b^2}{r^2} \right)^{\frac{1}{2}} dr \right] \tag{41}$$

The last integral is readily evaluated:

$$k \int_b^R \frac{(r^2 - b^2)^{\frac{1}{2}} \, dr}{r} = k \left[(r^2 - b^2)^{\frac{1}{2}} - b \arccos \frac{b}{r} \right] \Big|_b^R$$

$$= kR - \frac{kb\pi}{2} \tag{42}$$

Hence

$$\eta_\ell^{SC} = \lim_{R \to \infty} \left[\int_{r_c}^R (k_r - k) \, dr + k(R - r_c) - k(R - b\pi/2) \right]$$

$$= \int_{r_c}^\infty (k_r - k) \, dr - k(r_c - b\pi/2)$$

$$= \int_r^\infty (k_r - k) \, dr - kr_c + \left(\ell + \tfrac{1}{2} \right) \pi/2 \tag{43}$$

which is the standard form for the semiclassical phase shift as given in many quantum mechanics texts[5–11].

The deflection angle χ may also be expressed in a form similar to that for η_ℓ^{SC} by writing

$$\chi = \lim_{R\to\infty}\left[\left(\pi - 2\int_{\substack{\text{actual}\\\text{path}}} d\theta\right) - \left(\pi - 2\int_{\substack{V=0\\\text{path}}} d\theta\right)\right]$$

$$= -2b\left[\int_{r_c}^\infty \left[1 - \frac{V(r)}{E} - \frac{b^2}{r^2}\right]^{-\frac{1}{2}} \frac{dr}{r^2} - \int_b^\infty \left(1 - \frac{b^2}{r^2}\right)^{-\frac{1}{2}} \frac{dr}{r^2}\right] \quad (44)$$

Equation (44) may seem a peculiar way to express χ, and it might be wondered why bother. After all, the last integral in Eq. (44) is

$$\int_b^\infty \frac{dr}{r(r^2 - b^2)^{\frac{1}{2}}} = \left[\frac{1}{b}\arccos\frac{b}{r}\right]\Bigg|_b^\infty$$

$$= -\frac{1}{b}\left(\frac{\pi}{2}\right) \quad (45)$$

so that Eq. (44) reduces to Eq. (14). However, Eq. (44) allows us to deduce a simple and useful connection between the classical deflection angle χ and the rate of change of the semiclassical phase shift with angular momentum, $d\eta_\ell^{SC}/d\ell$. Recall that the general rule for differentiating a definite integral is

$$\frac{d}{dx}\int_{a(x)}^{b(x)} f(x,y)\,dy = \frac{db(x)}{dx}f[b(x),y] - \frac{da(x)}{dx}f[a(x),y] + \int_{a(x)}^{b(x)} \frac{\partial f(x,y)}{\partial x}dy$$

$$(46)$$

By differenting Eq. (40) with respect to ℓ and using Eq. (46) and the relation $(\partial b/\partial\ell)_E = b/k$, we find that

$$\frac{d\eta_\ell^{SC}}{d\ell} = \frac{1}{2}\chi \quad (47)$$

Next let us examine the differential cross section $I(\chi)$. The mathematical identity

$$\delta(1 - \cos\chi) = \frac{1}{2}\sum_{\ell=0}^\infty (2\ell + 1)P_\ell(\cos\chi) \quad (48)$$

is readily verified. If we exclude the singular point $\chi = 0$, which contributes negligibly to the total flux, we can rewrite Eq. (32) as

$$f(\chi) = \frac{\lambda}{2i}\sum_{\ell=0}^\infty (2\ell + 1)\exp(2i\eta_\ell)P_\ell(\cos\chi) \quad (49)$$

To obtain the classical limit, we consider the situation where the potential varies slowly over distances comparable to the de Broglie wavelength and that many ℓ values contribute to the scattering at a given angle so that the major contribution arises from large ℓ values. Then for $\chi \neq 0$,

$$f(\chi) = -\lambda \sum_{\ell=0}^{\infty} \left[\frac{(\ell + \frac{1}{2})}{2\pi \sin \chi} \right]^{\frac{1}{2}} \left[\exp(i\phi^+) - \exp(i\phi^-) \right] \tag{50}$$

where

$$\phi^{\pm} = 2\eta_{\ell} \pm (\ell + \tfrac{1}{2})\chi \pm \tfrac{1}{4}\pi \tag{51}$$

F. Derive Eq. (50). *Hint*: Reread Note [8] of Chapter 1.

In Eq. (50) the exponential factors are rapidly oscillating functions of ℓ, since the phases $\phi^{\pm}$ are large. Thus the majority of terms cancel and this sum is determined mainly from the range of ℓ values for which either ϕ^+ or ϕ^- is an extremum. From Eq. (51) this implies that Eq. (50) is essentially determined only by phase shifts satisfying the relation

$$\frac{2\,d\eta_{\ell}}{d\ell} \pm \chi = 0 \tag{52}$$

where the plus sign is for $d\phi^+/d\ell = 0$ and the minus sign for $d\phi^-/d\ell = 0$. Comparison of Eq. (52) with Eq. (47) demonstrates that only these phase shifts η_{ℓ} corresponding to the impact parameter b contribute to the differential cross section $I(\chi)$ in the classical limit. Conversely, the condition for classical scattering at a given deflection angle χ is that the values of ℓ be large for which Eq. (52) applies.

We conclude by examining one of the most striking phenomena in scattering, namely, *resonance scattering*, in which the total cross section shows an abrupt change as a function of energy. As shown in Eq. (34), each partial wave ℓ contributes to the total cross section a term

$$\sigma_{\ell} = \frac{4\pi}{k^2}(2\ell + 1)\sin^2 \eta_{\ell} \tag{53}$$

Generally the phase shifts η_{ℓ} are slowly varying functions of the energy. A resonance occurs for some partial wave ℓ when η_{ℓ} changes rapidly over a small energy range, and we write

$$\eta_{\ell}(E) = \eta_{\text{bg}} + \eta_{\text{res}} \tag{54}$$

where η_{bg} is the background phase shift and η_{res} is the resonance phase shift.

What makes a resonance occur? To understand this we turn our attention to Eq. (31), the asymptotic form of the wave function. For a given ℓ the ratio $S_{\ell}(k)$

of the outgoing wave amplitude to the incoming wave amplitude at energy $E = \hbar^2 k^2 / 2\mu$ is

$$S_\ell(k) = e^{2 i \eta_\ell(k)} \tag{55}$$

up to a constant factor. Here we have emphasized the dependence of the phase shift on energy by writing $\eta_\ell = \eta_\ell(k)$. The wave number k is a real variable. However, by the process of analytic continuation, $\eta_\ell(k)$ can be regarded as a function of a complex variable k. Suppose there exists an imaginary value of k, that is

$$k = i\kappa \tag{56}$$

where κ is real and positive definite such that

$$S_\ell(\kappa) = e^{2 i \eta_\ell(\kappa)} = 0 \tag{57}$$

Of course, this implies that the energy is negative, and

$$\eta_\ell(\kappa) = i\infty \tag{58}$$

that is, the asymptotic behavior of the wave function will go as $\exp(-kr + \ell\pi/2)/kr$, and there is no outgoing wave amplitude, only an exponentially dying form. This is exactly the condition of a bound state. Thus the zeros of S_ℓ correspond to bound states. Conversely, when $k = -i\kappa$, S_ℓ will grow without bound, and it follows that the poles (singularities in the complex plane) of S_ℓ correspond to scattering resonances. Close to a resonance

$$S_\ell(k) = e^{2 i \eta_\ell}$$

$$= e^{2 i \eta_{\text{bg}}} \frac{e^{i \eta_{\text{res}}}}{e^{-i \eta_{\text{res}}}}$$

$$= e^{2 i \eta_{\text{bg}}} \frac{1 + i \tan \eta_{\text{res}}}{1 - i \tan \eta_{\text{res}}} \tag{59}$$

The simplest way of causing S_ℓ to grow without bound near E_0 is to make it have a simple pole, that is, to set

$$\eta_{\text{res}} = \arctan \left[\frac{\Gamma/2}{E_0 - E} \right] \tag{60}$$

where the meaning of $\Gamma/2$ will become apparent but for the present may be assumed to be a constant independent of energy. Then Eq. (59) becomes

$$S_\ell(k) = \frac{(E_0 - E) + i(\Gamma/2)}{(E_0 - E) - i(\Gamma/2)} e^{2 i \eta_{\text{bg}}} \tag{61}$$

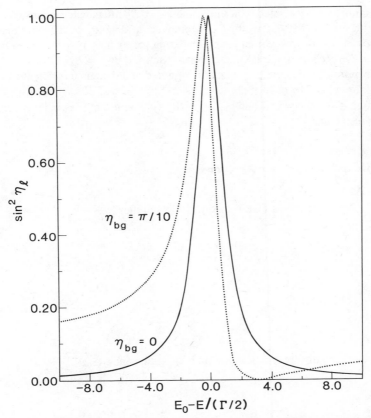

FIGURE 5   Line profile of an isolated scattering resonance for two different background phase shifts, $\eta_{bg} = 0$ and $\eta_{bg} = \pi/10$. Note that when $\eta_{bg} \neq 0$ it is possible for the cross section to vanish at a particular energy (the so-called Cooper minimum).

and $S_\ell(k)$ has a pole at $E = E_0 - i(\Gamma/2)$.

When η_{bg} can be neglected, substitution of Eq. (60) into Eq. (53) yields the result

$$\sigma_\ell = \frac{4\pi}{k^2}(2\ell + 1)\frac{(\Gamma/2)^2}{(E_0 - E)^2 + (\Gamma/2)^2} \qquad (62)$$

where we have made use of the identity

$$\arctan x = \arcsin \frac{x}{(x^2 + 1)^{\frac{1}{2}}} \qquad (63)$$

Equation (62) is called the *Breit–Wigner formula*[12] for an isolated scattering resonance. It describes a bell-shaped curve with a half-width at half-maximum of $\Gamma/2$. This suggests that the occurrence of a resonance at E_0 in the partial cross section σ_ℓ is associated with the formation of a virtual state (metastable bound state) with energy E_0 and width Γ. Moreover, in analogy to a bound state, the time dependence of this virtual state (scattering resonance) is given by $\exp[-i \times (E_0 - i\Gamma/2)t/\hbar]$ so that it is characterized by a lifetime $\tau = \hbar/\Gamma$.

In the more general case, η_{bg} cannot be ignored and there is interference between the background and resonance phase shifts. Figure 5 shows plots of $\sin^2 \eta_\ell(E)$ as a function of E for $\eta_{bg} = 0.0$ and $\eta_{bg} = \pi/10$, assuming that Γ is independent of E over the resonance.

G. Show that the partial cross section is then given by

$$\sigma_\ell = \frac{4\pi}{k^2}(2\ell + 1) \sin^2 \left[\eta_{bg} + \arctan\left(\frac{\Gamma/2}{E_0 - E} \right) \right]$$

$$= \frac{4\pi}{k^2}(2\ell + 1) \frac{(\varepsilon + q)^2}{1 + \varepsilon^2} \sin^2 \eta_{bg} \tag{64}$$

where

$$\varepsilon = \frac{E_0 - E}{\Gamma/2} \tag{65}$$

is the energy difference measured in half-widths and

$$q = \cot \eta_{bg} \tag{66}$$

is called the *line profile index*.

It has been shown by Fano[13] and Fano and Cooper[14] that absorption lines in photoionization may have profiles as illustrated above. Again this arises from interference between the continuum states produced by nonresonant photon absorption and the continuum state produced by the decay of a resonance.

One of the commonest situations that cause the appearance of resonances is an effective potential made up of an attractive part at small distances and a repulsive centrifugal barrier at long distances; as shown in Figure 6. For energies below the maximum in the centrifugal barrier, there would be bound states inside the attractive part of the potential if tunneling could be ignored. However, the presence of quantum mechanical tunneling permits particles "trapped" inside the attractive part of the potential to escape to infinity, and the tunneling rate depends on the height and thickness of the barrier. Conversely, particles incident on the potential at energy close to the virtual state energy are able to penetrate inside the attractive barrier. This behavior explains why resonances generally become narrower as ℓ increases. Larger ℓ values cause bigger centrifugal barriers, thus suppressing tunneling.

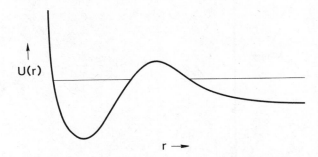

FIGURE 6   Plot of the potential energy including the centrifugal potential versus separation distance.

NOTES AND REFERENCES

1. J. O. Hirschfelder, R. F. Curtiss, and R. B. Bird, *Molecular Theory of Gases and Liquids*, Wiley, New York, 1954.

2. G. C. Maitland, M. Rigby, E. B. Smith, and W. A. Wakeham, *Intramolecular Forces*, Clarendon Press, Oxford, 1981.

3. R. B. Bernstein, *Chemical Dynamics via Molecular Beam and Laser Techniques*, Clarendon Press, Oxford, 1982, Chapter 5; *Adv. Chem. Phys.*, **10**, 75 (1966).

4. J. P. Toennies, "Molecular Beam Scattering Experiments on Elastic, Inelastic, and Reactive Collision," in *Physical Chemistry*, Vol. VIA, H. Eyring, D. Henderson, W. Jost, eds., Academic Press, New York, 1974, Chapter 5.

5. M. S. Child, *Molecular Collision Theory*, Academic Press, New York, 1974.

6. H. S. W. Mott and N. F. Massey, *The Theory of Atomic Collisions*, 3rd ed., Clarendon Press, Oxford, 1965.

7. R. G. Newton, *Scattering Theory of Waves and Particles*, McGraw-Hill, New York, 1966.

8. L. I. Schiff, *Quantum Mechanics*, 3rd ed., McGraw-Hill, New York, 1968.

9. L. D. Landau and E. M. Lifshitz, *Quantum Mechanics*, Addison-Wesley, Reading, MA, 1958.

10. J. R. Taylor, *Scattering Theory*, Wiley, New York, 1972.

11. U. Fano and A. R. P. Rau, *Atomic Collisions and Spectra*, Academic Press, Orlando, 1986.

12. G. Breit and E. Wigner, *Phys. Rev.*, **49**, 519 (1936).

13. U. Fano, *Phys. Rev.*, **124**, 1866 (1961).

14. U. Fano and J. W. Cooper, *Phys. Rev.*, **137**, A1364 (1965); *Rev. Mod. Phys.*, **40**, 441 (1968).

Chapter 2

COUPLING OF TWO ANGULAR MOMENTUM VECTORS

2.1 CLEBSCH–GORDAN COEFFICIENTS

In classical mechanics, the total angular momentum $\mathbf{j}$ of a system composed of two parts, one with angular momenta $\mathbf{j}_1$ and the other with $\mathbf{j}_2$, is given by the vector addition of $\mathbf{j}_1$ and $\mathbf{j}_2$:

$$\mathbf{j} = \mathbf{j}_1 + \mathbf{j}_2 \tag{2.1}$$

Actually, it is a simple matter to show that the sum of two angular momenta is also an angular momentum in the quantum mechanical sense in that it satisfies the commutation rules of Eq. (1.16). For example

$$[j_x, j_y] = [j_{1x} + j_{2x}, j_{1y} + j_{2y}]$$

$$= [j_{1x}, j_{1y}] + [j_{1x}, j_{2y}] + [j_{2x}, j_{1y}] + [j_{2x}, j_{2y}]$$

$$= ij_{1z} + 0 + 0 + ij_{2z} = ij_z \tag{2.2}$$

where the middle two commutators vanish because angular momenta in different spaces commute.

There are two useful points of view in describing such a compound system. One complete set of commuting angular momentum operators is $\mathbf{j}_1^2$, j_{1z}, $\mathbf{j}_2^2$, and j_{2z}. The states $|j_1 m_1, j_2 m_2\rangle \equiv |j_1 m_1\rangle |j_2 m_2\rangle$ are simultaneous eigenfunctions of these operators:

$$\mathbf{j}_1^2 |j_1 m_1, j_2 m_2\rangle = j_1(j_1 + 1) |j_1 m_1, j_2 m_2\rangle$$

$$j_{1z} |j_1 m_1, j_2 m_2\rangle = m_1 |j_1 m_1, j_2 m_2\rangle$$

$$\mathbf{j}_2^2 |j_1 m_1, j_2 m_2\rangle = j_2(j_2 + 1) |j_1 m_1, j_2 m_2\rangle$$

$$j_{2z} |j_1 m_1, j_2 m_2\rangle = m_2 |j_1 m_1, j_2 m_2\rangle \tag{2.3}$$

Here the $|j_1 m_1, j_2 m_2\rangle$ states are called the *uncoupled representation* and span a space of dimension $(2j_1 + 1)(2j_2 + 1)$.

It is easily verified that another complete set of commuting angular momentum operators is $\mathbf{j}_1^2$, $\mathbf{j}_2^2$, $\mathbf{j}^2 = (\mathbf{j}_1 + \mathbf{j}_2)^2$, and $j_z = j_{1z} + j_{2z}$. The states $|jm\rangle$, which are simultaneously eigenfunctions of these operators, are called the *coupled representation* and span a space of dimension $2j + 1$ for each j value. It is also common to write $|j_1 j_2 jm\rangle$ or $|(j_1 j_2)jm\rangle$ for $|jm\rangle$, but we prefer to use in general the more compact notation. The relations corresponding to Eq. (2.3) are

$$\mathbf{j}_1^2 |jm\rangle = j_1(j_1 + 1) |jm\rangle$$

$$\mathbf{j}_2^2 |jm\rangle = j_2(j_2 + 1) |jm\rangle$$

$$\mathbf{j}^2 |jm\rangle = j(j + 1) |jm\rangle$$

$$j_z |jm\rangle = m |jm\rangle \tag{2.4}$$

Note that each set of commuting angular momentum operators contains the same number of observables. Hence these two descriptions are equivalent and the two representations are connected by a unitary transformation[1]

$$|jm\rangle = \sum_{m_1, m_2} C(j_1 j_2 j; m_1 m_2 m) |j_1 m_1, j_2 m_2\rangle \tag{2.5}$$

or by the inverse transformation

$$|j_1 m_1, j_2 m_2\rangle = \sum_{j, m} C(j_1 j_2 j; m_1 m_2 m) |jm\rangle \tag{2.6}$$

where the elements of the unitary transformation, called *Clebsch–Gordan coefficients*, are chosen to be real. We make the formal identification

$$C(j_1 j_2 j; m_1 m_2 m) \equiv \langle j_1 m_1, j_2 m_2 | jm\rangle \equiv \langle jm | j_1 m_1, j_2 m_2\rangle \tag{2.7}$$

The goal of the remainder of this section as well as the next two sections is to elucidate the nature of the Clebsch–Gordan coefficients, which are also referred to in the literature as *vector coupling coefficients*, *vector addition coefficients*, and *Wigner coefficients*.

TABLE 2.1   Matrix of Clebsch–Gordan Coefficients[a]

	⟨3,1|	⟨3,0|	⟨2,1|	⋯	⟨−2,−1|	⟨−3,0|	⟨−3,−1|
|4 4⟩	⟨3,1|4 4⟩	0	0				
|4 3⟩	0	⟨3,0|4 3⟩	⟨2,1|4 3⟩				
|3 3⟩	0	⟨3,0|3 3⟩	⟨2,1|3 3⟩				
⋮	0	0	0	⋱	0	0	0
|3 −3⟩					0 ⟨−2,−1|3 −3⟩	⟨−3,0|3 −3⟩	0
|4 −3⟩					0 ⟨−2,−1|4 −3⟩	⟨−3,0|4 −3⟩	0
|4 −4⟩					0	0	0 ⟨−3,−1|4 −4⟩

[a] The $\langle j_1 m_1, j_2 m_2 | j m \rangle$ symbol is abbreviated here as $\langle m_1, m_2 | j m \rangle$. The case $j_1 = 3, j_2 = 1$ is illustrated. In general, $\mathbf{C}$ is a $(2j_1 + 1)(2j_2 + 1) \times (2j_1 + 1)(2j_2 + 1)$ unitary (orthogonal) matrix of (real) elements. Equations (2.8) and (2.9) express the relation $\mathbf{CC}^T = \mathbf{I}$ where $\mathbf{C}^T = \mathbf{C}^{-1}$. Most of the elements vanish because the triangle condition $m = m_1 + m_2$ is not satisfied. By grouping elements with the same m value, $\mathbf{C}$ is put into block diagonal form, as above.

The orthonormality of $|jm\rangle$ and $|j_1 m_1, j_2 m_2\rangle$ leads to the orthogonality relations

$$\sum_{m_1, m_2} \langle jm | j_1 m_1, j_2 m_2 \rangle \langle j_1 m_1, j_2 m_2 | j'm' \rangle = \delta_{j,j'} \delta_{m,m'} \tag{2.8}$$

and

$$\sum_{j,m} \langle j_1 m_1, j_2 m_2 | jm \rangle \langle jm | j_1 m_1', j_2 m_2' \rangle = \delta_{m_1, m_1'} \delta_{m_2, m_2'} \tag{2.9}$$

where we have chosen to use the bracket notation [see Eq. (2.7)] for the Clebsch–Gordan coefficients[2]. These orthogonality relations express the unitary nature of the matrix of Clebsch–Gordan coefficients (see Table 2.1), namely, the scalar product of any two column vectors or row vectors of this matrix vanishes and the scalar product of any vector with itself is unity.

The Clebsch–Gordan coefficients vanish unless the so-called triangle condition is satisfied, namely

$$m = m_1 + m_2 \tag{2.10}$$

and

$$|j_1 + j_2| \geq j \geq |j_1 - j_2| \tag{2.11}$$

that is, when coupling the angular momentum states $|j_1 m_1\rangle$ and $|j_2 m_2\rangle$ the magnetic quantum numbers m_1 and m_2 add algebraically while the angular momenta $\mathbf{j}_1$ and $\mathbf{j}_2$ add vectorially.

The derivation of Eq. (2.10) is straightforward. We apply $j_z = j_{1z} + j_{2z}$ to Eq. (2.5):

$$m|jm\rangle = \sum_{m_1,m_2} (m_1 + m_2) \langle j_1 m_1, j_2 m_2 | jm \rangle |j_1 m_1, j_2 m_2 \rangle$$

$$= \sum_{m_1',m_2'} m \langle j_1 m_1', j_2 m_2' | jm \rangle |j_1 m_1', j_2 m_2' \rangle \qquad (2.12)$$

By equating the coefficients of like terms of $|j_1 m_1, j_2 m_2\rangle$ we obtain the condition

$$(m - m_1 - m_2) \langle j_1 m_1, j_2 m_2 | jm \rangle = 0 \qquad (2.13)$$

It follows that either $m = m_1 + m_2$ or $\langle j_1 m_1, j_2 m_2 | jm \rangle$ vanishes. As a consequence of Eq. (2.10), the sum over m_1 and m_2 in Eq. (2.5) is constrained by the condition $m_1 + m_2 = m$ so that one of the two indices is redundant. Similarly, the sum over m in Eqs. (2.6) and (2.9) is also superfluous.

The derivation of Eq. (2.11) is more involved and often ignored in elementary texts. However, the ideas presented in this proof are useful in counting the states of atoms and diatomic molecules. We know that for a given j, $-j \le m \le j$ and $m_{max} = j$. Since $m = m_1 + m_2$, the maximum value of m for all j is $m_{max} = j_1 + j_2$, that is, the maximum value of $m_1 + m_2$. This must also be the maximum value of j, that is,

$$j_{max} = j_1 + j_2 \qquad (2.14)$$

for otherwise there would be values of m larger than m_{max}. Consider how the coupled representation $|jm\rangle$ is related to the uncoupled representation $|j_1 m_1, j_2 m_2\rangle$ (see Table 2.2). For the coupled states with $m = m_{max}$, there is only one such state $|j_{max} = j_1 + j_2, m_{max} = j_1 + j_2\rangle$, and it gives rise to only one uncoupled state $|j_1 j_1, j_2 j_2\rangle$. The next smaller m value is $j_1 + j_2 - 1$, and it occurs only in the coupled states $|j_1 + j_2, j_1 + j_2 - 1\rangle$ and $|j_1 + j_2 - 1, j_1 + j_2 - 1\rangle$. Combinations of these two coupled states give rise to the uncoupled states $|j_1 j_1 - 1, j_2 j_2\rangle$ and $|j_1 j_1, j_2 j_2 - 1\rangle$. Similarly, linear combinations of the three coupled states with $m = j_1 + j_2 - 2$, namely, $|j_1 + j_2, j_1 + j_2 - 2\rangle$, $|j_1 + j_2 - 1, j_1 + j_2 - 2\rangle$, and $|j_1 + j_2 - 2, j_1 + j_2 - 2\rangle$, give the three uncoupled states $|j_1 j_1, j_2 j_2 - 2\rangle$, $|j_1 j_1 - 1, j_2 j_2 - 1\rangle$, and $|j_1 j_1 - 2, j_2 j_2\rangle$. Thus each time m is reduced by one unit, one of the coupled states that occurs belongs to a j value of one less unit, and the m value of this state is at its maximum value. The lower limit j_{min} is reached when the number of coupled states (of which there are $2j + 1$ for each j value) has been matched against all the uncoupled states [of which there are $(2j_1 + 1)(2j_2 + 1)$], that is,

$$\sum_{j_{min}}^{j_{max}} (2j + 1) = (2j_1 + 1)(2j_2 + 1) \qquad (2.15)$$

TABLE 2.2   Relation between Coupled and Uncoupled Representations ($j_1 \geq j_2$)

Coupled Representation	Uncoupled Representation
$m = m_{max}$	
$\lvert j_1 + j_2, j_1 + j_2 \rangle$	$\lvert j_1 j_1, j_2 j_2 \rangle$
$m = m_{max} - 1$	
$\lvert j_1 + j_2, j_1 + j_2 - 1 \rangle$	$\lvert j_1 j_1, j_2 j_2 - 1 \rangle$
$\lvert j_1 + j_2 - 1, j_1 + j_2 - 1 \rangle$	$\lvert j_1 j_1 - 1, j_2 j_2 \rangle$
$m = m_{max} - 2$	
$\lvert j_1 + j_2, j_1 + j_2 - 2 \rangle$	$\lvert j_1 j_1, j_2 j_2 - 2 \rangle$
$\lvert j_1 + j_2 - 1, j_1 + j_2 - 2 \rangle$	$\lvert j_1 j_1 - 1, j_2 j_2 - 1 \rangle$
$\lvert j_1 + j_2 - 2, j_1 + j_2 - 2 \rangle$	$\lvert j_1 j_1 - 2, j_2 j_2 \rangle$
$m = m_{max} - 3$	
$\lvert j_1 + j_2, j_1 + j_2 - 3 \rangle$	$\lvert j_1 j_1, j_2 j_2 - 3 \rangle$
$\lvert j_1 + j_2 - 1, j_1 + j_2 - 3 \rangle$	$\lvert j_1 j_1 - 1, j_2 j_2 - 2 \rangle$
$\lvert j_1 + j_2 - 2, j_1 + j_2 - 3 \rangle$	$\lvert j_1 j_1 - 2, j_2 j_2 - 1 \rangle$
$\lvert j_1 + j_2 - 3, j_1 + j_2 - 3 \rangle$	$\lvert j_1 j_1 - 3, j_2 j_2 \rangle$
$m = m_{max} - 4$	
$\vdots$	$\vdots$

To evaluate Eq. (2.15) we use the following identities concerning the sums of integers or half-integers: For integral j

$$\sum_{j_{min}}^{j_{max}} j = \sum_{1}^{j_{max}} j - \sum_{1}^{j_{min}-1} j$$

$$= \tfrac{1}{2} j_{max}(j_{max} + 1) - \tfrac{1}{2}(j_{min} - 1) j_{min}$$

$$= \tfrac{1}{2} [\, j_{max}(j_{max} + 1) - j_{min}(j_{min} - 1)\,] \qquad (2.16)$$

The same result is readily shown to hold also for half-integral j:

$$\sum_{j_{min}}^{j_{max}} j = \sum_{1/2}^{j_{max}} j - \sum_{1/2}^{j_{min}-1} j$$

$$= \left[\tfrac{1}{2} j_{max}(j_{max} + 1) + \tfrac{1}{8} \right] - \left[\tfrac{1}{2}(j_{min} - 1)j_{min} + \tfrac{1}{8} \right]$$

$$= \tfrac{1}{2} \left[j_{max}(j_{max} + 1) - j_{min}(j_{min} - 1) \right] \qquad (2.17)$$

We also need the sum

$$\sum_{j_{min}}^{j_{max}} 1 = j_{max} - (j_{min} - 1) = j_{max} - j_{min} + 1 \qquad (2.18)$$

which also is valid for either integral or half-integral j. Equation (2.15) becomes

$$j_{max}(j_{max} + 1) - j_{min}(j_{min} - 1) + (j_{max} - j_{min} + 1) = (2j_1 + 1)(2j_2 + 1) \quad (2.19)$$

Substitution of $j_1 + j_2$ for j_{max} yields

$$j_{min}^2 = j_1^2 + j_2^2 - 2j_1 j_2 = (j_1 - j_2)^2 \qquad (2.20)$$

Since $j_{min} \geq 0$, it follows that

$$j_{min} = |j_1 - j_2| \qquad (2.21)$$

Equation (2.21), together with Eq. (2.14), establishes Eq. (2.11) for integral or half-integral j[3].

2.2 CLEBSCH–GORDAN COEFFICIENTS AND 3-j SYMBOLS: SYMMETRY PROPERTIES AND EXPLICIT VALUES

Recursion relations between the Clebsch–Gordan coefficients may be derived by applying the operator identity $j_\pm = j_{1\pm} + j_{2\pm}$ to both sides of Eq. (2.5). With the use of Eqs. (1.17) and (1.28) the left-hand side of Eq. (2.5) becomes

$$j_\pm |jm\rangle = [j(j + 1) - m(m \pm 1)]^{\frac{1}{2}} |jm \pm 1\rangle$$

$$= [j(j + 1) - m(m \pm 1)]^{\frac{1}{2}}$$

$$\times \sum_{m_1', m_2'} \langle j_1 m_1', j_2 m_2' | jm \pm 1 \rangle |j_1 m_1', j_2 m_2'\rangle \qquad (2.22)$$

while the right-hand side becomes

$$(j_{1\pm} + j_{2\pm}) \sum_{m_1, m_2} \langle j_1 m_1, j_2 m_2 | jm \rangle |j_1 m_1, j_2 m_2\rangle$$

$$= \sum_{m_1,m_2} \left\{ [j_1(j_1 + 1) - m_1(m_1 \pm 1)]^{\frac{1}{2}} \langle j_1 m_1, j_2 m_2 | jm \rangle | j_1 m_1 \pm 1, j_2 m_2 \rangle \right.$$

$$\left. + [j_2(j_2 + 1) - m_2(m_2 \pm 1)]^{\frac{1}{2}} \langle j_1 m_1, j_2 m_2 | jm \rangle | j_1 m_1, j_2 m_2 \pm 1 \rangle \right\}$$

$$(2.23)$$

By equating like coefficients of $|j_1 m_1, j_2 m_2 \rangle$ in the preceding two expansions [in Eqs. (2.22) and (2.23)], we obtain the relation

$$[j(j + 1) - m(m \pm 1)]^{\frac{1}{2}} \langle j_1 m_1, j_2 m_2 | jm \pm 1 \rangle$$

$$= [j_1(j_1 + 1) - m_1(m_1 \mp 1)]^{\frac{1}{2}} \langle j_1 m_1 \mp 1, j_2 m_2 | jm \rangle$$

$$+ [j_2(j_2 + 1) - m_2(m_2 \mp 1)]^{\frac{1}{2}} \langle j_1 m_1, j_2 m_2 \mp 1 | jm \rangle \quad (2.24)$$

The left-hand side of Eq. (2.24) vanishes for $m = j$ if we take the upper sign; this permits us to determine the various $\langle j_1 m_1, j_2 m_2 | jj \rangle$ subject to the orthonormality conditions given in Eqs. (2.8) and (2.9) and subject to an overall phase. The latter may be fixed by the convention that $\langle j_1 j_1, j_2 j - j_1 | jj \rangle$ is always real and positive[4]. The lower sign in Eq. (2.24) then relates $\langle j_1 m_1, j_2 m_2 | jj - 1 \rangle$ to the $\langle j_1 m_1', j_2 m_2' | jj \rangle$ so that one can work out all the Clebsch–Gordan coefficients by starting with $m = j$. Equation (2.24), which is based on the phase convention Eq. (1.28), implies that all the Clebsch–Gordan coefficients are real.

In this manner explicit expressions for the Clebsch–Gordan coefficients may be derived. In particular, after much algebra Racah[4] showed that

$$\langle j_1 m_1, j_2 m_2 | j_3 m_3 \rangle$$

$$= \delta_{m_1+m_2,m_3} \left[(2j_3 + 1) \frac{(s - 2j_3)!(s - 2j_2)!(s - 2j_1)!}{(s + 1)!} \right.$$

$$\left. \times (j_1 + m_1)!(j_1 - m_1)!(j_2 + m_2)!(j_2 - m_2)!(j_3 + m_3)!(j_3 - m_3)! \right]^{\frac{1}{2}}$$

$$\times \sum_{\nu} (-1)^{\nu} / [\nu!(j_1 + j_2 - j_3 - \nu)!(j_1 - m_1 - \nu)!(j_2 + m_2 - \nu)!$$

$$\times (j_3 - j_2 + m_1 + \nu)!(j_3 - j_1 - m_2 + \nu)!] \quad (2.25)$$

where $s = j_1 + j_2 + j_3$ and the index ν ranges over all integral values for which the factorial arguments are nonnegative[5].

From Eq. (2.25) the following symmetry relations are readily verified:

$$\langle j_1 m_1, j_2 m_2 | j_3 m_3 \rangle = (-1)^{j_1+j_2-j_3} \langle j_1 -m_1, j_2 -m_2 | j_3 -m_3 \rangle$$

$$= (-1)^{j_1+j_2-j_3} \langle j_2 m_2, j_1 m_1 | j_3 m_3 \rangle$$

$$= (-1)^{j_1-m_1} \left[\frac{2 j_3 + 1}{2 j_2 + 1} \right]^{\frac{1}{2}} \langle j_1 m_1, j_3 -m_3 | j_2 -m_2 \rangle$$

$$= (-1)^{j_2+m_2} \left[\frac{2 j_3 + 1}{2 j_1 + 1} \right]^{\frac{1}{2}} \langle j_3 -m_3, j_2 m_2 | j_1 -m_1 \rangle$$

$$= (-1)^{j_1-m_1} \left[\frac{2 j_3 + 1}{2 j_2 + 1} \right]^{\frac{1}{2}} \langle j_3 m_3, j_1 -m_1 | j_2 m_2 \rangle$$

$$= (-1)^{j_2+m_2} \left[\frac{2 j_3 + 1}{2 j_1 + 1} \right]^{\frac{1}{2}} \langle j_2 -m_2, j_3 m_3 | j_1 m_1 \rangle$$

$$(2.26)$$

The relations given in Eq. (2.26) take on their most symmetric form if we rewrite the Clebsch–Gordan coefficients as Wigner 3-j symbols, defined by

$$\begin{pmatrix} j_1 & j_2 & j_3 \\ m_1 & m_2 & m_3 \end{pmatrix} \equiv (-1)^{j_1-j_2-m_3} (2 j_3 + 1)^{-\frac{1}{2}} \langle j_1 m_1, j_2 m_2 | j_3 -m_3 \rangle \quad (2.27)$$

$$\langle j_1 m_1, j_2 m_2 | j_3 m_3 \rangle \equiv (-1)^{j_1-j_2+m_3} (2 j_3 + 1)^{\frac{1}{2}} \begin{pmatrix} j_1 & j_2 & j_3 \\ m_1 & m_2 & -m_3 \end{pmatrix} \quad (2.28)$$

Note particularly the minus sign appearing in the Clebsch–Gordan coefficient in Eq. (2.27) and the minus sign appearing in the 3-j symbol in Eq. (2.28). The 3-j symbols have the property that an even permutation of the columns leaves the numerical value unchanged

$$\begin{pmatrix} j_1 & j_2 & j_3 \\ m_1 & m_2 & m_3 \end{pmatrix} = \begin{pmatrix} j_2 & j_3 & j_1 \\ m_2 & m_3 & m_1 \end{pmatrix} = \begin{pmatrix} j_3 & j_1 & j_2 \\ m_3 & m_1 & m_2 \end{pmatrix} \quad (2.29)$$

while an odd permutation is equivalent to multiplication by $(-1)^{j_1+j_2+j_3}$

$$\begin{pmatrix} j_1 & j_2 & j_3 \\ m_1 & m_2 & m_3 \end{pmatrix} = (-1)^{j_1+j_2+j_3} \begin{pmatrix} j_2 & j_1 & j_3 \\ m_2 & m_1 & m_3 \end{pmatrix}$$

$$= (-1)^{j_1+j_2+j_3} \begin{pmatrix} j_1 & j_3 & j_2 \\ m_1 & m_3 & m_2 \end{pmatrix}$$

$$= (-1)^{j_1+j_2+j_3} \begin{pmatrix} j_3 & j_2 & j_1 \\ m_3 & m_2 & m_1 \end{pmatrix} \quad (2.30)$$

as is the replacement of the bottom row by the negative of all its arguments:

$$\begin{pmatrix} j_1 & j_2 & j_3 \\ m_1 & m_2 & m_3 \end{pmatrix} = (-1)^{j_1+j_2+j_3} \begin{pmatrix} j_1 & j_2 & j_3 \\ -m_1 & -m_2 & -m_3 \end{pmatrix} \qquad (2.31)$$

The orthogonality properties of the 3-j symbols are not as convenient as that of the Clebsch–Gordan coefficients:

$$\sum_{m_1,m_2} \begin{pmatrix} j_1 & j_2 & j_3 \\ m_1 & m_2 & m_3 \end{pmatrix} \begin{pmatrix} j_1 & j_2 & j_3' \\ m_1 & m_2 & m_3' \end{pmatrix} = (2j_3 + 1)^{-1} \delta_{j_3,j_3'} \delta_{m_3,m_3'} \qquad (2.32)$$

and

$$\sum_{j_3,m_3} (2j_3 + 1) \begin{pmatrix} j_1 & j_2 & j_3 \\ m_1 & m_2 & m_3 \end{pmatrix} \begin{pmatrix} j_1 & j_2 & j_3 \\ m_1' & m_2' & m_3 \end{pmatrix} = \delta_{m_1,m_1'} \delta_{m_2,m_2'} \qquad (2.33)$$

Nevertheless, the 3-j symbols are extensively used.

The monograph, *the 3-j and 6-j symbols*, by Rotenberg et al.[6], permits the numerical evaluation of all 3-j coefficients in steps of $\frac{1}{2}$ up to any argument equal to 8. Attention is also called to the existence of efficient algorithms for the exact[7] and semiclassical[8] evaluation of 3-j symbols. These are particularly useful for large quantum number arguments. In Tables 2.4 and 2.5 we list for reference purposes algebraic expressions for the Clebsch–Gordan coefficients and 3-j symbols when one of the angular momenta is 0, $\frac{1}{2}$, 1, $\frac{3}{2}$ or 2. We will have reason to make frequent use of these tables[9] since they give alegebraic expressions for almost all Clebsch–Gordan coefficients commonly encountered. However, often the numerical value is what matters in many practical problems, and for this purpose we list in the Appendix a simple (unoptimized) FORTRAN program for evaluating 3-j symbols.

2.3   CLEBSCH–GORDAN COEFFICIENTS AND 3-j SYMBOLS: GEOMETRIC INTERPRETATION

We look to the vector model as a means of providing an interpretation of the Clebsch–Gordan coefficients and 3-j symbols. According to this picture the coupled state $|jm\rangle$ is represented by two vectors $\mathbf{j}_1$ and $\mathbf{j}_2$ precessing in phase about their resultant $\mathbf{j}$, which, in turn, precesses about the z axis (Figure 2.1). The precession of $\mathbf{j}_1$ and $\mathbf{j}_2$ about $\mathbf{j}$ introduces an indeterminancy in their z components, m_1 and m_2, although at every instant $m = m_1 + m_2$ is fixed. The Clebsch–Gordan coefficient $\langle j_1 m_1, j_2 m_2 | jm \rangle$ thus represents the probability amplitude that the coupled state $|jm\rangle$ will be found having its component parts $\mathbf{j}_1$ and $\mathbf{j}_2$ making the projection m_1 and m_2 for j_{1z} and j_{2z}, respectively. The corresponding 3-j symbol is the probability amplitude divided by the factor $(2j + 1)^{1/2}$.

In terms of the vector model, the uncoupled state $|j_1 m_1, j_2 m_2\rangle$ is represented by two vectors $\mathbf{j}_1$ and $\mathbf{j}_2$ precessing independently about the z axis and making the

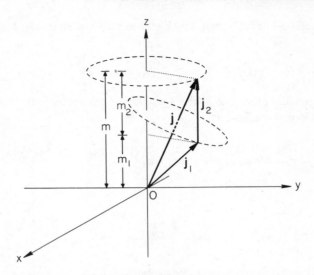

FIGURE 2.1　Vector model representation of the coupled state $|jm\rangle$.

projections m_1 and m_2, respectively (Figure 2.2). The Clebsch–Gordan coefficient $\langle j_1 m_1, j_2 m_2 | jm \rangle$ can thus be interpreted as giving the probability amplitude that the uncoupled state $|j_1 m_1, j_2 m_2\rangle$ at any instant couples together to form a resultant state $|jm\rangle$ of length $[j(j+1)]^{1/2}$. The squares of the Clebsch–Gordan coefficients represent probabilities and as such are proper fractions whose values range from 0 to 1.

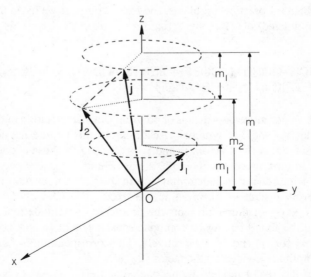

FIGURE 2.2　Vector model representation of the uncoupled state $|j_1 m_1, j_2 m_2\rangle$.

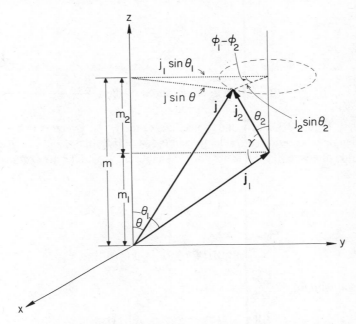

FIGURE 2.3 Alternative vector model representation of the uncoupled state $|j_1 m_1, j_2 m_2\rangle$.

It is instructive to use the vector model to approximate the value of $\langle j_1 m_1, j_2 m_2 | j m \rangle^2$. Let us return to Figure 2.2. Because of the cylindrical symmetry of the diagram as a whole, we may fix the orientation of $\mathbf{j}_1$ arbitrarily and consider $\mathbf{j}_2$ to precess uniformly about the z axis, as shown in Figure 2.3. We denote the polar angles of $\mathbf{j}_1$ and $\mathbf{j}_2$ by (θ_1, ϕ_1) and (θ_2, ϕ_2), respectively. Let us denote by γ the included angle between $\mathbf{j}_1$ and $\mathbf{j}_2$. Then

$$\mathbf{j}_1 \cdot \mathbf{j}_2 = |\mathbf{j}_1||\mathbf{j}_2| \cos \gamma$$

$$= |\mathbf{j}_1||\mathbf{j}_2| \left(\cos \theta_1 \cos \theta_2 + \sin \theta_1 \sin \theta_2 \cos(\phi_1 - \phi_2) \right) \qquad (2.34)$$

where $\phi_1 - \phi_2$ is the dihedral angle between the $j_1 z$ and $j_2 z$ planes. It is precisely this angle that changes uniformly as $\mathbf{j}_2$ precesses about the z axis.

Let us differentiate Eq. (2.34) with respect to time to obtain the rate of change of the length of $\mathbf{j}$. From the law of cosines

$$|\mathbf{j}|^2 = |\mathbf{j}_1|^2 + |\mathbf{j}_2|^2 - 2|\mathbf{j}_1||\mathbf{j}_2| \cos \gamma$$

$$= |\mathbf{j}_1|^2 + |\mathbf{j}_2|^2 - 2\mathbf{j}_1 \cdot \mathbf{j}_2 \qquad (2.35)$$

so that

$$\frac{d}{dt}(\mathbf{j}_1 \cdot \mathbf{j}_2) = -\frac{1}{2} \frac{d}{dt} \left[|\mathbf{j}|^2 - |\mathbf{j}_1|^2 - |\mathbf{j}_2|^2 \right]$$

$$= -|\mathbf{j}|\frac{d|\mathbf{j}|}{dt} \tag{2.36}$$

Hence

$$|\mathbf{j}|\frac{d|\mathbf{j}|}{dt} = |\mathbf{j}_1||\mathbf{j}_2|\sin\theta_1\sin\theta_2\sin(\phi_1 - \phi_2)\frac{d(\phi_1 - \phi_2)}{dt} \tag{2.37}$$

However, as Figure 2.3 shows, the projections of the $\mathbf{j}_1, \mathbf{j}_2, \mathbf{j}$ vectors in the xy plane form a triangle whose sides are $s_1 = |\mathbf{j}_1|\sin\theta_1$, $s_2 = |\mathbf{j}_2|\sin\theta_2$, and $s = |\mathbf{j}|\sin\theta$ and whose angle opposite $|\mathbf{j}|\sin\theta$ is the dihedral angle $\phi_1 - \phi_2$. The area of this triangle is given by

$$A_{xy}(\mathbf{j}_1, \mathbf{j}_2, \mathbf{j}) = \frac{1}{2}|s_1 \times s_2|$$

$$= \frac{1}{2}|\mathbf{j}_1|\sin\theta_1|\mathbf{j}_2|\sin\theta_2\sin(\phi_1 - \phi_2) \tag{2.38}$$

Comparison with Eq. (2.37) shows that

$$\frac{d|\mathbf{j}|}{dt} = 2\frac{d(\phi_1 - \phi_2)}{dt}\frac{A_{xy}(\mathbf{j}_1, \mathbf{j}_2, \mathbf{j})}{|\mathbf{j}|} \tag{2.39}$$

We normalize the rate by setting $d(\phi_1 - \phi_2)/dt = 2\pi$, that is, one revolution per unit time. Then the probability density $P(j)$ that the resultant of $\mathbf{j}_1$ and $\mathbf{j}_2$ is a vector of length $|\mathbf{j}|$ is given by the dwell time, that is, by $2(d|\mathbf{j}|/dt)^{-1}$ where the factor of two arises because $|\mathbf{j}|$ occurs twice as the dihedral angle $\phi_1 - \phi_2$ runs from 0 to 2π:

$$P(j) = \frac{|\mathbf{j}|}{2\pi A_{xy}(\mathbf{j}_1, \mathbf{j}_2, \mathbf{j})} \tag{2.40}$$

Hence, according to the vector model[10], $\langle j_1 m_1, j_2 m_2|j m\rangle^2$ is proportional to $|\mathbf{j}|/\pi$ and inversely proportional to twice the area of the $(\mathbf{j}_1, \mathbf{j}_2, \mathbf{j})$ triangle projected onto the xy plane. This may be expressed in terms of the square of the corresponding 3-j symbol as

$$\begin{pmatrix} j_1 & j_2 & j \\ m_1 & m_2 & -m \end{pmatrix}^2 = \frac{|\mathbf{j}|}{2\pi(2j+1)A_{xy}(\mathbf{j}_1, \mathbf{j}_2, \mathbf{j})}$$

$$\simeq \frac{1}{4\pi A_{xy}(\mathbf{j}_1, \mathbf{j}_2, \mathbf{j})} \tag{2.41}$$

where $|\mathbf{j}|$ has been approximated by $j + \frac{1}{2}$ for large j values. Equation (2.41) gives a geometric interpretation of the square of the 3-j symbol in the limit of large angular momenta.

Let us make Eq. (2.41) quantitative by referring again to Figure 2.3. From the law of cosines

$$\sin(\phi_1 - \phi_2) = \left[1 - \cos^2(\phi_1 - \phi_2)\right]^{\frac{1}{2}}$$

$$= \left[1 - \left(\frac{|\mathbf{j}|^2 \sin^2 \theta - |\mathbf{j_1}|^2 \sin^2 \theta_1 - |\mathbf{j_2}|^2 \sin^2 \theta_2}{2|\mathbf{j_1}| \sin \theta_1 |\mathbf{j_2}| \sin \theta_2}\right)^2\right]^{\frac{1}{2}} \qquad (2.42)$$

Substitution of Eq. (2.42) into Eq. (2.38) yields for the area the expression

$$A_{xy}(\mathbf{j_1}, \mathbf{j_2}, \mathbf{j}) = \tfrac{1}{4}\left[(a + b + c)^2 - 2(a^2 + b^2 + c^2)\right]^{\frac{1}{2}} \qquad (2.43)$$

where

$$a = j_1(j_1 + 1) - m_1^2$$

$$b = j_2(j_2 + 1) - m_2^2$$

$$c = j(j + 1) - m^2 \qquad (2.44)$$

so that Eq. (2.41) becomes

$$\begin{pmatrix} j_1 & j_2 & j \\ m_1 & m_2 & -m \end{pmatrix}^2 \simeq \frac{1}{\pi\left[(a + b + c)^2 - 2(a^2 + b^2 + c^2)\right]^{\frac{1}{2}}} \qquad (2.45)$$

Note that when $(a + b + c)^2 - 2(a^2 + b^2 + c^2)$ is negative, the 3-j symbol (and corresponding Clebsch–Gordan coefficient) is classically forbidden and the classical probability is zero.

Table 2.3 is a numerical test of the validity of Eq. (2.45). The behavior is typical. Equation (2.45) does not approximate well the squares of individual 3-j symbols, but it does provide an approximation for the average taken over a few neighboring values[11]. The reason for this is that the 3-j symbols and corresponding Clebsch–Gordan coefficients exhibit the usual quantum behavior, namely, rapid oscillations in the region classically allowed by the vector model and exponential decrease in the classically forbidden region. Nevertheless, the vector model does provide a good insight into the nature of 3-j symbols and Clebsch–Gordan coefficients. Regarded in this manner, these hostile-appearing symbols bristling with six different arguments may even turn out to be friendly companions in this journey through angular momentum land.

TABLE 2.3

Square of 3-j Symbol	Exact Value	Vector Model Estimate
$\begin{pmatrix} 4 & 4 & 8 \\ 1 & 0 & -1 \end{pmatrix}^2$	0.0202	0.0143
$\begin{pmatrix} 4 & 4 & 7 \\ 1 & 0 & -1 \end{pmatrix}^2$	0.0054	0.0090
$\begin{pmatrix} 4 & 4 & 6 \\ 1 & 0 & -1 \end{pmatrix}^2$	0.0082	0.0082
$\begin{pmatrix} 4 & 4 & 5 \\ 1 & 0 & -1 \end{pmatrix}^2$	0.0105	0.0084
$\begin{pmatrix} 4 & 4 & 4 \\ 1 & 0 & -1 \end{pmatrix}^2$	0.0045	0.0095
$\begin{pmatrix} 4 & 4 & 3 \\ 1 & 0 & -1 \end{pmatrix}^2$	0.0195	0.0117
$\begin{pmatrix} 4 & 4 & 2 \\ 1 & 0 & -1 \end{pmatrix}^2$	0.0022	0.0167
$\begin{pmatrix} 4 & 4 & 1 \\ 1 & 0 & -1 \end{pmatrix}^2$	0.0556	0.0365

TABLE 2.4   Algebraic Expressions for Some Commonly Occurring Clebsch–Gordan Coefficients $\langle j_1 m_1, j_2 m_2 | jm \rangle^a$

$$\text{A. } j_2 = 0$$

$$\langle j_1 \; m_1, 0 \; 0 | j \; m \rangle \qquad\qquad = \delta_{m_1 m} \delta_{j_1 j}$$

$$\text{B. } j_2 = \tfrac{1}{2}$$

$$\langle j_1 \; m - \tfrac{1}{2}, \tfrac{1}{2} \; \tfrac{1}{2} | j_1 + \tfrac{1}{2} \; m \rangle \;\; = \left[\frac{(j_1 + m + \tfrac{1}{2})}{(2j_1 + 1)} \right]^{\frac{1}{2}}$$

$$\langle j_1 \; m - \tfrac{1}{2}, \tfrac{1}{2} \; \tfrac{1}{2} | j_1 - \tfrac{1}{2} \; m \rangle \;\; = - \left[\frac{(j_1 - m + \tfrac{1}{2})}{(2j_1 + 1)} \right]^{\frac{1}{2}}$$

$$\langle j_1 \; m + \tfrac{1}{2}, \tfrac{1}{2} \; -\tfrac{1}{2} | j_1 + \tfrac{1}{2} \; m \rangle = \left[\frac{(j_1 - m + \tfrac{1}{2})}{(2j_1 + 1)} \right]^{\frac{1}{2}}$$

$$\langle j_1 \; m + \tfrac{1}{2}, \tfrac{1}{2} \; -\tfrac{1}{2} | j_1 - \tfrac{1}{2} \; m \rangle = \left[\frac{(j_1 + m + \tfrac{1}{2})}{(2j_1 + 1)} \right]^{\frac{1}{2}}$$

$$\text{C. } j_2 = 1$$

$$\langle j_1 \; m - 1, 1 \; 1 | j_1 + 1 \; m \rangle \qquad = \left[\frac{(j_1 + m)(j_1 + m + 1)}{(2j_1 + 1)(2j_1 + 2)} \right]^{\frac{1}{2}}$$

$$\langle j_1 \; m - 1, 1 \; 1 | j_1 \; m \rangle \qquad = - \left[\frac{(j_1 + m)(j_1 - m + 1)}{2j_1(j_1 + 1)} \right]^{\frac{1}{2}}$$

$$\langle j_1 \; m - 1, 1 \; 1 | j_1 - 1 \; m \rangle \qquad = \left[\frac{(j_1 - m)(j_1 - m + 1)}{2j_1(2j_1 + 1)} \right]^{\frac{1}{2}}$$

$$\langle j_1 \; m, 1 \; 0 | j_1 + 1 \; m \rangle \qquad = \left[\frac{(j_1 - m + 1)(j_1 + m + 1)}{(2j_1 + 1)(j_1 + 1)} \right]^{\frac{1}{2}}$$

$$\langle j_1 \; m, 1 \; 0 | j_1 \; m \rangle \qquad = \frac{m}{[j_1(j_1 + 1)]^{\frac{1}{2}}}$$

$$\langle j_1 \; m, 1 \; 0 | j_1 - 1 \; m \rangle \qquad = - \left[\frac{(j_1 - m)(j_1 + m)}{j_1(2j_1 + 1)} \right]^{\frac{1}{2}}$$

$$\langle j_1 \; m + 1, 1 \; -1 | j_1 + 1 \; m \rangle \;\; = \left[\frac{(j_1 - m)(j_1 - m + 1)}{(2j_1 + 1)(2j_1 + 2)} \right]^{\frac{1}{2}}$$

$$\langle j_1 \; m + 1, 1 \; -1 | j_1 \; m \rangle \qquad = \left[\frac{(j_1 - m)(j_1 + m + 1)}{2j_1(j_1 + 1)} \right]^{\frac{1}{2}}$$

$$\langle j_1 \; m + 1, 1 \; -1 | j_1 - 1 \; m \rangle \;\; = \left[\frac{(j_1 + m + 1)(j_1 + m)}{2j_1(2j_1 + 1)} \right]^{\frac{1}{2}}$$

TABLE 2.4 (continued)

D. $j_2 = \frac{3}{2}$

$$\langle j_1 \; m - \tfrac{3}{2}, \tfrac{3}{2} \; \tfrac{3}{2} | j_1 + \tfrac{3}{2} \; m \rangle \;=\; \left[\frac{(j_1 + m - \tfrac{1}{2})(j_1 + m + \tfrac{1}{2})(j_1 + m + \tfrac{3}{2})}{(2j_1 + 1)(2j_1 + 2)(2j_1 + 3)} \right]^{\frac{1}{2}}$$

$$\langle j_1 \; m - \tfrac{3}{2}, \tfrac{3}{2} \; \tfrac{3}{2} | j_1 + \tfrac{1}{2} \; m \rangle \;=\; -\left[\frac{3(j_1 + m - \tfrac{1}{2})(j_1 + m + \tfrac{1}{2})(j_1 - m + \tfrac{3}{2})}{2j_1(2j_1 + 1)(2j_1 + 3)} \right]^{\frac{1}{2}}$$

$$\langle j_1 \; m - \tfrac{3}{2}, \tfrac{3}{2} \; \tfrac{3}{2} | j_1 - \tfrac{1}{2} \; m \rangle \;=\; \left[\frac{3(j_1 + m - \tfrac{1}{2})(j_1 - m + \tfrac{1}{2})(j_1 - m + \tfrac{3}{2})}{(2j_1 - 1)(2j_1 + 1)(2j_1 + 2)} \right]^{\frac{1}{2}}$$

$$\langle j_1 \; m - \tfrac{3}{2}, \tfrac{3}{2} \; \tfrac{3}{2} | j_1 - \tfrac{3}{2} \; m \rangle \;=\; -\left[\frac{(j_1 - m - \tfrac{1}{2})(j_1 - m + \tfrac{1}{2})(j_1 - m + \tfrac{3}{2})}{(2j_1 - 1)(2j_1)(2j_1 + 1)} \right]^{\frac{1}{2}}$$

$$\langle j_1 \; m - \tfrac{1}{2}, \tfrac{3}{2} \; \tfrac{1}{2} | j_1 + \tfrac{3}{2} \; m \rangle \;=\; \left[\frac{3(j_1 + m + \tfrac{1}{2})(j_1 + m + \tfrac{3}{2})(j_1 - m + \tfrac{3}{2})}{(2j_1 + 1)(2j_1 + 2)(2j_1 + 3)} \right]^{\frac{1}{2}}$$

$$\langle j_1 \; m - \tfrac{1}{2}, \tfrac{3}{2} \; \tfrac{1}{2} | j_1 + \tfrac{1}{2} \; m \rangle \;=\; -(j_1 - 3m + \tfrac{3}{2})\left[\frac{j_1 + m + \tfrac{1}{2}}{2j_1(2j_1 + 1)(2j_1 + 3)} \right]^{\frac{1}{2}}$$

$$\langle j_1 \; m - \tfrac{1}{2}, \tfrac{3}{2} \; \tfrac{1}{2} | j_1 - \tfrac{1}{2} \; m \rangle$$
$$=\; -(j_1 + 3m - \tfrac{1}{2})\left[\frac{j_1 - m + \tfrac{1}{2}}{(2j_1 - 1)(2j_1 + 1)(2j_1 + 2)} \right]^{\frac{1}{2}}$$

$$\langle j_1 \; m - \tfrac{1}{2}, \tfrac{3}{2} \; \tfrac{1}{2} | j_1 - \tfrac{3}{2} \; m \rangle \;=\; \left[\frac{3(j_1 + m - \tfrac{1}{2})(j_1 - m - \tfrac{1}{2})(j_1 - m + \tfrac{1}{2})}{(2j_1 - 1)(2j_1)(2j_1 + 1)} \right]^{\frac{1}{2}}$$

$$\langle j_1 \; m + \tfrac{1}{2}, \tfrac{3}{2} \; -\tfrac{1}{2} | j_1 + \tfrac{3}{2} \; m \rangle \;=\; \left[\frac{3(j_1 + m + \tfrac{3}{2})(j_1 - m + \tfrac{1}{2})(j_1 - m + \tfrac{3}{2})}{(2j_1 + 1)(2j_1 + 2)(2j_1 + 3)} \right]^{\frac{1}{2}}$$

$$\langle j_1 \; m + \tfrac{1}{2}, \tfrac{3}{2} \; -\tfrac{1}{2} | j_1 + \tfrac{1}{2} \; m \rangle \;=\; (j_1 + 3m + \tfrac{3}{2})\left[\frac{j_1 - m + \tfrac{1}{2}}{2j_1(2j_1 + 1)(2j_1 + 3)} \right]^{\frac{1}{2}}$$

$$\langle j_1 \; m + \tfrac{1}{2}, \tfrac{3}{2} \; -\tfrac{1}{2} | j_1 - \tfrac{1}{2} \; m \rangle$$
$$=\; -(j_1 - 3m - \tfrac{1}{2})\left[\frac{j_1 + m + \tfrac{1}{2}}{(2j_1 - 1)(2j_1 + 1)(2j_1 + 2)} \right]^{\frac{1}{2}}$$

$$\langle j_1 \; m + \tfrac{1}{2}, \tfrac{3}{2} \; -\tfrac{1}{2} | j_1 - \tfrac{3}{2} \; m \rangle \;=\; -\left[\frac{3(j_1 + m - \tfrac{1}{2})(j_1 + m + \tfrac{1}{2})(j_1 - m - \tfrac{1}{2})}{(2j_1 - 1)(2j_1)(2j_1 + 1)} \right]^{\frac{1}{2}}$$

TABLE 2.4 (continued)

$$\langle j_1 \; m + \tfrac{3}{2}, \tfrac{3}{2} \; -\tfrac{3}{2} | j_1 + \tfrac{3}{2} \; m \rangle = \left[\frac{(j_1 - m - \tfrac{1}{2})(j_1 - m + \tfrac{1}{2})(j_1 - m + \tfrac{3}{2})}{(2j_1 + 1)(2j_1 + 2)(2j_1 + 3)} \right]^{\tfrac{1}{2}}$$

$$\langle j_1 \; m + \tfrac{3}{2}, \tfrac{3}{2} \; -\tfrac{3}{2} | j_1 + \tfrac{1}{2} \; m \rangle = \left[\frac{3(j_1 + m + \tfrac{3}{2})(j_1 - m - \tfrac{1}{2})(j_1 - m + \tfrac{1}{2})}{2j_1(2j_1 + 1)(2j_1 + 3)} \right]^{\tfrac{1}{2}}$$

$$\langle j_1 \; m + \tfrac{3}{2}, \tfrac{3}{2} \; -\tfrac{3}{2} | j_1 - \tfrac{1}{2} \; m \rangle = \left[\frac{3(j_1 + m + \tfrac{1}{2})(j_1 + m + \tfrac{3}{2})(j_1 - m - \tfrac{1}{2})}{(2j_1 - 1)(2j_1 + 1)(2j_1 + 2)} \right]^{\tfrac{1}{2}}$$

$$\langle j_1 \; m + \tfrac{3}{2}, \tfrac{3}{2} \; -\tfrac{3}{2} | j_1 - \tfrac{3}{2} \; m \rangle = \left[\frac{(j_1 + m - \tfrac{1}{2})(j_1 + m + \tfrac{1}{2})(j_1 + m + \tfrac{3}{2})}{(2j_1 - 1)(2j_1)(2j_1 + 1)} \right]^{\tfrac{1}{2}}$$

E. $j_2 = 2$

$$\langle j_1 \; m - 2, 2 \; 2 | j_1 + 2 \; m \rangle$$
$$= \left[\frac{(j_1 + m - 1)(j_1 + m)(j_1 + m + 1)(j_1 + m + 2)}{(2j_1 + 1)(2j_1 + 2)(2j_1 + 3)(2j_1 + 4)} \right]^{\tfrac{1}{2}}$$

$$\langle j_1 \; m - 2, 2 \; 2 | j_1 + 1 \; m \rangle$$
$$= - \left[\frac{(j_1 + m - 1)(j_1 + m)(j_1 + m + 1)(j_1 - m + 2)}{(j_1 + 1)(j_1 + 2)(2j_1)(2j_1 + 1)} \right]^{\tfrac{1}{2}}$$

$$\langle j_1 \; m - 2, 2 \; 2 | j_1 \; m \rangle = \left[\frac{3(j_1 + m - 1)(j_1 + m)(j_1 - m + 1)(j_1 - m + 2)}{(j_1 + 1)(2j_1 - 1)(2j_1)(2j_1 + 3)} \right]^{\tfrac{1}{2}}$$

$$\langle j_1 \; m - 2, 2 \; 2 | j_1 - 1 \; m \rangle$$
$$= \left[\frac{(j_1 + m - 1)(j_1 - m)(j_1 - m + 1)(j_1 - m + 2)}{j_1(j_1 + 1)(2j_1 - 2)(2j_1 + 1)} \right]^{\tfrac{1}{2}}$$

$$\langle j_1 \; m - 2, 2 \; 2 | j_1 - 2 \; m \rangle$$
$$= \left[\frac{(j_1 - m - 1)(j_1 - m)(j_1 - m + 1)(j_1 - m + 2)}{(2j_1 - 2)(2j_1 - 1)(2j_1)(2j_1 + 1)]} \right]^{\tfrac{1}{2}}$$

$$\langle j_1 \; m - 1, 2 \; 1 | j_1 + 2 \; m \rangle = \left[\frac{(j_1 - m + 2)(j_1 + m + 2)(j_1 + m + 1)(j_1 + m)}{(j_1 + 1)(j_1 + 2)(2j_1 + 1)(2j_1 + 3)} \right]^{\tfrac{1}{2}}$$

$$\langle j_1 \; m - 1, 2 \; 1 | j_1 + 1 \; m \rangle$$
$$= -(j_1 - 2m + 2) \left[\frac{(j_1 + m)(j_1 + m + 1)}{(j_1 + 1)(j_1 + 2)(2j_1)(2j_1 + 1)} \right]^{\tfrac{1}{2}}$$

$$\langle j_1 \; m - 1, 2 \; 1 | j_1 \; m \rangle = (1 - 2m) \left[\frac{3(j_1 - m + 1)(j_1 + m)}{j_1(2j_1 - 1)(2j_1 + 2)(2j_1 + 3)} \right]^{\tfrac{1}{2}}$$

TABLE 2.4 (continued)

$\langle j_1 \; m-1, 2 \; 1 | j_1 - 1 \; m \rangle$

$$= (j_1 + 2m - 1) \left[\frac{(j_1 - m)(j_1 - m + 1)}{(j_1 - 1)(j_1)(2j_1 + 1)(2j_1 + 2)} \right]^{\frac{1}{2}}$$

$\langle j_1 \; m-1, 2 \; 1 | j_1 - 2 \; m \rangle$

$$= - \left[\frac{(j_1 - m - 1)(j_1 - m)(j_1 - m + 1)(j_1 + m - 1)}{(j_1 - 1)(j_1)(2j_1 - 1)(2j_1 + 1)} \right]^{\frac{1}{2}}$$

$\langle j_1 \; m, 2 \; 0 | j_1 + 2 \; m \rangle$

$$= \left[\frac{3(j_1 - m + 1)(j_1 - m + 2)(j_1 + m + 1)(j_1 + m + 2)}{(j_1 + 2)(2j_1 + 1)(2j_1 + 2)(2j_1 + 3)} \right]^{\frac{1}{2}}$$

$\langle j_1 \; m, 2 \; 0 | j_1 + 1 \; m \rangle \quad = m \left[\dfrac{3(j_1 - m + 1)(j_1 + m + 1)}{j_1(j_1 + 1)(j_1 + 2)(2j_1 + 1)} \right]^{\frac{1}{2}}$

$\langle j_1 \; m, 2 \; 0 | j_1 m \rangle \quad = \dfrac{3m^2 - j_1(j_1 + 1)}{[j_1(j_1 + 1)(2j_1 - 1)(2j_1 + 3)]^{\frac{1}{2}}}$

$\langle j_1 \; m, 2 \; 0 | j_1 - 1 \; m \rangle \quad = -m \left[\dfrac{3(j_1 - m)(j_1 + m)}{(j_1 - 1)(j_1)(j_1 + 1)(2j_1 + 1)} \right]^{\frac{1}{2}}$

$\langle j_1 \; m, 2 \; 0 | j_1 - 2 \; m \rangle \quad = \left[\dfrac{3(j_1 - m - 1)(j_1 - m)(j_1 + m - 1)(j_1 + m)}{j_1(2j_1 - 2)(2j_1 - 1)(2j_1 + 1)} \right]^{\frac{1}{2}}$

$\langle j_1 \; m+1, 2 \; -1 | j_1 + 2 \; m \rangle$

$$= \left[\frac{(j_1 - m)(j_1 - m + 1)(j_1 - m + 2)(j_1 + m + 2)}{(j_1 + 1)(j_1 + 2)(2j_1 + 1)(2j_1 + 3)} \right]^{\frac{1}{2}}$$

$\langle j_1 \; m+1, 2 \; -1 | j_1 + 1 \; m \rangle = (j_1 + 2m + 2) \left[\dfrac{(j_1 - m)(j_1 - m + 1)}{j_1(j_1 + 2)(2j_1 + 1)(2j_1 + 2)} \right]^{\frac{1}{2}}$

$\langle j_1 \; m+1, 2 \; -1 | j_1 \; m \rangle \quad = (2m + 1) \left[\dfrac{3(j_1 - m)(j_1 + m + 1)}{j_1(2j_1 - 1)(2j_1 + 2)(2j_1 + 3)} \right]^{\frac{1}{2}}$

$\langle j_1 \; m+1, 2 \; -1 | j_1 - 1 \; m \rangle$

$$= -(j_1 - 2m - 1) \left[\frac{(j_1 + m)(j_1 + m + 1)}{(j_1 - 1)(j_1)(2j_1 + 1)(2j_1 + 2)} \right]^{\frac{1}{2}}$$

$\langle j_1 \; m+1, 2 \; -1 | j_1 - 2 \; m \rangle$

$$= - \left[\frac{(j_1 - m - 1)(j_1 + m - 1)(j_1 + m)(j_1 + m + 1)}{(j_1 - 1)(j_1)(2j_1 - 1)(2j_1 + 1)} \right]^{\frac{1}{2}}$$

$\langle j_1 \; m+2, 2 \; -2 | j_1 + 2 \; m \rangle$

$$= \left[\frac{(j_1 - m - 1)(j_1 - m)(j_1 - m + 1)(j_1 - m + 2)}{(2j_1 + 1)(2j_1 + 2)(2j_1 + 3)(2j_1 + 4)} \right]^{\frac{1}{2}}$$

TABLE 2.4 (continued)

$\langle j_1 \; m + 2, 2 \; -2 | j_1 + 1 \; m \rangle$

$$= \left[\frac{(j_1 - m - 1)(j_1 - m)(j_1 - m + 1)(j_1 + m + 2)}{j_1(j_1 + 1)(2j_1 + 1)(2j_1 + 4)} \right]^{\frac{1}{2}}$$

$\langle j_1 \; m + 2, 2 \; -2 | j_1 \; m \rangle$

$$= \left[\frac{3(j_1 - m - 1)(j_1 - m)(j_1 + m + 1)(j_1 + m + 2)}{j_1(2j_1 - 1)(2j_1 + 2)(2j_1 + 3)} \right]^{\frac{1}{2}}$$

$\langle j_1 \; m + 2, 2 \; -2 | j_1 - 1 \; m \rangle$

$$= \left[\frac{(j_1 - m - 1)(j_1 + m)(j_1 + m + 1)(j_1 + m + 2)}{(j_1 - 1)(j_1)(2j_1 + 1)(2j_1 + 2)} \right]^{\frac{1}{2}}$$

$\langle j_1 \; m + 2, 2 \; -2 | j_1 - 2 \; m \rangle$

$$= \left[\frac{(j_1 + m - 1)(j_1 + m)(j_1 + m + 1)(j_1 + m + 2)}{(2j_1 - 2)(2j_1 - 1)(2j_1)(2j_1 + 1)} \right]^{\frac{1}{2}}$$

[a]Symmetry relations [see Equation (2.26)] may be used in conjunction with this table to evaluate all nonvanishing Clebsch–Gordan coefficients with one angular momentum argument equal to 0, $\frac{1}{2}$, 1, $\frac{3}{2}$ or 2.

TABLE 2.5   Algebraic Expressions for Some Commonly Occurring 3-j Symbols $\begin{pmatrix} j_1 & j_2 & j_3 \\ m_1 & m_2 & m_3 \end{pmatrix}^a$

A. $j_3 = 0$

$$\begin{pmatrix} j & j & 0 \\ m & -m & 0 \end{pmatrix} = (-1)^{j-m}(2j+1)^{-\frac{1}{2}}$$

B. $j_3 = \frac{1}{2}$

$$\begin{pmatrix} j+\frac{1}{2} & j & \frac{1}{2} \\ m & -m-\frac{1}{2} & \frac{1}{2} \end{pmatrix} = (-1)^{j-m-\frac{1}{2}}\left[\frac{j-m+\frac{1}{2}}{(2j+2)(2j+1)}\right]^{\frac{1}{2}}$$

C. $j_3 = 1$

$$\begin{pmatrix} j+1 & j & 1 \\ m & -m-1 & 1 \end{pmatrix} = (-1)^{j-m-1}\left[\frac{(j-m)(j-m+1)}{(2j+3)(2j+2)(2j+1)}\right]^{\frac{1}{2}}$$

$$\begin{pmatrix} j+1 & j & 1 \\ m & -m & 0 \end{pmatrix} = (-1)^{j-m-1}\left[\frac{2(j+m+1)(j-m+1)}{(2j+3)(2j+2)(2j+1)}\right]^{\frac{1}{2}}$$

$$\begin{pmatrix} j & j & 1 \\ m & -m-1 & 1 \end{pmatrix} = (-1)^{j-m}\left[\frac{2(j-m)(j+m+1)}{(2j+2)(2j+1)(2j)}\right]^{\frac{1}{2}}$$

$$\begin{pmatrix} j & j & 1 \\ m & -m & 0 \end{pmatrix} = (-1)^{j-m}\frac{2m}{[(2j+2)(2j+1)(2j)]^{\frac{1}{2}}}$$

D. $j_3 = \frac{3}{2}$

$$\begin{pmatrix} j+\frac{3}{2} & j & \frac{3}{2} \\ m & -m-\frac{3}{2} & \frac{3}{2} \end{pmatrix}$$
$$= (-1)^{j-m+\frac{1}{2}}A_{\frac{3}{2}}(2j+4)[(j-m-\tfrac{1}{2})(j-m+\tfrac{1}{2})(j-m+\tfrac{3}{2})]^{\frac{1}{2}}$$

$$\begin{pmatrix} j+\frac{3}{2} & j & \frac{3}{2} \\ m & -m-\frac{1}{2} & \frac{1}{2} \end{pmatrix}$$
$$= (-1)^{j-m+\frac{1}{2}}A_{\frac{3}{2}}(2j+4)[3(j-m+\tfrac{1}{2})(j-m+\tfrac{3}{2})(j+m+\tfrac{3}{2})]^{\frac{1}{2}}$$

$$\begin{pmatrix} j+\frac{1}{2} & j & \frac{3}{2} \\ m & -m-\frac{3}{2} & \frac{3}{2} \end{pmatrix}$$
$$= (-1)^{j-m-\frac{1}{2}}A_{\frac{3}{2}}(2j+3)[3(j-m-\tfrac{1}{2})(j-m+\tfrac{1}{2})(j+m+\tfrac{3}{2})]^{\frac{1}{2}}$$

$$\begin{pmatrix} j+\frac{1}{2} & j & \frac{3}{2} \\ m & -m-\frac{1}{2} & \frac{1}{2} \end{pmatrix}$$
$$= (-1)^{j-m-\frac{1}{2}}A_{\frac{3}{2}}(2j+3)(j+3m+\tfrac{3}{2})[j-m+\tfrac{1}{2}]^{\frac{1}{2}}$$

where $A_{\frac{3}{2}}(X) = [X(X-1)(X-2)(X-3)]^{-\frac{1}{2}}$

TABLE 2.5 (continued)

E. $j_3 = 2$

$$\begin{pmatrix} j+2 & j & 2 \\ m & -m-2 & 2 \end{pmatrix}$$
$$= (-1)^{j-m} A_2 (2j+5) [(j-m-1)(j-m)(j-m+1)(j-m+2)]^{\frac{1}{2}}$$

$$\begin{pmatrix} j+2 & j & 2 \\ m & -m-1 & 1 \end{pmatrix}$$
$$= (-1)^{j-m} A_2 (2j+5) [4(j+m+2)(j-m+2)(j-m+1)(j-m)]^{\frac{1}{2}}$$

$$\begin{pmatrix} j+2 & j & 2 \\ m & -m & 0 \end{pmatrix}$$
$$= (-1)^{j-m} A_2 (2j+5) [6(j+m+2)(j+m+1)(j-m+2)(j-m+1)]^{\frac{1}{2}}$$

$$\begin{pmatrix} j+1 & j & 2 \\ m & -m-2 & 2 \end{pmatrix}$$
$$= (-1)^{j-m+1} A_2 (2j+4) [4(j-m-1)(j-m)(j-m+1)(j+m+2)]^{\frac{1}{2}}$$

$$\begin{pmatrix} j+1 & j & 2 \\ m & -m-1 & 1 \end{pmatrix}$$
$$= (-1)^{j-m+1} A_2 (2j+4) 2(j+2m+2) [(j-m+1)(j-m)]^{\frac{1}{2}}$$

$$\begin{pmatrix} j+1 & j & 2 \\ m & -m & 0 \end{pmatrix}$$
$$= (-1)^{j-m+1} A_2 (2j+4) 2m [6(j+m+1)(j-m+1)]^{\frac{1}{2}}$$

$$\begin{pmatrix} j & j & 2 \\ m & -m-2 & 2 \end{pmatrix}$$
$$= (-1)^{j-m} A_2 (2j+3) [6(j-m-1)(j-m)(j+m+1)(j+m+2)]^{\frac{1}{2}}$$

$$\begin{pmatrix} j & j & 2 \\ m & -m-1 & 1 \end{pmatrix}$$
$$= (-1)^{j-m} A_2 (2j+3) (2m+1) [6(j+m+1)(j-m)]^{\frac{1}{2}}$$

$$\begin{pmatrix} j & j & 2 \\ m & -m & 0 \end{pmatrix} = (-1)^{j-m} A_2 (2j+3) 2 [3m^2 - j(j+1)]$$

where $A_2(X) = [X(X-1)(X-2)(X-3)(X-4)]^{-\frac{1}{2}}$

---

[a]Symmetry relations [see Equations (2.29)–(2.31)] may be used in conjunction with this table to evaluate all nonvanishing 3-j symbols with one angular momentum argument equal to $0, \frac{1}{2}, 1, \frac{3}{2}$, or 2.

NOTES AND REFERENCES

1. In the language of group theory, the set of functions $|j_1 m_1\rangle$ span a $(2j_1 + 1) \times (2j_1 + 1)$ representation $\mathbf{D}^{j_1}$ of $\mathbf{j}_1^2$ and j_{1z}, while the $|j_2 m_2\rangle$ span a $(2j_2 + 1) \times (2j_2 + 1)$ representation $\mathbf{D}^{j_2}$ of $\mathbf{j}_2^2$ and j_{2z}. Then the direct product of the two sets, $|j_1 m_1, j_2 m_2\rangle$, spans the direct product representation $\mathbf{D}^{j_1} \otimes \mathbf{D}^{j_2}$, which is in general a reducible representation. The group theory content of Eq. (2.6) is that $\mathbf{D}^{j_1} \otimes \mathbf{D}^{j_2}$ can be decomposed into the irreducible representations $\mathbf{D}^{j}$, where j ranges from $j_1 + j_2$ to $|j_1 - j_2|$ in unit steps:

$$\mathbf{D}^{j_1} \otimes \mathbf{D}^{j_2} = \mathbf{D}^{j_1+j_2} + \mathbf{D}^{j_1+j_2-1} + \cdots + \mathbf{D}^{|j_1-j_2|}$$

2. There are numerous notations for the Clebsch–Gordan coefficients. Because $m = m_1 + m_2$ (as will be proved in the text), it is common to drop m and write $C(j_1 j_2 j; m_1 m_2)$, as advocated by M. E. Rose, *Elementary Theory of Angular Momentum* (Wiley, New York, 1957). We prefer the bracket notation $\langle j_1 m_1, j_2 m_2 | j m \rangle$ because it makes very explicit the coupling (decoupling) of the angular momentum arguments involved. A. R. Edmonds, *Angular Momentum in Quantum Mechanics* (Princeton University Press, Princeton, NJ, 1974) uses the notation $\langle j_1 m_1 j_2 m_2 | j_1 j_2 j m \rangle$; D. M. Brink and G. R. Satchler, *Angular Momentum* (Oxford University Press, Oxford, 1979) employ $\langle j_1 j_2 m_1 m_2 | j m \rangle$ and sometimes just $\langle m_1 m_2 | j m \rangle$; while L. C. Biedenharn and J. D. Louck, *Angular Momentum in Quantum Physics* (Addison-Wesley, Reading, MA, 1981) use $C_{m_1 m_2 m}^{j_1 j_2 j}$. For a list of other notations, see Edmonds, p. 52, and Biedenharn and Louck, pp. 150–152.

3. Please note that this treatment takes no account of the Pauli exclusion principle and hence cannot be used for equivalent electrons (and other fermions).

4. See G. Racah, *Phys. Rev.*, **62**, 438 (1942). This is equivalent to the phase convention adopted by E. U. Condon and G. H. Shortley, *Theory of Atomic Spectra* (Cambridge University Press, 1935).

5. An explicit formula for the Clebsch–Gordan coefficients was first derived by E. P. Wigner in 1931 (see E. P. Wigner, *Group Theory* (Academic Press, New York, 1959), pp. 188–191. A different but equivalent expression was obtained by B. L. Van der Waerden, *Der Gruppentheoretische Methode in der Quantenmechanik* (Springer, Berlin, 1932), pp. 68–71. Racah's derivation (see reference [3]) is the first not involving group theoretic arguments. Other derivations have been given by S. D. Majumdar, *Progr. Theor. Phys.*, **20**, 798 (1958), J. Schwinger in L. C. Biedenharn and H. Van Dam, *Quantum Theory of Angular Momentum* (Academic Press, New York, 1965), pp. 241–246, and R. T. Sharp, *Am. J. Phys.*, **28**, 116 (1960).

6. M. Rotenberg, R. Bivins, N. Metropolis, and J. K. Wooten, Jr., *the 3-j and 6-j symbols* (The Technology Press, MIT, Cambridge, MA, 1959). See also *Landolt–Börnstein, Numerical Data and Functional Relationships in Science and Technology*, K.-H. Hellwege, ed., (Springer-Verlag, Berlin, 1968) Group I, Vol. 3, "Numerical Tables for Angular Correlation Computations in α-, β-, and γ- Spectroscopy: $3\text{-}j$, $6\text{-}j$, $9\text{-}j$ Symbols, F- and Γ-Coefficients."

7. K. Schulten and R. G. Gordon, *J. Math. Phys.*, **16**, 1961 (1975); *Computer Phys. Commun.*, **11**, 269 (1976).

8. K. Schulten and R. G. Gordon, *J. Math. Phys.*, **16**, 1971 (1975).

9. Tables of Clebsch–Gordan coefficients or 3-j symbols in algebraic form are given by Edmonds, pp. 126–127 for angular momentum arguments $\frac{3}{2}$ and 2, by R. Saito and M. Morita, *Progr. Theor. Phys.*, **13**, 540 (1955) for $\frac{5}{2}$, and by D. L. Falkoff, G. S. Colladay, and R. E. Sells, *Can. J. Phys.*, **30**, 253 (1952) for 3. This information is also collected and displayed in L. C. Biedenharn and J. D. Louck, *Angular Momentum in Quantum Physics* (Addison-Wesley, Reading, MA, 1981), pp. 635–643.

10. This result was first obtained by E. P. Wigner, *Group Theory* (Academic Press, New York, 1959), pp. 351–353.

11. For further investigations of the correspondence between the quantum behavior for large angular momenta and the classical limit, see P. J. Brussaard and H. A. Tolhoek, *Physica*, **23**, 955 (1957); G. Ponzano and T. Regge, in *Spectroscopic and Group Theoretic Methods in Physics* (North-Holland, Amsterdam, 1968), pp. 1–58; W. H. Miller, *Adv. Chem. Phys.*, **25**, 69 (1974); and K. Schulten and R. G. Gordon, *J. Math. Phys.*, **16**, 1961 (1975).

APPLICATION 2
THE WIGNER–WITMER RULES

E. Wigner and E. E. Witmer, Z. *Physik*, **51**, 859 (1928) were the first to derive the number of different electronic states (terms) of a diatomic molecule formed by bringing together two atoms 1 and 2 having orbital angular momenta L_1 and L_2 and spin angular momenta S_1 and S_2. This matter is of interest when describing the nature of the compound state when two atoms collide or the nature of the atomic fragments when a diatomic molecule dissociates.

An atomic term is written as ^{2S+1}L, where $2S + 1$ is the multiplicity expressing the degeneracy associated with the total spin S (which has $2S + 1$ values of M_S) and the total angular momentum L takes the values 0, 1, 2, ... and is denoted by the capital Latin letters S, P, D, ...

For a (nonrotating) diatomic molecule, the total orbital angular momentum of the electrons is not a good quantum number since the molecule does not possess spherical symmetry and the orbital angular momentum operator does not commute with the total molecular Hamiltonian. However, the projection of the orbital angular momentum on the internuclear axis is conserved since the diatomic molecule has axial symmetry about a line passing through the two nuclei. The electronic terms of a diatomic molecule are classified according to the value of this projection. The absolute value of the projected orbital angular momentum along the axis of the molecule is denoted by Λ. It takes on the values 0, 1, 2, ... Terms with different Λ are denoted by the capital Greek letters Σ, Π, Δ, ... corresponding to the capital Latin letters S, P, D, ... for atomic terms with $L = 0, 1, 2, ...$ To this, the multiplicity is affixed on the left as a superscript, that is, $^{2S+1}\Lambda$, just as for the atomic term ^{2S+1}L.

The electronic wave function for a many-electron atom can be written as the product of one-electron atomic orbitals, or more properly, as the antisymmetrized linear combination of such products. Recall that the sum of angular momenta is also an angular momentum. Hence the atomic term ^{2S+1}L has a spin part that transforms as $|SM_S\rangle$ and a space part that transforms as $|LM_L\rangle \equiv Y_{LM_L}$.

Let us consider first the case where the molecule consists of two different atoms. Then $\Lambda = |M_{L_1} + M_{L_2}|$. Let $L_>$ and $L_<$ stand for the greater and lesser value of the (L_1, L_2) pair, respectively. Let $S_>$ and $S_<$ have a similar meaning.

A. Show that there are

$$(2S_< + 1)(2L_< + 1)$$

Σ terms (called Σ *states*) of different multiplicity.

B. Show that there are a total of

$$(2S_< + 1)(2L_< + 1)(L_> + 1)$$

electronic terms where Λ ranges from 0 to $L_1 + L_2$ and S ranges from $|S_1 - S_2|$ to $S_1 + S_2$, that is, from $S_> - S_<$ to $S_> + S_<$.

In addition to rotation through any angle about the internuclear axis, a diatomic molecule also has reflection symmetry in any plane that contains this axis. On reflection the sense of the precession of $\mathbf{L}$ about the z axis (taken to coincide with the internuclear axis) is reversed and M changes sign. Diatomic molecules with $\Lambda \neq 0$ are doubly degenerate in the absence of rotation. To each value of the energy there are two states that differ in the direction of the projection of the orbital angular momentum on the internuclear axis. Hence the reflection symmetry leads to no additional classification for nonrotating $\Lambda \neq 0$ states. However, for Σ states ($\Lambda = 0$), the state of the molecule is not changed on reflection, so that Σ states are nondegenerate. The wave function of a Σ state can be multiplied by only a constant phase factor p. Since double reflection in the same plane restores the system to its original condition, $p^2 = 1$, and it follows that $p = \pm 1$. Σ states whose wave functions are unaltered by reflection are called Σ^+ *states*; those that change sign are Σ^- *states*. Recall that for each multiplicity there are $(2L_< + 1)$ Σ states. The problem we next confront is how many of these are Σ^+ states and how many are Σ^- states.

We may use the $|L_1 M_{L_1}, L_2 M_{L_2}\rangle$ uncoupled representation as basis functions, where $M_{L_1} + M_{L_2} = 0$. For $M = |M_{L_1}| = |M_{L_2}| \neq 0$, the basis functions can be written as symmetric and antisymmetric products under reflection:

$$|\Sigma^+; M\rangle = |L_1 M, L_2 - M\rangle + |L_1 - M, L_2 M\rangle$$

$$|\Sigma^-; M\rangle = |L_1 M, L_2 - M\rangle - |L_1 - M, L_2 M\rangle$$

Clearly there are $L_<$ reflection functions of the form $|\Sigma^+; M\rangle$ and $L_<$ of $|\Sigma^-; M\rangle$.

The energies of these nondegenerate Σ states are found by solving a set of $(2L_< + 1)$ secular equations. The use of the reflection basis set $|\Sigma^\pm; M\rangle$ factors the secular determinant into two blocks. Depending on the reflection symmetry of $|L_1 0, L_2 0\rangle$, this will lead to a $L_< + 1$ by $L_< + 1$ secular determinant for the energies of the Σ states of one reflection symmetry and a $L_<$ by $L_<$ secular determinant for the other. Hence we need only establish how $|L_1 0, L_2 0\rangle$ behaves under reflection to complete this problem.

Consider a coordinate system with its origin at the center of atom 1 and its z axis along the internuclear axis, as shown in Figure 1. It is an easy matter to convince oneself that a reflection in the xz plane, $\sigma(xz)$, is equivalent to an inversion through the origin, i, followed by a $180°$ rotation about the y axis, $C_2(y)$. This may be established by sketching what happens to some representative points. Then

$$\sigma(xz)|L_1 0, L_2 0\rangle = C_2(y)i|L_1 0, L_2 0\rangle$$

Under inversion, an atomic wave function remains unchanged or changes sign depending on whether the parity p of the atomic term is even or odd, respectively.

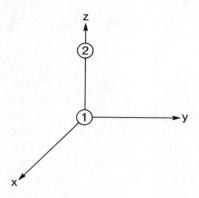

FIGURE 1 Coordinate system for a diatomic
molecule.

Thus the wave function is multiplied by the factor p whose value is determined as
follows.

C. Under inversion $x \rightarrow -x$, $y \rightarrow -y$, and $z \rightarrow -z$. Show that $\phi \rightarrow \pi + \phi$ and
$\theta \rightarrow \pi - \theta$.

D. Use the preceding result to prove that the spherical harmonics under inversion
obey the relation

$$iY_{\ell m}(\theta, \phi) = (-1)^{\ell} Y_{\ell m}(\theta, \phi)$$

from which it is concluded that the parity of the state of a particle with angular
momentum ℓ is independent of m. *Hint*: Show that under this transformation
$\theta \rightarrow \pi - \theta$, and $\cos \theta \rightarrow -\cos \theta$.

E. For an atomic system, present arguments to show that

$$p = (-1)^{\sum_i \ell_i}$$

where ℓ_i is the orbital angular momentum quantum number of the ith electron
and the sum runs over all electrons.

From this equation we have

$$\sigma(xz) |L_1 0, L_2 0\rangle = p_1 p_2 C_2(y) |L_1 0, L_2 0\rangle$$

But the rotation by $180°$ about the y axis causes the transformation $x \rightarrow -x$, $y \rightarrow y$,
$z \rightarrow -z$.

F. Prove that

$$C_2(y) |L0\rangle = (-1)^{L} |L0\rangle$$

Hint: Use the fact that $|L0\rangle$ is proportional to $P_L(\cos \theta)$ to derive the above.

Thus $|L_1 0, L_2 0\rangle$ is multiplied by the factor $p_1 p_2 (-1)^{L_1 + L_2}$ on reflection, and depending on whether this factor is +1 or −1 there will be one more Σ^+ or Σ^- term, respectively, for each multiplicity.

Consider now a molecule consisting of the same atoms (the so-called homonuclear diatomic). There now exists a new symmetry, inversion of the electronic wave function through the midpoint of the internuclear axis. Those wave functions remaining unchanged are called *gerade*, while those changing sign are called *ungerade* states. A subscript g or u is attached to the right of the term symbol. If the two atoms are in different electronic states (one excited and the other not, or one neutral and the other ionized), the total number of possible terms is doubled, since each term can be either gerade or ungerade. If both atoms are in the same electronic state, the total number of states is the same as for a molecule composed from different atoms.

We must then sort out which states belong to gerade and ungerade symmetry. The sorting process is involved [see P. Pechukas and R. N. Zare, *Am. J. Phys.*, **40**, 1687 (1972)]. The results are as follows:

If Λ is odd, the number of $^{2S+1}\Lambda_g$ states equals the number of $^{2S+1}\Lambda_u$ states.

If Λ is even, the number of $^{2S+1}\Lambda_g$ states is one greater or lesser than the number of $^{2S+1}\Lambda_u$ states. The "extra" $^{2S+1}\Lambda$ state is g or u as S is even or odd.

If S is even, all Σ^+ states are g and all Σ^- states are u.

If S is odd, all Σ^+ states are u and all Σ^- states are g.

Note that the g, u designations refer to the electronic wave function. Thus, for example, this classification is used to describe the electronic states of $^{14}N_2$ as well as $^{14}N^{15}N$.

G. The ground state of the hydrogen atom is

$$H: 1s\ ^2S$$

and that of the oxygen atom is

$$O: (1s)^2 (2s)^2 (2p)^4\ ^3P$$

Find all the electronic states of H_2, OH, and O_2 that can be built up by bringing together their respective ground-state atoms.

APPLICATION 3
THE ROTATIONAL ENERGY LEVELS OF A $^2\Sigma$ FREE RADICAL

To a first approximation we may imagine a diatomic molecule as a three-dimensional rigid rotor. The Hamiltonian $\mathcal{H}$ for the total system would be simply

$$\mathcal{H} = \mathcal{H}_0 + B_v \mathbf{N}^2 \tag{1}$$

where $\mathcal{H}_0$ represents all those terms that fix the origin of the vibrational state v, B_v is the rotational constant of the molecule in the vibrational state v, and $\mathbf{N}^2$ is the square of the rotational angular momentum. The wave functions that diagonalize $\mathcal{H}$ may be written as

$$|vN M_N\rangle = |v\rangle |N M_N\rangle = |v\rangle Y_{N M_N}(\theta, \phi) \tag{2}$$

where $|v\rangle$ is the vibrational wave function and $Y_{N M_N}$ is a spherical harmonic. We take

$$\mathcal{H}_0 |v\rangle = \nu_0 |v\rangle \tag{3}$$

where ν_0 is the band origin and

$$\mathbf{N}^2 |N M_N\rangle = N(N + 1) |N M_N\rangle \tag{4}$$

Hence the energy levels are given by

$$E(v, N) = \langle vN M_N| \mathcal{H} |vN M_N\rangle = \nu_0 + B_v N(N + 1) \tag{5}$$

Actually, this is a fine starting point for a $^1\Sigma$ diatomic molecule. In a $^2\Sigma$ molecule, however, we must include the presence of nonzero electronic spin $\mathbf{S}$, arising from an unpaired electron. We add to the Hamiltonian a phenomenological term $\gamma_v \mathbf{N} \cdot \mathbf{S}$ to represent the effective interaction between $\mathbf{N}$ and $\mathbf{S}$, where γ_v is called the *spin–rotation constant* for the molecule in the vibrational state v. We also include (to first order) the nonrigidity of the molecule by introducing into the Hamiltonian the vibration–rotation term $-D_v \mathbf{N}^4$, where D_v is called the *centrifugal distortion constant* for the molecule in the vibrational state v. The term $-D_v \mathbf{N}^4$ expresses, in part, the fact that in a nonrigid molecule the moment of inertia changes with the rate of rotation. Thus the total Hamiltonian becomes

$$\mathcal{H} = \mathcal{H}_0 + B_v \mathbf{N}^2 - D_v \mathbf{N}^4 + \gamma_v \mathbf{N} \cdot \mathbf{S} \tag{6}$$

A. Use the coupled representation $|NSJM\rangle$ to calculate explicit expressions in terms of the rotational quantum number N for the energies of the rotational levels $J = N + \frac{1}{2}$ and $J = N - \frac{1}{2}$. Hence show that levels with the same value of N are split (doubled) by spin–rotation splitting by the amount $\gamma_v(N + \frac{1}{2})$, which grows as N increases. *Hint*: Construct the 2×2 secular determinant, using the $\left|N\frac{1}{2}\ N + \frac{1}{2}M\right\rangle\ \left|N\frac{1}{2}\ N - \frac{1}{2}M\right\rangle$ basis functions. Use the relation $2\,\mathbf{N}\cdot\mathbf{S} = \mathbf{J}^2 - \mathbf{N}^2 - \mathbf{S}^2$ to evaluate the matrix elements of $\mathbf{N}\cdot\mathbf{S}$.

B. Repeat part A but this time use the uncoupled representation $|N M_N, S M_S\rangle$. *Hint*: Work specifically with $\left|N M_N, \frac{1}{2}\frac{1}{2}\right\rangle$ and $\left|N M_N + 1, \frac{1}{2} -\frac{1}{2}\right\rangle$ Use the identity

$$\mathbf{N}\cdot\mathbf{S} = N_x S_x + N_y S_y + N_z S_z$$

$$= \frac{1}{2}(N_+ S_- + N_- S_+) + N_z S_z \tag{7}$$

to evaluate the matrix elements of $\mathbf{N}\cdot\mathbf{S}$ appearing in the 2×2 secular determinant with these basis functions.

Consider next the Zeeman effect in $^2\Sigma$ molecules. Because $^2\Sigma$ molecules have an unpaired electron spin, they are paramagnetic. Application of a magnetic field $\mathbf{H}$ splits the M degeneracy of each J level (the Zeeman effect). The energy of interaction between the magnetic dipole $\boldsymbol{\mu}_S$ associated with the electron spin $\mathbf{S}$ and the external field is

$$\mathcal{H}_z = -\boldsymbol{\mu}_S \cdot \mathbf{H} \tag{8}$$

Each unit of spin momentum produces slightly more than two electronic Bohr magnetons, that is,

$$\boldsymbol{\mu}_S = -g_S \mu_0 \mathbf{S} \tag{9}$$

where $g_S = 2.002$ and μ_0 is the electronic Bohr magneton (~ 1.4 MHz/G). We choose the field direction to be along the z axis. Hence we write

$$\mathcal{H}_z = 2.002\,\mu_0\,H S_z \tag{10}$$

If the field strength is weak, first-order perturbation theory may be used to calculate the energy of each magnetic sublevel M of the level J:

$$\Delta E_z(M) = \langle NSJM|\mathcal{H}_z|NSJM\rangle \tag{11}$$

C. Evaluate $\Delta E_z(M)$ by rewriting $|NSJM\rangle$ in terms of the $|N M_N, S M_S\rangle$ and then calculating the matrix elements of $\mathcal{H}_z$. Show that

$$\Delta E_z(M) = \pm g_S \mu_0 M H/(2N + 1) \tag{12}$$

where the upper sign is for $J = N + \frac{1}{2}$ and the lower sign is for $J = N - \frac{1}{2}$.

At larger field strengths, it is necessary to add $\mathcal{H}_z$ to the molecular Hamiltonian and evaluate the resulting new 2×2 secular determinant for its roots (eigenenergies).

D. Work in the coupled representation to obtain an explicit expression for the energies of the magnetic sublevels of a $^2\Sigma$ molecule in a magnetic field H. Show that for small field strengths these expressions approach the results of first-order perturbation theory in which the energies of the magnetic sublevels vary linearly with H.

As the field strength increases, the energies of the magnetic sublevels deviate from linearity in the field strength (onset of the molecular Paschen–Back effect) and at still larger field strength the unpaired electron of the free radical decouples from the molecule and acts as a free electron.

E. Plot a graph showing the variation of the $^2\Sigma$ energy levels with magnetic field strength for the lowest eight levels ($N = 0, 1$). Use $B_v = 1 \text{ cm}^{-1}$, $D_v = 1 \times 10^{-6} \text{ cm}^{-1}$, and $\gamma_v = 1 \times 10^{-2} \text{ cm}^{-1}$.

Chapter 3

TRANSFORMATION UNDER ROTATION

3.1 ANGULAR MOMENTUM OPERATORS AS GENERATORS OF INFINITESIMAL ROTATIONS

We come to a matter both simple and profound. The symmetries of a physical system and its conservation laws are related to one another. To bring out this concept, let us look at the action of a linear transformation $\mathbf{U}$ on the eigenvectors $|\psi_j\rangle$ of the system, which changes its representation into a new set of basis functions $|\psi_j'\rangle$:

$$|\psi_j'\rangle = \mathbf{U}\,|\psi_j\rangle \tag{3.1}$$

We have

$$\langle\psi_i'|\psi_j'\rangle = \langle\psi_i|\,\mathbf{U}^\dagger\mathbf{U}\,|\psi_j\rangle \tag{3.2}$$

where the matrix representing $\mathbf{U}^\dagger$ is obtained from the matrix representing $\mathbf{U}$ by taking its transpose complex conjugate. Equation (3.2) shows that in the $\{\langle\psi'|\}$ vector space the lengths of the vectors and the angles between any two of them will be preserved (will be the same as in the $\{\langle\psi|\}$ vector space) if

$$\mathbf{U}^\dagger = \mathbf{U}^{-1} \tag{3.3}$$

Transformations whose matrices satisfy this condition are said to be unitary, for $\mathbf{U}^\dagger\mathbf{U}$ equals the unit matrix $\mathbf{I}$ (the identity matrix).

Equation (3.1) shows how the representation of a vector changes under the unitary transformation $\mathbf{U}$. It is convenient to know how to relate the matrices representing the same operator $\mathbf{Q}$ in the two different representations. In the "old" representation

$$Q_{ij} = \langle\psi_i|\,\mathbf{Q}\,|\psi_j\rangle \tag{3.4}$$

73

Introducing the identity $U^\dagger U = I$, we may rewrite Eq. (3.4) as

$$Q_{ij} = \langle \psi_i | \, U^\dagger U Q U^\dagger U \, | \psi_j \rangle$$

$$= \langle \psi_i' | \, U Q U^\dagger \, | \psi_j' \rangle$$

$$= Q_{ij}' \tag{3.5}$$

Thus a unitary transformation U transforms an operator Q into

$$Q' = U Q U^\dagger = U Q U^{-1} \tag{3.6}$$

Suppose that the operator U defining a unitary transformation approaches the identity matrix I as the change ε in some variable caused by the transformation becomes smaller and smaller. Then U may be represented as

$$U = I + i\varepsilon S \tag{3.7}$$

where ε is an infinitesimal real quantity and the operator S is called the *generator of the infinitesimal transformation*[1, 2]. For the condition $U^\dagger U = I$ to obtain, it is necessary that

$$S^\dagger = S \tag{3.8}$$

Operators that satisfy Eq. (3.8) are said to be Hermitian.

Two successive unitary transformations are still a unitary transformation. We apply n successive infinitesimal transformations, each having a change $\delta = \varepsilon/n$, so that to the same order in ε

$$U = [I + i\delta S]^n \tag{3.9}$$

We take the limit as n increases:

$$U = \lim_{\substack{\delta \to 0 \\ n \to \infty}} [I + i\delta S]^n = \lim_{n \to \infty} \left[I + i\frac{\varepsilon}{n} S \right]^n$$

$$= \exp(i\varepsilon S) \tag{3.10}$$

In Eq. (3.10) the exponential of an operator means no more than its power series expansion

$$\exp(Q) = I + Q + \frac{1}{2!}Q^2 + \frac{1}{3!}Q^3 + \cdots \tag{3.11}$$

Thus Eq. (3.7) is seen to be just the leading two terms of Eq. (3.10).

The operators U and S commute; thus

$$[\exp(i\varepsilon S), S] = 0 \tag{3.12}$$

which can be readily demonstrated by replacing $\exp(i\varepsilon S)$ by its power series expansion in S. The eigenfunctions of S are also eigenfunctions of U, and the eigenvalues of S are unchanged (conserved) by the transformation U. Hence the invariance of the system under the unitary transformation U implies a conservation law for the eigenvalues of the Hermitian operator S, which is the generator of the infinitesimal transformation[3].

We illustrate this result with three important examples. However, before we start we must distinguish carefully between *active* and *passive* transformations[4]. In the active sense, one transforms the physical system from one position to another; in the passive sense, one transforms the coordinate frame. Both descriptions, of course, are equivalent, but they lead to confusion unless it is made explicit what is being transformed and what is remaining fixed. As we shall see, transformation of the physical system in one "direction" is equivalent to transformation of the coordinate frame by an equal amount in the opposite "direction."

First, let us consider the transformation $T(t)$, which carries the physical system from time t_0 to time $t_0 + t$; hence the time coordinate from t_0 to $t' = t_0 - t$:

$$T(t)\,|t_0\rangle = |t'\rangle \tag{3.13}$$

The operator $T(t)$ is called the *time displacement operator*. As $t \to 0$, $T(t) \to I$. For an infinitesimal change in time we can replace $|t'\rangle$ by its Taylor's series expansion

$$|t'\rangle = |t_0\rangle + (-t)\frac{\partial}{\partial t}|t_0\rangle + \frac{1}{2!}(-t)^2\frac{\partial^2}{\partial t^2}|t_0\rangle + \cdots$$

$$= \exp\left(-t\frac{\partial}{\partial t}\right)|t_0\rangle \tag{3.14}$$

In Eq. (3.14) we denote the name of the time variable and the value of that variable by the same symbol t. This lazy but traditional notation may cause some confusion at first; please note in the power series expansion of $\exp(-t\partial/\partial t)$ that t is a constant (the amount of the time change) and commutes with $\partial/\partial t$. Comparison of Eq. (3.14) with Eq. (3.10) shows that

$$T(t) = \exp(it\mathcal{H}) \tag{3.15}$$

where $\mathcal{H} = i\partial/\partial t$ (in units of $\hbar = 1$) is recognized as the Hamiltonian of the system (for those systems whose Hamiltonians do not depend explicitly on time). Hence the symmetry of time invariance of the system implies the conservation of energy (the eigenvalues of $\mathcal{H}$).

In particular, if $|\psi(t_0)\rangle$ is an eigenstate of $\mathcal{H}$ with the eigenvalue E_0, then time evolution of the eigenstate (*not* the time variable) through a time interval t causes the eigenstate to transform as $|\psi'(t_0)\rangle = T(t)\,|\psi(t_0)\rangle = |\psi(t_0 - t)\rangle$. For $t = t_0$ it follows that $|\psi'(t)\rangle = |\psi(0)\rangle$. Thus $|\psi'(t_0)\rangle$ has the same value at time $t = t_0$ that

$|\psi(t_0)\rangle$ has at time $t_0 = 0$. Consequently, $|\psi'(t)\rangle$ can be obtained from $|\psi(t)\rangle$ by letting $|\psi(t)\rangle$ develop *backward* in time from t to 0:

$$|\psi(t)\rangle = \mathbf{T}(-t)\,|\psi(0)\rangle = \exp(-it\mathcal{H})\,|\psi(0)\rangle = \exp(-iE_0 t)\,|\psi(0)\rangle \quad (3.16)$$

Traditionally $\mathbf{U}(t) = \mathbf{T}(-t) = \mathbf{T}(t)^{\dagger}$ is called the *time evolution operator* or *time development operator*, which operates on the eigenstates of the system to cause them to advance in time. Comparison of $\mathbf{U}(t)$ with $\mathbf{T}(t)$ again shows the expected relation between active and passive transformations; a change in the active sense is equivalent to an equal and opposite change in the passive sense.

Next, consider the one-dimensional displacement operator $\mathbf{D}(z)$, which carries the system from z_0 to $z_0 + z$; hence the position coordinate z_0 into $z' = z_0 - z$:

$$\mathbf{D}(z)\,|z_0\rangle = |z'\rangle \quad (3.17)$$

Again $\mathbf{D}(0) = \mathbf{I}$, and for an infinitesimal displacement

$$|z'\rangle = |z_0\rangle + (-z)\frac{\partial}{\partial z}\,|z_0\rangle + \frac{1}{2!}(-z)^2 \frac{\partial^2}{\partial z^2}\,|z_0\rangle + \cdots$$

$$= \exp\left(-z\frac{\partial}{\partial z}\right)|z_0\rangle \quad (3.18)$$

We identify

$$\mathbf{D}(z) = \exp(-izp_z) \quad (3.19)$$

where $p_z = -i\partial/\partial z$ (in units of $\hbar = 1$) is the z component of the linear momentum operator. Thus p_z is the generator of an infinitesimal displacement along z, and the invariance of the system to such a change implies that the z component of the linear momentum is conserved.

As our last example, we consider a rotational transformation. We define a *positive* rotation as one that carries a right-handed screw in the positive direction (forward) along its axis. Any rotation $\mathbf{R}_n(a)$ can be specified by giving three parameters, two to fix its rotation axis $\hat{n}$ and one to fix its rotation angle a about $\hat{n}$. For convenience, let $\hat{n}$ be chosen to point along the z axis. The transformation $\mathbf{R}_z(\phi)$ then carries the system from ϕ_0 to $\phi_0 + \phi$ and hence the angle ϕ_0 to $\phi' = \phi_0 - \phi$ in the coordinate frame:

$$\mathbf{R}_z(\phi)\,|\phi_0\rangle = |\phi'\rangle \quad (3.20)$$

By the same line of argument as before

$$|\phi'\rangle = \exp\left(-\phi\frac{\partial}{\partial \phi}\right)|\phi_0\rangle \quad (3.21)$$

and we make the identification

$$\mathbf{R}_z(\phi) = \exp(-i\phi J_z) \tag{3.22}$$

where we have replaced L_z by J_z since the algebraic properties of the operators L_x, L_y, L_z are determined by their commutation relations, which are the same for the more general operators J_x, J_y, J_z. Thus the z component of the angular momentum operator is the generator of an infinitesimal rotation about z.

The choice of the z axis was for convenience. For an arbitrary rotation about $\hat{\mathbf{n}}$ by an angle a, we obtain

$$\mathbf{R}_n(a) = \exp(-ia\mathbf{J} \cdot \hat{\mathbf{n}}) \tag{3.23}$$

where

$$\mathbf{J} \cdot \hat{\mathbf{n}} = -i\frac{\partial}{\partial a} \tag{3.24}$$

If the system is invariant under rotation about the axis $\hat{\mathbf{n}}$, then the component of the angular momentum along $\hat{\mathbf{n}}$ is conserved.

3.2 PARAMETERIZATION OF ROTATIONS BY EULER ANGLES

As we have remarked, any finite rotation may be uniquely specified by giving the three components of the vector $a\hat{\mathbf{n}}$, where a is the angle of rotation and $\hat{\mathbf{n}}$ is a unit vector pointing along the axis of rotation. For many purposes, however, an equivalent but more useful description is achieved by introducing the three Euler angles ϕ, θ, and χ shown in Figure 3.1. Here the angles ϕ and θ are familiar spherical polar coordinates, where θ measures the angle from the Z axis to the z axis and ϕ measures the angle from the X axis to the projection of the z axis on the XY plane. The angle χ measures the angle from the line of nodes N, defined to be the intersection of the XY and xy planes, to the y axis. Thus χ is an azimuthal angle about the z axis just as ϕ is an azimuthal angle about the Z axis. Note that the line of nodes ON is perpendicular to both the z and Z axes.

The Euler angles ϕ, θ, and χ should be regarded as defining a prescription whereby the $F = XYZ$ frame (the space-fixed frame) may be made to coincide with the $g = xyz$ frame (the body-fixed frame) by three successive finite rotations (Figure 3.2):

1. A counterclockwise rotation ϕ about Z, the vertical axis. This carries the Y axis into the line of nodes N.

2. A counterclockwise rotation θ about the line of nodes N. This carries the Z axis into the z axis (the figure axis of the body).

3. A counterclockwise rotation χ about z, the figure axis. This carries the line of nodes N into the y axis.

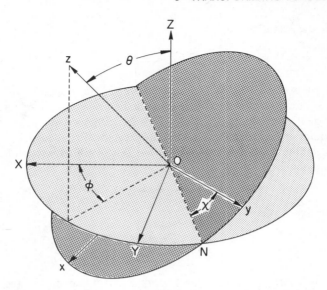

FIGURE 3.1   Euler angles ϕ, θ, and χ relating the space-fixed $F = XYZ$ and molecule-fixed $g = xyz$ frames.

Note that all rotations are positive. This definition of the Euler angles ϕ, θ, and χ, which are also frequently written as α, β, and γ, agrees with that of Rose[5], Edmonds[6], and Brink and Satchler[7]. According to Eq. (3.23), we may express the above prescription as

$$\mathbf{R}(\phi, \theta, \chi) = \exp(-i\chi\mathbf{J} \cdot \hat{\mathbf{n}}_\chi) \exp(-i\theta\mathbf{J} \cdot \hat{\mathbf{n}}_\theta) \exp(-i\phi\mathbf{J} \cdot \hat{\mathbf{n}}_\phi)$$

$$= \exp(-i\chi J_z) \exp(-i\theta J_N) \exp(-i\phi J_Z) \tag{3.25}$$

where the form of $\mathbf{R}$ is chosen to be an active rotation of the physical system[4, 8].

For many purposes Eq. (3.25) is rather awkward because it is a mixed expression involving angular momentum operators referred to both the F and g coordinate frames. We now show that $\mathbf{R}(\phi, \theta, \chi)$ is equivalent to first a rotation χ about the Z axis, then a rotation θ about the Y axis, and finally a rotation ϕ about the same Z axis:

$$\mathbf{R}(\phi, \theta, \chi) = \exp(-i\phi J_Z) \exp(-i\theta J_Y) \exp(-i\chi J_Z) \tag{3.26}$$

The proof of this rather surprising result makes extensive use of Eq. (3.6), namely, that a unitary transformation $\mathbf{U}$ carries an operator $\mathbf{Q}$ into $\mathbf{UQU}^{-1}$. Thus $\exp(-i\theta J_N)$ is the transform of $\exp(-i\theta J_Y)$ under the rotation $\exp(-i\phi J_Z)$, which carried the Y axis into the line of nodes N. Hence

$$\exp(-i\theta J_N) = \exp(-i\phi J_Z) \exp(-i\theta J_Y) \exp(i\phi J_Z) \tag{3.27}$$

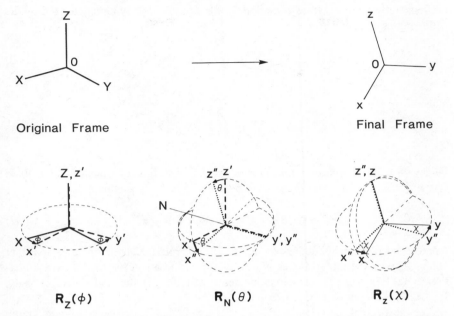

FIGURE 3.2   Transformation of the $F = XYZ$ frame to the $g = xyz$ frame having a common origin O by three successive rotations, first $R_Z(\phi)$, then $R_N(\theta)$, and finally $R_z(\chi)$.

Similarly, $\exp(-i\chi J_z)$ is the transform of $\exp(-i\chi J_Z)$ under the previous rotation $\exp(-i\theta J_N)$, which carried Z into z:

$$\exp(-i\chi J_z) = \exp(-i\theta J_N) \exp(-i\chi J_Z) \exp(i\theta J_N) \qquad (3.28)$$

Finally, we can always write the identity

$$\exp(-i\chi J_Z) = \exp(-i\phi J_Z) \exp(-i\chi J_Z) \exp(i\phi J_Z) \qquad (3.29)$$

since J_Z commutes with itself. If one likes, Eq. (3.29) may be regarded as expressing that $\exp(-i\chi J_Z)$ is also the transform of the first rotation by the angle ϕ about the Z axis. When Eqs. (3.27)–(3.29) are substituted into Eq. (3.25), Eq. (3.26) results. Equation (3.26) shows that the three Euler angle rotations may all be carried out in the *same* coordinate frame if the order of the rotations is reversed!

3.3   THE DIRECTION COSINE MATRIX ELEMENTS

Let **r** be an arbitrary vector having the Cartesian components $r_F = (r_X, r_Y, r_Z)$ in the space-fixed (laboratory) frame and the Cartesian components $r_g = (r_x, r_y, r_z)$ in the molecule-fixed (body) frame. These two equivalent descriptions of the vector **r** are related by a unitary transformation Φ that is conveniently parameterized by the Euler angles ϕ, θ, and χ:

$$r_F = \sum_g \Phi_{Fg}(\phi, \theta, \chi) r_g \tag{3.30}$$

and

$$r_g = \sum_F \Phi_{gF}(\phi, \theta, \chi) r_F \tag{3.31}$$

where Φ is called the *direction cosine matrix*. The unitary transformation Φ is conveniently expressed by the product of three Euler angle rotations

$$\Phi(\phi, \theta, \chi) = \mathbf{R}_z(\chi) \mathbf{R}_N(\theta) \mathbf{R}_Z(\phi) \tag{3.32}$$

where ϕ is a rotation about the original Z axis, θ is rotation about the new y axis that coincides with the line of nodes N, and χ is rotation about the final z axis. Equation (3.32) brings the F frame in coincidence with the g frame, and not surprisingly, Eq. (3.32) is identical to Eq. (3.25).

Let the original coordinate frame be denoted by XYZ, the coordinate frame after the rotation $\mathbf{R}_Z(\phi)$ by $x'y'z'$, the coordinate frame after the rotation $\mathbf{R}_N(\theta)$ by $x''y''z''$, and the final coordinate frame after the rotation $\mathbf{R}_z(\chi)$ by xyz (see Figure 3.2). Specifically,

$$\begin{bmatrix} x' \\ y' \\ z' \end{bmatrix} = \mathbf{R}_Z(\phi) \begin{bmatrix} X \\ Y \\ Z \end{bmatrix} = \begin{pmatrix} \cos\phi & \sin\phi & 0 \\ -\sin\phi & \cos\phi & 0 \\ 0 & 0 & 1 \end{pmatrix} \begin{bmatrix} X \\ Y \\ Z \end{bmatrix} \tag{3.33}$$

$$\begin{bmatrix} x'' \\ y'' \\ z'' \end{bmatrix} = \mathbf{R}_N(\theta) \begin{bmatrix} x' \\ y' \\ z' \end{bmatrix} = \begin{pmatrix} \cos\theta & 0 & -\sin\theta \\ 0 & 1 & 0 \\ \sin\theta & 0 & \cos\theta \end{pmatrix} \begin{bmatrix} x' \\ y' \\ z' \end{bmatrix} \tag{3.34}$$

and

$$\begin{bmatrix} x \\ y \\ z \end{bmatrix} = \mathbf{R}_z(\chi) \begin{bmatrix} x'' \\ y'' \\ z'' \end{bmatrix} = \begin{pmatrix} \cos\chi & \sin\chi & 0 \\ -\sin\chi & \cos\chi & 0 \\ 0 & 0 & 1 \end{pmatrix} \begin{bmatrix} x'' \\ y'' \\ z'' \end{bmatrix} \tag{3.35}$$

In Eq. (3.34), note that $x'' = x'\cos\theta - z'\sin\theta$, and so forth, because the "new" x'' coordinate is rotated out of the "old" $x'z'$ first quadrant, as shown in Figure 3.2. Equation (3.31) then takes the form

$$\begin{bmatrix} x \\ y \\ z \end{bmatrix} = \mathbf{R}_z(\chi) \mathbf{R}_N(\theta) \mathbf{R}_Z(\phi) \begin{bmatrix} X \\ Y \\ Z \end{bmatrix}$$

$$= \begin{pmatrix} \cos\chi & \sin\chi & 0 \\ -\sin\chi & \cos\chi & 0 \\ 0 & 0 & 1 \end{pmatrix} \begin{pmatrix} \cos\theta & 0 & -\sin\theta \\ 0 & 1 & 0 \\ \sin\theta & 0 & \cos\theta \end{pmatrix}$$

$$\times \begin{pmatrix} \cos\phi & \sin\phi & 0 \\ -\sin\phi & \cos\phi & 0 \\ 0 & 0 & 1 \end{pmatrix} \begin{bmatrix} X \\ Y \\ Z \end{bmatrix}$$

$$= \begin{bmatrix} c\phi\,c\theta\,c\chi - s\phi\,s\chi & s\phi\,c\theta\,c\chi + c\phi\,s\chi & -s\theta\,c\chi \\ -c\phi\,c\theta\,s\chi - s\phi\,c\chi & -s\phi\,c\theta\,s\chi + c\phi\,c\chi & s\theta\,s\chi \\ c\phi\,s\theta & s\phi\,s\theta & c\theta \end{bmatrix} \begin{bmatrix} X \\ Y \\ Z \end{bmatrix}$$

$$(3.36)$$

where c and s represent cos and sin respectively. Because the Φ_{gF} are the real elements of a unitary transformation, $\Phi_{gF}^{-1} = \Phi_{Fg}$. Then Eq. (3.30) may be rewritten as

$$\begin{bmatrix} X \\ Y \\ Z \end{bmatrix} = \begin{bmatrix} c\phi\,c\theta\,c\chi - s\phi\,s\chi & -c\phi\,c\theta\,s\chi - s\phi\,c\chi & c\phi\,s\theta \\ s\phi\,c\theta\,c\chi + c\phi\,s\chi & -s\phi\,c\theta\,s\chi + c\phi\,c\chi & s\phi\,s\theta \\ -s\theta\,c\chi & s\theta\,s\chi & c\theta \end{bmatrix} \begin{bmatrix} x \\ y \\ z \end{bmatrix}$$

$$(3.37)$$

The elements appearing in the 3×3 matrices of Eqs. (3.36) and (3.37) are simply the cosines of the angles between the various pairs of axes (see Figures 3.1 and 3.2), that is, the direction cosines Φ_{gF} and $\Phi_{Fg'}$, respectively. They have the property that

$$\sum_{g} \Phi_{Fg}\Phi_{F'g} = \delta_{FF'}$$

$$\sum_{F} \Phi_{Fg}\Phi_{Fg'} = \delta_{gg'} \qquad (3.38)$$

where the indices Fg on Φ_{Fg} denote the row and column, respectively. Equation (3.38) is nothing more than the familiar orthonormality conditions for the elements of a real unitary matrix.

3.4 REPRESENTATION OF THE ANGULAR MOMENTUM OPERATORS IN THE SPACE-FIXED AND MOLECULE-FIXED FRAMES

Previously we found that the projection of the angular momentum $\mathbf{J}$ on an arbitrary spatial axis $\hat{n}$ is the operator that generates an infinitesimal rotation about $\hat{n}$, that is, $\mathbf{J} \cdot \hat{n} = -i\partial/\partial a_n$, where a_n is the angle associated with a rotation about $\hat{n}$. By examining Figure 3.2 it is seen that the unit vectors $\hat{n}_\phi$, $\hat{n}_\theta$, and $\hat{n}_\chi$ associated with the three Euler angles can be expressed in terms of the Cartesian unit vectors $\hat{\mathbf{n}}_X$, $\hat{\mathbf{n}}_Y$, and $\hat{\mathbf{n}}_Z$ in the space-fixed frame by

$$\hat{\mathbf{n}}_\phi = \hat{\mathbf{n}}_Z$$

$$\hat{\mathbf{n}}_\theta = -\hat{\mathbf{n}}_X \sin\phi + \hat{\mathbf{n}}_Y \cos\phi$$

$$\hat{\mathbf{n}}_\chi = \hat{\mathbf{n}}_X \sin\theta \cos\phi + \hat{\mathbf{n}}_Y \sin\theta \sin\phi + \hat{\mathbf{n}}_Z \cos\theta \qquad (3.39)$$

When Eq. (3.24) is applied to Eq. (3.39), there results

$$-i\frac{\partial}{\partial\phi} = J_Z$$

$$-i\frac{\partial}{\partial\theta} = -\sin\phi J_X + \cos\phi J_Y$$

$$-i\frac{\partial}{\partial\chi} = \sin\theta \cos\phi J_X + \sin\theta \sin\phi J_Y + \cos\theta J_Z \qquad (3.40)$$

which may be solved algebraically to yield

$$J_X = -i\cos\phi\left[-\cot\theta\frac{\partial}{\partial\phi} + \frac{1}{\sin\theta}\frac{\partial}{\partial\chi}\right] + i\sin\phi\frac{\partial}{\partial\theta}$$

$$J_Y = -i\sin\phi\left[-\cot\theta\frac{\partial}{\partial\phi} + \frac{1}{\sin\theta}\frac{\partial}{\partial\chi}\right] - i\cos\phi\frac{\partial}{\partial\theta}$$

$$J_Z = -i\frac{\partial}{\partial\phi}$$

$$J_\pm = -i\exp(\pm i\phi)\left\{\left[-\cot\theta\frac{\partial}{\partial\phi} + \frac{1}{\sin\theta}\frac{\partial}{\partial\chi}\right] \pm i\frac{\partial}{\partial\theta}\right\}$$

$$\mathbf{J}^2 = -\frac{\partial^2}{\partial\theta^2} - \cot\theta\frac{\partial}{\partial\theta} - \frac{1}{\sin^2\theta}\left(\frac{\partial^2}{\partial\phi^2} + \frac{\partial^2}{\partial\chi^2} - 2\cos\theta\frac{\partial^2}{\partial\phi\partial\chi}\right) \qquad (3.41)$$

It is readily verified that J_X, J_Y, and J_Z satisfy the normal commutation rules, that is, $[J_X, J_Y] = iJ_Z$, and so on. These are the angular momentum operators in the Euler angle representation referred to the space-fixed frame.

To find the expression analogous to Eq. (3.41) in the molecule-fixed frame, we start once again with Eq. (3.39) but use the direction cosine transformation

$$\hat{\mathbf{n}}_F = \sum_g \Phi_{Fg}(\phi, \theta, \chi)\, \hat{\mathbf{n}}_g \qquad (3.42)$$

to replace the unit vectors in the space-fixed frame by the unit vectors in the molecule-fixed frame:

$$\hat{\mathbf{n}}_\phi = -\hat{\mathbf{n}}_x \sin\theta \cos\chi + \hat{\mathbf{n}}_y \sin\theta \sin\chi + \hat{\mathbf{n}}_z \cos\theta$$

$$\hat{\mathbf{n}}_\theta = \hat{\mathbf{n}}_x \sin\chi + \hat{\mathbf{n}}_y \cos\chi$$

$$\hat{\mathbf{n}}_\chi = \hat{\mathbf{n}}_z \tag{3.43}$$

Proceeding in the same manner as in Eqs. (3.40) and (3.41), we find

$$-i\frac{\partial}{\partial\phi} = -\sin\theta\cos\chi J_x + \sin\theta\sin\chi J_y + \cos\theta J_z$$

$$-i\frac{\partial}{\partial\theta} = \sin\chi J_x + \cos\chi J_y$$

$$-i\frac{\partial}{\partial\chi} = J_z \tag{3.44}$$

which may be solved algebraically to yield

$$J_x = -i\cos\chi\left[\cot\theta\frac{\partial}{\partial\chi} - \frac{1}{\sin\theta}\frac{\partial}{\partial\phi}\right] - i\sin\chi\frac{\partial}{\partial\theta}$$

$$J_y = i\sin\chi\left[\cot\theta\frac{\partial}{\partial\chi} - \frac{1}{\sin\theta}\frac{\partial}{\partial\phi}\right] - i\cos\chi\frac{\partial}{\partial\theta}$$

$$J_z = -i\frac{\partial}{\partial\chi}$$

$$J_\pm = i\exp(\mp i\chi)\left\{\left[-\cot\theta\frac{\partial}{\partial\chi} + \frac{1}{\sin\theta}\frac{\partial}{\partial\phi}\right] \mp i\frac{\partial}{\partial\theta}\right\}$$

$$\mathbf{J}^2 = -\frac{\partial^2}{\partial\theta^2} - \cot\theta\frac{\partial}{\partial\theta} - \frac{1}{\sin^2\theta}\left(\frac{\partial^2}{\partial\phi^2} + \frac{\partial^2}{\partial\chi^2} - 2\cos\theta\frac{\partial^2}{\partial\phi\partial\chi}\right) \tag{3.45}$$

If we examine the commutation relations of J_x, J_y, and J_z, we find, for example,

$$J_x J_y - J_y J_x = \cot^2\theta\frac{\partial}{\partial\chi} + \frac{d}{d\theta}(\cot\theta)\frac{\partial}{\partial\chi} = -\frac{\partial}{\partial\chi} = -iJ_z \tag{3.46}$$

and so forth. The commutation relations of J_x, J_y, and J_z are *anomalous*[9–11] in that they differ from the normal ones by a change in the sign of i.

This result is more than a mere curiosity and can lead to irreconcilable phase errors if simply ignored. To apply the angular momentum apparatus to molecular problems,

the molecular angular momentum operators must satisfy three conditions: (1) be
defined with respect to a common center, (2) have the same axes of quantization, and
(3) obey a consistent set of commutation rules. The last condition poses a dilemma.

There are many ways to resolve this paradox[11–18], but certainly one of the
simplest is to set the matrix elements of the raising and lowering operator of the
total angular momentum referred to the body-fixed frame $J_\pm = J_x \pm iJ_y$ so that it
behaves as $J_\mp = J_X \mp iJ_Y$ in the space-fixed frame[13–15]. Let $|J\Omega\rangle$ be the basis
set functions that diagonalize $\mathbf{J}^2$ and J_z:

$$\langle J'\Omega'| \mathbf{J}^2 |J\Omega\rangle = J(J + 1)\delta_{J'J}\delta_{\Omega',\Omega} \tag{3.47}$$

$$\langle J'\Omega'| J_z |J\Omega\rangle = \Omega\delta_{J'J}\delta_{\Omega',\Omega} \tag{3.48}$$

Then we insist that

$$\langle J'\Omega'| J_\pm |J\Omega\rangle = [J(J + 1) - \Omega(\Omega \mp 1)]^{\frac{1}{2}}\delta_{J'J}\delta_{\Omega',\Omega\mp1} \tag{3.49}$$

The rationale for this choice is as follows. The space-fixed $F = XYZ$ and
molecule-fixed $g = xyz$ frames rotate about one another, having only the common
origin O fixed. Thus an observer in the space-fixed frame sees the molecule-fixed
frame rotate in a sense opposite to that of an observer in the molecule-fixed frame
looking at the space-fixed frame. This is exactly the same as time reversal in classical
mechanics, where the sign of the time variable t is changed to $-t$. This operation does
not affect the value of any coordinate q, but time reversal does change the sign of the
velocity dq/dt and hence the signs of the linear and angular momenta. In particular,
the change in the sign of the angular momentum corresponds to a change in the sense
of rotation, that is, from clockwise to counterclockwise or vice versa.

What is the nature of the time reversal transformation in quantum mechanics? It
is insufficient to change only the time variable t into $-t$ for then the time-dependent
Schrödinger equation, $H\psi = (-\hbar/i)\partial\psi/\partial t$, would not be invariant under time
reversal for a time-independent Hamiltonian. In quantum mechanics, momenta are
represented by operators of the form $(\hbar/i)\partial/\partial q$ that do not explicitly contain the
time variable. However, in contrast to position coordinates, they do contain the pure
imaginary number i. This suggests a formalism in which time reversal corresponds
not only to substituting $-t$ for t but also taking the complex conjugate[19]. Let θ
denote the time reversal operator. Then in this formalism

$$\theta\mathbf{J}\theta^{-1} = \mathbf{J}^* = -\mathbf{J} \tag{3.50}$$

and in particular, $\theta J_z\theta^{-1} = -J_z$ and $\theta J_\pm\theta^{-1} = -J_\mp$. Note that the commutation
rule for the components of $\mathbf{J}$ changes sign when the sign of each component is
reversed. This suggests that all we need to do in referring $\mathbf{J}$ in the space-fixed frame
to the molecule-fixed frame is apply the time reversal operation to it, causing $J_\pm$ to
behave as $J_\mp$. Then the commutation rules for $\mathbf{J}$ referred to the molecule-fixed frame
are the same as those for the internal angular momentum operators (e.g., $\mathbf{L}$ and $\mathbf{S}$)

defined in this frame[20]. Alternatively, we can refer molecule-fixed operators to the space-fixed coordinate system, thereby removing the anomalous commutation properties. This approach is discussed in more detail in Chapter 5. In conclusion, the so-called anomalous commutation relations of J_x, J_y, and J_z are no anomaly, but a necessary consequence of time reversal symmetry.

3.5 THE ROTATION MATRICES

We return to the general problem of the transformation properties of angular momentum under rotation. First we note that an arbitrary rotation cannot change the value of J since $\mathbf{J}^2$ commutes with the rotation operator [defined in Eq. (3.23)]:

$$[\mathbf{R}_n(a), \mathbf{J}^2] = [\exp(-ia\mathbf{J} \cdot \hat{\mathbf{n}}), \mathbf{J}^2]$$

$$= \sum_\nu \frac{1}{\nu!}(-ia)^\nu \left[(\mathbf{J} \cdot \hat{\mathbf{n}})^\nu, \mathbf{J}^2\right] = 0 \tag{3.51}$$

Thus when a rotation acts on the eigenstate $|JM\rangle$ of $\mathbf{J}^2$ and J_Z, it can only transform $|JM\rangle$ into a linear combination of other M values:

$$\mathbf{R}(\phi, \theta, \chi)|JM\rangle = \sum_{M'} D^J_{M'M}(\phi, \theta, \chi)|JM'\rangle \tag{3.52}$$

where the expansion coefficients

$$D^J_{M'M}(\phi, \theta, \chi) = \langle JM'|\mathbf{R}(\phi, \theta, \chi)|JM\rangle \tag{3.53}$$

are the elements of a $(2J+1) \times (2J+1)$ unitary matrix for $\mathbf{R}$, called the *rotation matrix*[21].

Substitution of Eq. (3.26) reduces Eq. (3.53) to the form

$$D^J_{M'M}(\phi, \theta, \chi) = e^{-i\phi M'} d^J_{M'M}(\theta) e^{-i\chi M} \tag{3.54}$$

where

$$d^J_{M'M}(\theta) = \langle JM'|e^{-i\theta J_Y}|JM\rangle \tag{3.55}$$

In deriving Eq. (3.54) we made use of the identity

$$e^{-iaJ_Z}|JM\rangle = \sum_\nu \frac{1}{\nu!}(-ia)^\nu (J_Z)^\nu |JM\rangle$$

$$= \sum_\nu \frac{1}{\nu!}(-iaM)^\nu |JM\rangle$$

$$= e^{-iaM}|JM\rangle \tag{3.56}$$

A central problem to the development of the properties of the rotation matrices is the evaluation of $d^J_{M'M}(\theta)$. Equation (3.55) shows that the matrix $\mathbf{d}^J(\theta)$ can be calculated from the matrix representation of J_Y. As Wigner[21] first demonstrated, the $d^J_{M'M}(\theta)$ can be expressed as a finite polynomial in arguments of the half-angle $\theta/2$:

$$d^J_{M'M}(\theta) = \left[(J+M)!(J-M)!(J+M')!(J-M')!\right]^{\frac{1}{2}}$$

$$\times \sum_\nu \frac{(-1)^\nu}{(J-M'-\nu)!(J+M-\nu)!(\nu+M'-M)!\nu!}$$

$$\times \left[\cos\left(\frac{\theta}{2}\right)\right]^{2J+M-M'-2\nu} \left[-\sin\left(\frac{\theta}{2}\right)\right]^{M'-M+2\nu} \qquad (3.57)$$

where the sum over ν is for all integers for which the factorial arguments are nonnegative. One way in which this result may be derived is as follows.

As shown in Eq. (1.67), an arbitrary eigenfunction $|JM\rangle$ of $\mathbf{J}^2$ and J_z can be constructed in terms of the spin functions α and β by

$$|JM\rangle = \frac{(\alpha)^{J+M}(\beta)^{J-M}}{[(J+M)!(J-M)!]^{\frac{1}{2}}} \qquad (3.58)$$

Let us use α' and β' to denote the rotated spin functions. Then Eq. (3.52) may be recast as

$$\frac{(\alpha')^{J+M}(\beta')^{J-M}}{[(J+M)!(J-M)!]^{\frac{1}{2}}}$$

$$= \sum_{M'} D^J_{M'M}(R) \frac{(\alpha)^{J+M'}(\beta)^{J-M'}}{[(J+M')!(J-M')!]^{\frac{1}{2}}} \qquad (3.59)$$

In order to evaluate $D^J_{M'M}(R)$ we must find the values of α' and β' under the rotation R. We use Eq. (3.55) to write

$$d^J_{M'M}(\theta) = \langle JM'|e^{-i\theta J_Y}|JM\rangle = \langle JM'|\exp\left[\left(\frac{\theta}{2}\right)\begin{pmatrix} 0 & -1 \\ 1 & 0 \end{pmatrix}\right]|JM\rangle \quad (3.60)$$

where we have replaced J_Y by $\frac{1}{2}\sigma_Y$ [see Eq. (1.62)]. In the second line of Eq. (3.60), $\langle JM'|$ and $|JM\rangle$ are row and column vectors. Expressing the exponential of a matrix by its power series expansion we find that

$$e^{-i\theta J_Y} = I + \frac{\theta}{2}\begin{pmatrix} 0 & -1 \\ 1 & 0 \end{pmatrix} + \frac{1}{2!}\left(\frac{\theta}{2}\right)^2\begin{pmatrix} 0 & -1 \\ 1 & 0 \end{pmatrix}^2 + \cdots$$

$$= \left[1 - \frac{1}{2!}\left(\frac{\theta}{2}\right)^2 + \cdots\right]\begin{pmatrix} 1 & 0 \\ 0 & 1 \end{pmatrix}$$

$$+ \left[\frac{\theta}{2} - \frac{1}{3!}\left(\frac{\theta}{2}\right)^3 + \cdots\right]\begin{pmatrix} 0 & -1 \\ 1 & 0 \end{pmatrix}$$

$$= \begin{pmatrix} \cos\theta/2 & -\sin\theta/2 \\ \sin\theta/2 & \cos\theta/2 \end{pmatrix} \tag{3.61}$$

Thus the rotation matrix elements for $J = \frac{1}{2}$ are

$$D^{\frac{1}{2}}_{M'M}(\phi,\theta,\chi) = e^{-iM'\phi}d^{\frac{1}{2}}_{M'M}(\theta)e^{-iM\chi} \tag{3.62}$$

where the $d^{\frac{1}{2}}_{M'M}(\theta)$ is simply the $M'M$ element of the matrix shown in Eq. (3.61). For the special case $\phi = 0, \chi = 0$, the transformation of the spin functions is

$$\alpha' = \begin{pmatrix} \cos\theta/2 & -\sin\theta/2 \\ \sin\theta/2 & \cos\theta/2 \end{pmatrix}\begin{pmatrix} 1 \\ 0 \end{pmatrix} = \begin{pmatrix} \cos\theta/2 \\ \sin\theta/2 \end{pmatrix} = \alpha\cos\frac{\theta}{2} + \beta\sin\frac{\theta}{2} \tag{3.63}$$

and

$$\beta' = \begin{pmatrix} \cos\theta/2 & -\sin\theta/2 \\ \sin\theta/2 & \cos\theta/2 \end{pmatrix}\begin{pmatrix} 0 \\ 1 \end{pmatrix} = \begin{pmatrix} -\sin\theta/2 \\ \cos\theta/2 \end{pmatrix} = \beta\cos\frac{\theta}{2} + \alpha\sin\frac{\theta}{2} \tag{3.64}$$

On substituting Eqs. (3.63) and (3.64) into Eq. (3.59) and setting $\phi = 0$ and $\chi = 0$, we obtain the expression

$$[(J+M)!(J-M)!]^{-\frac{1}{2}}\left[\sum_i \frac{(J+M)!}{i!(J+M-i)!}\left(\alpha\cos\frac{\theta}{2}\right)^{J+M-i}\left(\beta\sin\frac{\theta}{2}\right)^i\right]$$

$$\times \left[\sum_j \frac{(J-M)!}{j!(J-M-j)!}\left(\beta\cos\frac{\theta}{2}\right)^{J-M-j}\left(-\alpha\sin\frac{\theta}{2}\right)^j\right]$$

$$= \sum_{M'}[(J+M')!(J-M')!]^{-\frac{1}{2}}d^J_{M'M}(\theta)\,(\alpha)^{J+M'}(\beta)^{J-M'}$$

$$\tag{3.65}$$

where we have made use of the binomial expansion

$$(a+b)^n = \sum_k n!/[k!(n-k)!]\,a^{n-k}b^k$$

We equate like powers of α and β by setting $M' = M + j - i$, that is, $j = M' - M + i$. Hence

$$\frac{d_{M'M}^J(\theta)}{[(J + M')!(J - M)!]^{\frac{1}{2}}}$$

$$= [(J + M)!(J - M)!]^{\frac{1}{2}}$$

$$\times \sum_i (-1)^{M'-M+i} \frac{\left(\cos \frac{\theta}{2}\right)^{2J+M-M'-2i} \left(\sin \frac{\theta}{2}\right)^{M'-M+2i}}{i!(M' - M + i)!(J + M - i)!(J - M' - i)!}$$

$$(3.66)$$

and Eq. (3.66) is readily rearranged to the form of Eq. (3.57) by replacing the index i with the index ν.

For reference purposes we list in Table 3.1 the matrix elements of $d_{M'M}^J(\theta)$ for $J = 0, \frac{1}{2}, 1, \frac{3}{2}$, and 2. Explicit expressions for $d_{M'M}^J(\theta)$ for $J = 3, 4, 5$, and 6 may be found elsewhere[22].

The $d_{M'M}^J$ satisfy a number of useful symmetry relations. Equation (3.57) is invariant under the substitution $M \rightarrow -M'$ and $M' \rightarrow -M$:

$$d_{M'M}^J(\theta) = d_{-M,-M'}^J(\theta) \tag{3.67}$$

Moreover, the replacement of θ by $-\theta$ in Eq. (3.57) yields

$$d_{M'M}^J(-\theta) = (-1)^{M'-M} d_{M'M}^J(\theta) \tag{3.68}$$

However, rotation by $-\theta$ is the inverse of rotation by θ, that is, $d_{M'M}^J(-\theta) = [d_{M'M}^J(\theta)]^{-1}$. The $d_{M'M}^J$ are real and are elements of a unitary transformation, specifically, $[d_{M'M}^J(\theta)]^{-1} = [d_{M'M}^J(\theta)]^\dagger = d_{MM'}^J(\theta)$. Thus Eq. (3.68) may also be written

$$d_{M'M}^J(-\theta) = d_{MM'}^J(\theta) \tag{3.69}$$

Combination of the preceding results gives

$$d_{M'M}^J(\theta) = (-1)^{M'-M} d_{MM'}^J(\theta) = (-1)^{M'-M} d_{-M'-M}^J(\theta) \tag{3.70}$$

Examination of Eq. (3.57) also shows that

$$d_{M'M}^J(0) = \delta_{M'M} \tag{3.71}$$

$$d_{M'M}^J(\pi) = (-1)^{J+M'} \delta_{M'-M} \tag{3.72}$$

TABLE 3.1   Algebraic Expressions for the $d^J_{M'M}(\theta)$ for $J = 0, \frac{1}{2}, 1, \frac{3}{2}$, and 2.

A. $J = 0$

$$d^0_{00}(\theta) \quad = 1$$

B. $J = \frac{1}{2}$

$$d^{\frac{1}{2}}_{\frac{1}{2}\frac{1}{2}}(\theta) \quad = d^{\frac{1}{2}}_{-\frac{1}{2},-\frac{1}{2}}(\theta) \; = \cos\left(\frac{\theta}{2}\right)$$

$$d^{\frac{1}{2}}_{\frac{1}{2},-\frac{1}{2}}(\theta) = -d^{\frac{1}{2}}_{-\frac{1}{2},\frac{1}{2}}(\theta) = -\sin\left(\frac{\theta}{2}\right)$$

C. $J = 1$

$$d^1_{11}(\theta) \quad = d^1_{-1,-1}(\theta) \; = \cos^2\left(\frac{\theta}{2}\right) \quad = \tfrac{1}{2}(1 + \cos\theta)$$

$$d^1_{1,-1}(\theta) \quad = d^1_{-1,1}(\theta) \; = \sin^2\left(\frac{\theta}{2}\right) \quad = \tfrac{1}{2}(1 - \cos\theta)$$

$$d^1_{01}(\theta) \quad = d^1_{-1,0}(\theta) \; = -d^1_{0,-1}(\theta) = -d^1_{10}(\theta) \qquad = \tfrac{1}{\sqrt{2}}\sin\theta$$

$$d^1_{00}(\theta) \quad = \cos\theta$$

D. $J = \frac{3}{2}$

$$d^{\frac{3}{2}}_{\frac{3}{2}\frac{3}{2}}(\theta) \quad = d^{\frac{3}{2}}_{-\frac{3}{2}-\frac{3}{2}}(\theta) \; = \cos^3\left(\frac{\theta}{2}\right)$$

$$d^{\frac{3}{2}}_{\frac{3}{2}\frac{1}{2}}(\theta) \quad = d^{\frac{3}{2}}_{-\frac{1}{2}-\frac{3}{2}}(\theta) \; = -d^{\frac{3}{2}}_{\frac{1}{2}\frac{3}{2}}(\theta) \quad = -d^{\frac{3}{2}}_{-\frac{3}{2}-\frac{1}{2}}(\theta) \; = -\sqrt{3}\cos^2\left(\frac{\theta}{2}\right)\sin\left(\frac{\theta}{2}\right)$$

$$d^{\frac{3}{2}}_{\frac{3}{2}-\frac{1}{2}}(\theta) = d^{\frac{3}{2}}_{-\frac{1}{2}\frac{3}{2}}(\theta) \; = d^{\frac{3}{2}}_{\frac{1}{2}-\frac{3}{2}}(\theta) \quad = d^{\frac{3}{2}}_{-\frac{3}{2}\frac{1}{2}}(\theta) \; = \sqrt{3}\cos\left(\frac{\theta}{2}\right)\sin^2\left(\frac{\theta}{2}\right)$$

$$d^{\frac{3}{2}}_{\frac{3}{2}-\frac{3}{2}}(\theta) = -d^{\frac{3}{2}}_{-\frac{3}{2}\frac{3}{2}}(\theta) \; = -\sin^3\left(\frac{\theta}{2}\right)$$

$$d^{\frac{3}{2}}_{\frac{1}{2}\frac{1}{2}}(\theta) \quad = d^{\frac{3}{2}}_{-\frac{1}{2}-\frac{1}{2}}(\theta) \; = \cos\left(\frac{\theta}{2}\right)\left[3\cos^2\left(\frac{\theta}{2}\right) - 2\right]$$

$$d^{\frac{3}{2}}_{\frac{1}{2}-\frac{1}{2}}(\theta) = d^{\frac{3}{2}}_{-\frac{1}{2}\frac{1}{2}}(\theta) \; = \sin\left(\frac{\theta}{2}\right)\left[3\sin^2\left(\frac{\theta}{2}\right) - 2\right]$$

E. $J = 2$

$$d^2_{22}(\theta) \quad = d^2_{-2-2}(\theta) \; = \cos^4\left(\frac{\theta}{2}\right)$$

$$d^2_{21}(\theta) \quad = -d^2_{12}(\theta) \quad = -d^2_{-2-1}(\theta) = -d^2_{-1-2}(\theta) \; = -\tfrac{1}{2}\sin\theta(1 + \cos\theta)$$

$$d^2_{20}(\theta) \quad = d^2_{02}(\theta) \quad = d^2_{-20}(\theta) \quad = d^2_{0-2}(\theta) \quad = \sqrt{\tfrac{3}{8}}\sin^2\theta$$

$$d^2_{2-1}(\theta) \quad = d^2_{1-2}(\theta) \quad = -d^2_{-21}(\theta) \; = -d^2_{-12}(\theta) \; = \tfrac{1}{2}\sin\theta(\cos\theta - 1)$$

$$d^2_{2-2}(\theta) \quad = d^2_{-22}(\theta) \quad = \sin^4\left(\frac{\theta}{2}\right)$$

$$d^2_{11}(\theta) \quad = d^2_{-1-1}(\theta) \; = \tfrac{1}{2}(2\cos\theta - 1)(\cos\theta + 1)$$

$$d^2_{1-1}(\theta) \quad = d^2_{-11}(\theta) \quad = \tfrac{1}{2}(2\cos\theta + 1)(1 - \cos\theta)$$

$$d^2_{10}(\theta) \quad = d^2_{0-1}(\theta) \quad = -d^2_{01}(\theta) \quad = -d^2_{-10}(\theta) \; = \sqrt{\tfrac{3}{2}}\sin\theta\cos\theta$$

$$d^2_{00}(\theta) \quad = \tfrac{1}{2}(3\cos^2\theta - 1)$$

and

$$d_{M'M}^{J}(-\pi) = (-1)^{J-M'}\delta_{M'-M} \tag{3.73}$$

Because successive rotations correspond to multiplication of the appropriate rotation matrices, we find

$$d_{M'M}^{J}(\pi + \theta) = \sum_{M''} d_{M'M''}^{J}(\pi) \, d_{M''M}^{J}(\theta)$$

$$= (-1)^{J+M'} d_{-M'\,M}^{J}(\theta) \tag{3.74}$$

and

$$d_{M'M}^{J}(\pi - \theta) = \sum_{M''} d_{M'M''}^{J}(\pi) \, d_{M''M}^{J}(-\theta)$$

$$= (-1)^{J+M'} d_{-M'\,M}^{J}(-\theta)$$

$$= (-1)^{J+M'} d_{M\,-M'}^{J}(\theta) \tag{3.75}$$

In particular when $\theta = \pi$, Eq. (3.74) yields the result

$$d_{M'M}^{J}(2\pi) = (-1)^{J+M'} d_{-M'\,M}^{J}(\pi)$$

$$= (-1)^{2J}\delta_{M'M}$$

$$= (-1)^{2J} d_{M'M}^{J}(0) \tag{3.76}$$

Hence for integral J a rotation by $\theta = 2\pi$ is the same as a rotation by $\theta = 0$, in agreement with our physical intuition for bodies in three-dimensional space. However, for half-integral J, there is a change of sign in going from $\theta = 0$ to $\theta = 2\pi$, and a rotation by $\theta = 4\pi$ is required to return the system to its initial state. Thus in the range $(0, 2\pi)$ the rotation matrices for spin angular momenta are double-valued, the two values differing by a phase factor of -1. This is another manifestation of the nonintuitive behavior of spin angular momenta. Provided care is taken to make a consistent phase choice throughout a calculation, this ambiguity, however, does not affect observable quantities.

Finally, because the rotation matrices are unitary, they satisfy the sum rules

$$\sum_{M'} \left[D_{M'M}^{J}(R)\right]^{\dagger} D_{M'N}^{J}(R) = \sum_{M'} d_{MM'}^{J}(\theta) \, d_{M'N}^{J}(\theta) = \delta_{MN} \tag{3.77}$$

and

$$\sum_{M} \left[D_{M'M}^{J}(R)\right]^{\dagger} D_{N'M}^{J}(R) = \sum_{M} d_{MM'}^{J}(\theta) \, d_{N'M}^{J}(\theta) = \delta_{M'N'} \tag{3.78}$$

where the argument R represents the Euler angles ϕ, θ, and χ, and † is the transpose complex conjugate.

3.6 THE ROTATION MATRICES: GEOMETRICAL INTERPRETATION

We can readily interpret Eq. (3.52) once again in terms of the vector model[23]. The state $|JM\rangle$ is represented by a vector of length $\sqrt{J(J+1)}$ making a fixed projection M on the Z axis. The indeterminancy in the values of J_X and J_Y is represented in the vector model by making $\mathbf{J}$ precess uniformly about the Z axis. Suppose that the rotation $\mathbf{R}(\phi, \theta, \chi)$ transforms the system from the "old" XYZ frame to the "new" frame $X'Y'Z'$ with a common origin O. Then the OZ' axis will be inclined at an angle θ with respect to the OZ axis, and the probability for finding $\mathbf{J}$ making the projection M' on the OZ' axis is just

$$\left| D_{M'M}^{J}(\phi, \theta, \chi) \right|^2 = \left[d_{M'M}^{J}(\theta) \right]^2 \tag{3.79}$$

Thus $D_{M'M}^{J}(\phi, \theta, \chi)$ represents the probability amplitude that an angular momentum vector $\mathbf{J}$ making a projection M on the OZ axis will be found making a projection M' on the OZ' axis that is related to the original frame by the rotation $\mathbf{R}(\phi, \theta, \chi)$. The rotation $\mathbf{R}$ includes the two Euler angles ϕ and χ representing azimuthal rotation about the old and new Z axes, respectively. Equation (3.54) shows that these angles are associated with the phase factors $e^{-iM\phi}$ and $e^{-iM'\chi}$, which can be seen from Eq. (3.79) to play no role in the probability interpretation described above. On the basis of the vector model, we should expect to find the spread in the values of M' between the two dotted projection lines shown in Figure 3.3. However, the vector model is not an exact description. The effect of quantum indeterminancy of J_X and J_Y is to allow values of M' to fall outside these limits. The probability $[d_{M'M}^{J}(\theta)]^2$ can be shown to die exponentially in these classically forbidden regions and to oscillate in the classically allowed region[23]. The number of oscillations increases with J, and, as expected, the envelope of oscillations approaches the classical limit predicted by the vector model.

This geometric interpretation of the rotation matrices allows us to assess at once without detailed calculation the relative absolute magnitudes of the $d_{M'M}^{J}(\theta)$ for any pair of M, M' for a given J as a function of the angle θ. An instructive example concerns molecular beam magnetic resonance[24–26]. A schematic of the experimental setup is shown in Figure 3.4. If an atom or molecule with a magnetic moment $\boldsymbol{\mu} = g\mu_0 \mathbf{J}$ (where μ_0 is the electronic Bohr magneton, $\mathbf{J}$ the total angular momentum, and g a proportionality constant called the g factor) passes through an inhomogeneous magnetic field whose gradient is perpendicular to the direction of motion, the particle experiences a force proportional to the component of $\boldsymbol{\mu}$ projected along the direction of the gradient. Let us choose the Z axis to be along the direction of the magnetic field gradient. Then a beam will be split into $2J + 1$ components on emerging from the inhomogeneous field, each corresponding to the expectation value of J_Z, that is, M. Thus in the classic Stern–Gerlach experiment, a spin $\frac{1}{2}$

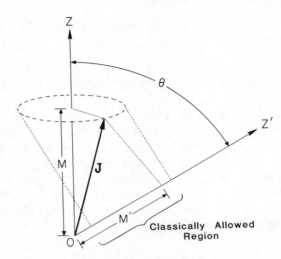

FIGURE 3.3   Vector model picture for the rotation of the state function $|JM\rangle$ from Z to Z' by the angle θ.

system on passing through the inhomogeneous field is deflected along the trajectories symmetrically displaced above and below the initial direction and produces two symmetric spots when the particles impinge on a screen perpendicular to their initial direction of motion. Each spot corresponds to particles in the state $|\frac{1}{2} \frac{1}{2}\rangle$ or $|\frac{1}{2} -\frac{1}{2}\rangle$. In the experiment shown in Figure 3.4, the first Stern–Gerlach magnet, labeled A, is used as a state selector to prepare a beam of (noninteracting) particles in a particular state $|JM\rangle$; all other particles with $M' \neq M$ are removed by beam stops. The particles next pass through the so-called interaction region having a homogeneous magnetic field along Z. In the interaction region external perturbations may be applied, such as a radiofrequency (rf) field to induce changes in M. The particles pass through a second inhomogeneous magnetic field, labeled B, which acts as an M state analyzer and then are detected (counted) by various means, often by the use of an electron impact ionization mass spectrometer. In the interaction region as the homogeneous magnetic field strength is changed for a fixed rf frequency, or alternatively as the rf frequency is swept for a fixed magnetic field strength, a different count rate is observed when the energy difference between neighboring M levels matches the energy of the rf photon (the resonance condition). In this manner an rf spectrum of the beam species is obtained by using a molecular beam magnetic resonance spectrometer[24–27].

Let us assume at first that the beam is unaffected in its journey through the interaction region, that is, that no external perturbations are present in the C region. We wish to know the transmission probability when the B magnet is inclined at angle θ with respect to the A magnet (double Stern–Gerlach experiment). Let the detector be set to intercept only those particles in the state $|JM'\rangle$. We wish to determine how many particles are incident on the detector, that is, what fraction of the

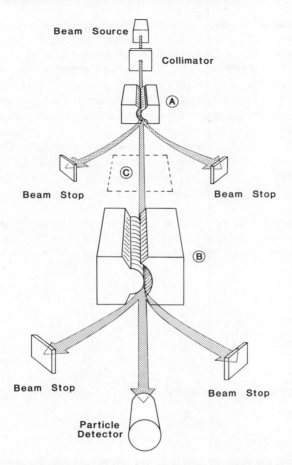

FIGURE 3.4 Schematic of a molecular beam magnetic resonance spectrometer. Two Stern–Gerlach (inhomogeneous) magnets A and B are separated by an interaction region C in which some external perturbation such as a radiofrequency (rf) field may be used to induce transitions. Magnet A and its beam stops serve to prepare the state $|JM\rangle$ of a beam of noninteracting particles; magnet B, beam stops, and detector serve to analyze particles in the state $|JM'\rangle$. For simplicity, a $J = 1$ system is pictured.

particles in the state $|JM\rangle$ find themselves later on passing through magnet B in the state $|JM'\rangle$. Clearly, the answer to this question is given simply by $|\langle JM|JM'\rangle|^2 = [d^J_{M'M}(\theta)]^2$. Thus knowledge of the rotation matrices is equivalent to understanding the outcome of a double Stern–Gerlach experiment. Moreover, the vector model picture for the rotation matrices allows us to make qualitative predictions.

Let us consider some special cases in the double Stern–Gerlach experiment. When $\theta = 0°$, Eq. (3.71) shows that $[d^J_{M'M}(0)]^2 = \delta_{M'M}$; all particles appear at the same M value. This agrees with the vector model and is a consequence of the fact that $|JM\rangle$ is an eigenstate of J_Z and a "remeasurement" of J_Z cannot alter this state. As θ begins to increase, other $M' \neq M$ states begin to appear, but those with the same sign as M are the more populated (for $M \neq 0$). At $\theta = 90°$, many M' states are present but the population in $|JM'\rangle$ is the same as that in $|J-M'\rangle$. Again this result follows at once from the vector model or may be shown to be a consequence of Eq. (3.75) with θ set equal to $\pi/2$. As θ becomes larger than $90°$, M' states having a sign opposite to M become the more populated (for $M \neq 0$). At $\theta = 180°$, the vector model predicts that no particles with $M' = M$ are transmitted for $M \neq 0$; that is, the polarizer–analyzer pair of Stern–Gerlach magnets are "crossed." The same result is seen to be a consequence of Eqs. (3.72) and (3.73), which show that $[d^J_{M'M}(\pm\pi)]^2 = \delta_{M'-M}$; all particles appear at the value of M' of the same magnitude but opposite sign when the direction of the inhomogeneous magnetic field is reversed.

Let us suppose that the perturbation in the interaction region C has the effect of only altering the populations of the individual $|JM\rangle$ states. Then we represent the new state $|JM_C\rangle$ in terms of the $|JM\rangle$ basis functions as

$$|JM_C\rangle = \sum_M f_M |JM\rangle \tag{3.80}$$

where the f_M coefficients are real and have the property that

$$\sum_M f_M^2 = 1 \tag{3.81}$$

The detector signal is then proportional to

$$S \propto \sum_M f_M^2 [d^J_{M'M}(\theta)]^2 \tag{3.82}$$

For example, with the polarizer and analyzer magnets crossed, a fraction f_M^2 of the original beam will be transmitted and the detector will respond only to transitions in the interaction region in which M is changed into $M_C = -M$. In particular, let us imagine that the rf field was replaced by another Stern–Gerlach magnet whose gradient direction is at an angle θ_C with respect to magnet A. Then

$$|JM_C\rangle = \sum_M D^J_{M_CM}(\phi_C, \theta_C, \chi_C) |JM\rangle \tag{3.83}$$

and

$$S \propto \sum_M [d^J_{M_CM}(\theta_C)]^2 [d^J_{M'M}(\theta)]^2 \tag{3.84}$$

Equation (3.84) shows that particles will pass through the crossed polarizer–analyzer combination provided the gradient direction of the intermediate Stern–Gerlach magnet is parallel to neither magnet A nor magnet B[28]. Again this conclusion is anticipated on the basis of the geometric interpretation of the rotation matrices. Thus magnetic resonance spectroscopy and a triple Stern–Gerlach experiment have many similarities to each other.

3.7   THE SPHERICAL HARMONIC ADDITION THEOREM

An extremely useful relation between the coordinates of particles i and j is expressed by the spherical harmonic addition theorem:

$$P_L(\cos\theta_{ij}) = \frac{4\pi}{2L+1} \sum_M Y_{LM}^*(\theta_i,\phi_i) Y_{LM}(\theta_j,\phi_j) \tag{3.85}$$

where (θ_i,ϕ_i) and (θ_j,ϕ_j) are the spherical polar coordinates of i and j, respectively, and θ_{ij} is the angle between the position vectors to these two particles (Figure 3.5). In particular, for the case $L = 1$, Eq. (3.85) reduces to the familiar expression

$$\cos\theta_{ij} = \cos\theta_i\cos\theta_j + \sin\theta_i\sin\theta_j\cos(\phi_i - \phi_j) \tag{3.86}$$

We may derive Eq. (3.85) by specializing Eq. (3.52) to the case of integral J, which we denote by the orbital angular momentum L. Then in the spherical polar coordinate representation

$$\mathbf{R}Y_{LM}(\theta_i,\phi_i) = \sum_{M'} D_{M'M}^L(\phi,\theta,\chi)\, Y_{LM'}(\theta_i,\phi_i)$$

$$= Y_{LM}(\theta_i',\phi_i') \tag{3.87}$$

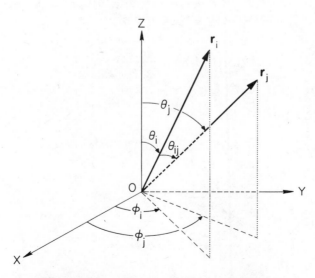

FIGURE 3.5   Spherical coordinates of particles i and j.

where the rotation $\mathbf{R}(\phi, \theta, \chi)$ causes the primed and unprimed frames to coincide. Let us consider the quantity

$$Q = \sum_M Y_{LM}^*(\theta_i, \phi_i) Y_{LM}(\theta_j, \phi_j) \tag{3.88}$$

under the rotational transformation $\mathbf{R}$:

$$Q' = \sum_M \left[\mathbf{R} Y_{LM}^*(\theta_i, \phi_i) \right] \left[\mathbf{R} Y_{LM}(\theta_j, \phi_j) \right]$$

$$= \sum_M \sum_{M'} \sum_{M''} \left[D_{M'M}^{L*}(\phi, \theta, \chi) \right]$$

$$\times \left[D_{M''M}^L(\phi, \theta, \chi) \right] Y_{LM'}^*(\theta_i, \phi_i) Y_{LM''}(\theta_j, \phi_j)$$

$$= \sum_{M'} \sum_{M''} \delta_{M'M''} Y_{LM'}^*(\theta_i, \phi_i) Y_{LM''}(\theta_j, \phi_j)$$

$$= \sum_{M'} Y_{LM'}^*(\theta_i, \phi_i) Y_{LM'}(\theta_j, \phi_j)$$

$$= Q \tag{3.89}$$

where we have used Eqs. (3.78) and (3.87). Thus we conclude that Q is invariant to rotation, that is, is a scalar quantity. We can select therefore any set of axes in Figure 3.5. We choose $\theta_i = 0$, ϕ_i arbitrary, $\theta_j = \theta_{ij}$, and ϕ_j arbitrary. From the properties of spherical harmonics [see Eq. (1.55)]

$$Y_{LM}(0, \phi_i) = \left(\frac{2L+1}{4\pi} \right)^{\frac{1}{2}} \delta_{M,0} \tag{3.90}$$

and

$$Q = \sum_M \left(\frac{2L+1}{4\pi} \right)^{\frac{1}{2}} \delta_{M0} Y_{LM}(\theta_{ij}, \phi_j)$$

$$= \left(\frac{2L+1}{4\pi} \right)^{\frac{1}{2}} Y_{L0}(\theta_{ij}, \phi_j)$$

$$= \left(\frac{2L+1}{4\pi} \right) P_L(\cos \theta_{ij}) \tag{3.91}$$

which completes the proof of the spherical harmonic addition theorem.

We specialize Eq. (3.87) to the case where $M = 0$. We find

$$P_L(\cos \theta_i') = \left(\frac{2L+1}{4\pi}\right)^{-\frac{1}{2}} \sum_{M'} D_{M'0}^L (\phi, \theta, \chi) \, Y_{LM'}(\theta_i, \phi_i) \qquad (3.92)$$

Reference to Eq. (3.85) yields the identification

$$D_{M0}^L (\phi, \theta, \chi) = \left(\frac{4\pi}{2L+1}\right)^{\frac{1}{2}} Y_{LM}^*(\theta, \phi) \qquad (3.93)$$

which is the same as

$$Y_{LM}(\theta, \phi) = \left(\frac{2L+1}{4\pi}\right)^{\frac{1}{2}} D_{M0}^{L*} (\phi, \theta, \chi) \qquad (3.94)$$

In particular,

$$D_{00}^L (\phi, \theta, \chi) = P_L(\cos \theta) \qquad (3.95)$$

Equation (3.93) may also be written

$$d_{M0}^L(\theta) e^{-iM\phi} = \left(\frac{4\pi}{2L+1}\right)^{\frac{1}{2}} Y_{LM}^*(\theta, \phi)$$

$$= (-1)^M \left[\frac{(L-M)!}{(L+M)!}\right]^{\frac{1}{2}} P_L^M(\cos \theta) e^{-iM\phi} \qquad (3.96)$$

It follows that

$$D_{0M}^L (\phi, \theta, \chi) = d_{0M}^L(\theta) e^{-iM\chi} = (-1)^M d_{M0}^L(\theta) e^{-iM\chi}$$

$$= (-1)^M \left(\frac{4\pi}{2L+1}\right)^{\frac{1}{2}} Y_{LM}^*(\theta, \chi)$$

$$= \left(\frac{4\pi}{2L+1}\right)^{\frac{1}{2}} Y_{L-M}(\theta, \chi) \qquad (3.97)$$

where we have used Eq. (3.93) and the relation [see Eq. (1.56)]

$$Y_{LM}^* = (-1)^M Y_{L-M} \qquad (3.98)$$

As we saw in Eq. (3.86), the $L = 1$ case has great practical importance. In particular, the rotation matrix for $L = 1$ occurs so often that it deserves some special consideration. For $L = 1$, Eq. (3.87) becomes

$$Y_{1M}(\theta_i', \phi_i') = \sum_{M'} D_{M'M}^1 (\phi, \theta, \chi) \, Y_{1M'}(\theta_i, \phi_i) \qquad (3.99)$$

The spherical harmonics Y_{1M} of order one have the explicit form in Cartesian components

$$Y_{11}(\theta, \phi) = \frac{1}{r}\sqrt{\frac{3}{4\pi}}\left[-\frac{1}{\sqrt{2}}(x + iy)\right]$$

$$Y_{10}(\theta, \phi) = \frac{1}{r}\sqrt{\frac{3}{4\pi}}\,[z]$$

$$Y_{1\,-1}(\theta, \phi) = \frac{1}{r}\sqrt{\frac{3}{4\pi}}\left[\frac{1}{\sqrt{2}}(x - iy)\right] \tag{3.100}$$

The factor $r^{-1}(3/4\pi)^{\frac{1}{2}}$ appearing in Eq. (3.100) is invariant under rotation so that the transformation properties of $Y_{1M}(\theta, \phi)$ are determined by the linear combination of Cartesian components indicated in square brackets. This suggests expressing an arbitrary vector $\mathbf{r}$ in this so-called *spherical basis*. For the space-fixed frame we define r_M^F by

$$r_{\pm 1}^F = \mp\frac{1}{\sqrt{2}}(r_X \pm ir_Y), \qquad\qquad r_0^F = r_Z \tag{3.101}$$

and for the molecule-fixed frame by

$$r_{\pm 1}^g = \mp\frac{1}{\sqrt{2}}(r_x \pm ir_y), \qquad\qquad r_0^g = r_z \tag{3.102}$$

The two descriptions are related by

$$r_M^F = \sum_{M'} D_{MM'}^1(\phi, \theta, \chi)\, r_{M'}^g \tag{3.103}$$

Note that Eq. (3.103) differs from Eq. (3.99) in the form of the rotational transformation. Comparing Eq. (3.103) to Eq. (3.30), we conclude that $D_{MM'}^1(\phi, \theta, \chi)$ and $\Phi_{Fg}(\phi, \theta, \chi)$ are related by a unitary transformation, which is the transformation from a spherical basis to a Cartesian basis[29]. Thus the rotation matrix elements for $L = 1$ are no more or less familiar than the direction cosine matrix elements; either can be written as a linear combination of the other.

3.8 THE CLEBSCH–GORDAN SERIES AND ITS INVERSE

The general properties of the rotation matrices may be further developed by considering the connection between the uncoupled $|J_1 M_1\rangle|J_2 M_2\rangle$ and coupled $|J_3 M_3\rangle$

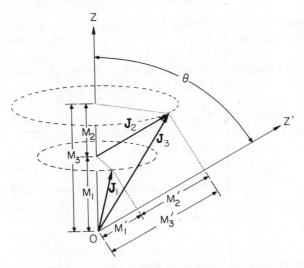

FIGURE 3.6   Vector model picture of the Clebsch–Gordan series. The angular momenta J_1 and J_2 precess independently, making the projections M_1 and M_2 on the Z axis.

representations under a rotational transformation. Applying a rotation $\mathbf{R}$ to both sides of Eq. (2.6) and using Eq. (3.52), we obtain the relation

$$\sum_{M_1'}\sum_{M_2'} D^{J_1}_{M_1'M_1}(R)\, D^{J_2}_{M_2'M_2}(R)\, |J_1\,M_1'\rangle\, |J_2\,M_2'\rangle$$

$$= \sum_{J_3}\sum_{M_3'} \langle J_1\,M_1, J_2\,M_2|J_3\,M_3\rangle\, D^{J_3}_{M_3'M_3}(R)\, |J_3\,M_3'\rangle \qquad (3.104)$$

We multiply both sides of Eq. (3.104) by $\langle J_1\,M_1'|\,\langle J_2\,M_2'|$ and obtain

$$
\boxed{
\begin{aligned}
&D^{J_1}_{M_1'M_1}(R)\,D^{J_2}_{M_2'M_2}(R)\\[2mm]
&= \sum_{J_3} \langle J_1\,M_1, J_2\,M_2|J_3\,M_3\rangle \langle J_1\,M_1', J_2\,M_2'|J_3\,M_3'\rangle\, D^{J_3}_{M_3'M_3}(R)
\end{aligned}
}
$$

$$(3.105)$$

which is called the *Clebsch–Gordan series*. If desired, this can be given a simple probability interpretation (see Figure 3.6) in terms of the vector model pictures shown in Figures 1.1, 1.2, 1.3, and 2.2. Suppose that the two angular momenta $\mathbf{J}_1$ and $\mathbf{J}_2$ independently precess about the Z axis with projections M_1

and M_2. Consider a rotation through angular coordinates R. The probability amplitude that the projections of $\mathbf{J}_1$ and $\mathbf{J}_2$ in the new frame are M_1' and M_2' is given by $D_{M_1' M_1}^{J_1}(R) D_{M_2' M_2}^{J_2}(R)$. Let $\mathbf{J}_1$ and $\mathbf{J}_2$ couple to form $\mathbf{J}_3$. The probability amplitude that a particular value of J_3 is formed in the old coordinate frame is $\langle J_1 M_1, J_2 M_2 | J_3 M_3 \rangle$, the probability amplitude that the same value of J_3 results in the new coordinate frame is $\langle J_1 M_1', J_2 M_2' | J_3 M_3' \rangle$, and the probability amplitude that $\mathbf{J}_3$ having the projection $M_1 + M_2 = M_3$ in the old frame makes the projection $M_1' + M_2' = M_3'$ in the new frame is $D_{M_3' M_3}^{J_3}(R)$. Combining these probabilities and summing over all possible $\mathbf{J}$ values that can be the resultant of $\mathbf{J}_1$ and $\mathbf{J}_2$, we have $\sum_{J_3} \langle J_1 M_1, J_2 M_2 | J_3 M_3 \rangle \langle J_1 M_1', J_2 M_2' | J_3 M_3' \rangle D_{M_3' M_3}^{J_3}(R)$. This may be equated to $D_{M_1' M_1}^{J_1}(R) D_{M_2' M_2}^{J_2}(R)$ to obtain the Clebsch–Gordan series. Thus we see that in spite of the formidable appearance of Eq. (3.105) it has the geometrical interpretation of relating the way the uncoupled and coupled representations transform under rotation.

An alternative expression is obtained by applying a rotation by $\mathbf{R}$ to both sides of Eq. (2.5). Using the same procedure, we find

$$
D_{M_3' M_3}^{J_3}(R) = \sum_{M_1, M_1'} \langle J_1 M_1, J_2 M_2 | J_3 M_3 \rangle \langle J_1 M_1', J_2 M_2' | J_3 M_3' \rangle
$$

$$
\times D_{M_1' M_1}^{J_1}(R) D_{M_2' M_2}^{J_2}(R) \tag{3.106}
$$

which is called the *inverse Clebsch–Gordan series*.

3.9 INTEGRALS OVER PRODUCTS OF ROTATION MATRICES

Equations (3.105) and (3.106) allow us to evaluate the integral over two rotation matrices having the same arguments where the solid angle element $d\Omega$ is $d\phi \sin\theta \, d\theta \, d\chi$:

$$
\int_0^{2\pi} d\phi \int_0^{\pi} \sin\theta \, d\theta \int_0^{2\pi} d\chi \; D_{M_1' M_1}^{J_1 *}(R) D_{M_2' M_2}^{J_2}(R)
$$

$$
= \int d\Omega \; (-1)^{M_1' - M_1} D_{-M_1' -M_1}^{J}(R) D_{M_2' M_2}^{J_2}(R)
$$

$$
= \sum_{J_3} (-1)^{M_1' - M_1} \langle J_1 -M_1, J_2 M_2 | J_3, -M_1 + M_2 \rangle
$$

$$
\times \langle J_1 -M_1', J_2 M_2' | J_3, -M_1' + M_2' \rangle
$$

$$\times \int d\Omega\, D^{J_3}_{-M'_1+M'_2,-M_1+M_2}(R) \qquad (3.107)$$

The integration over the angles ϕ and χ are readily carried out:

$$\int_0^{2\pi} d\phi\, e^{-i(M'_2-M'_1)\phi} = 2\pi\, \delta_{M'_1 M'_2}$$

$$\int_0^{2\pi} d\chi\, e^{-i(M_2-M_1)\chi} = 2\pi\, \delta_{M_1 M_2} \qquad (3.108)$$

The integration over the angle θ becomes

$$\int_0^\pi \sin\theta d\theta\, d^{J_3}_{00}(\theta) = \int_0^\pi \sin\theta d\theta\, P_{J_3}(\cos\theta)$$

$$= 2\,\delta_{J_3 0} \qquad (3.109)$$

where we have used the orthogonality properties of the Legendre polynomials [see Eq. (1.59)]. Thus Eq. (3.107) becomes

$$\int d\Omega\, D^{J_1^*}_{M'_1 M_1}(R)\, D^{J_2}_{M'_2 M_2}(R)$$

$$= 8\pi^2 (-1)^{M'_1-M_1} \langle J_1-M_1, J_2 M_1 | 00 \rangle$$

$$\times \langle J_1-M'_1, J_2 M'_1 | 00 \rangle \qquad (3.110)$$

However, according to Eq. (2.28) and with the help of Table 2.5,

$$\langle J_1-M_1, J_2 M_1 | 00 \rangle = (-1)^{J_1-J_2} \begin{pmatrix} J_1 & J_2 & 0 \\ -M_1 & M_1 & 0 \end{pmatrix}$$

$$= (-1)^{J_1-M_1}(2J_1+1)^{-\frac{1}{2}}\delta_{J_1 J_2} \qquad (3.111)$$

and similarly

$$\langle J_1-M'_1, J_2 M'_1 | 00 \rangle = (-1)^{J_1-M'_1}(2J_1+1)^{-\frac{1}{2}}\delta_{J_1 J_2} \qquad (3.112)$$

Hence we conclude that[30]

$$\int d\Omega\, D^{J_1^*}_{M'_1 M_1}(R)\, D^{J_2}_{M'_2 M_2}(R) = \frac{8\pi^2}{2J_1+1}\delta_{J_1 J_2}\delta_{M'_1 M'_2}\delta_{M_1 M_2} \qquad (3.113)$$

and a normalized rotation matrix is $[(2J + 1)/8\pi^2]^{\frac{1}{2}} D^J_{M'M}(R)$. In Eq. (3.113) the phase is unity since $(-1)^{2(J_1-M_1)} = 1$ regardless of whether J_1 is integral or half-integral.

With the aid of the Clebsch–Gordan series and Eq. (3.113), we evaluate the integral over a triple product of rotation matrices having the same angular arguments:

$$\int d\Omega\, D^{J_3^*}_{M_3'M_3}(R)\, D^{J_2}_{M_2'M_2}(R)\, D^{J_1}_{M_1'M_1}(R)$$

$$= \sum_J \langle J_1 M_1, J_2 M_2 | J, M_1 + M_2 \rangle \langle J_1 M_1', J_2 M_2' | J, M_1' + M_2' \rangle$$

$$\times \int d\Omega\, D^{J_3^*}_{M_3'M_3}(R)\, D^J_{M_1'+M_2', M_1+M_2}(R)$$

$$= \sum_J \langle J_1 M_1, J_2 M_2 | J, M_1 + M_2 \rangle \langle J_1 M_1', J_2 M_2' | J, M_1' + M_2' \rangle$$

$$\times \frac{8\pi^2}{2J_3 + 1} \delta_{M_1+M_2, M_3} \delta_{M_1'+M_2', M_3'} \delta_{J_3, J}$$

$$= \frac{8\pi^2}{2J_3 + 1} \langle J_1 M_1, J_2 M_2 | J_3 M_3 \rangle \langle J_1 M_1', J_2 M_2' | J_3 M_3' \rangle \qquad (3.114)$$

This integral occurs in various applications. We shall encounter it again in the calculation of rotational line strengths. A special case of Eq. (3.114), for example, is the integral of three spherical harmonics

$$\int d\Omega\, Y^*_{L_3 M_3}(\theta, \phi) Y_{L_2 M_2}(\theta, \phi) Y_{L_1 M_1}(\theta, \phi)$$

$$= \left[\frac{(2L_1 + 1)(2L_2 + 1)}{4\pi(2L_3 + 1)} \right]^{\frac{1}{2}} \langle L_1 M_1, L_2 M_2 | L_3 M_3 \rangle \langle L_1 0, L_2 0 | L_3 0 \rangle$$

$$(3.115)$$

where here $d\Omega = \sin\theta\, d\theta\, d\phi$. Equation (3.115) may be derived from Eq. (3.114) by setting to zero the indices M_1, M_2, and M_3 and using Eq. (3.93). This integral is needed for the calculation of the atomic structure multiplet splittings in Russell–Saunders coupling[31]. If $L_2 = 1$, this expression also occurs in the calculation of matrix elements for electric dipole radiation between angular momentum states for spinless particles.

The use of 3-j symbols in place of Clebsch–Gordan coefficients allows us to rewrite the results presented in this section. In particular, it is readily shown that

$$
D^{J_1}_{M_1' M_1}(R)\, D^{J_2}_{M_2' M_2}(R) = \sum_{J_3} (2J_3 + 1) \begin{pmatrix} J_1 & J_2 & J_3 \\ M_1' & M_2' & M_3' \end{pmatrix}
$$

$$
\times \begin{pmatrix} J_1 & J_2 & J_3 \\ M_1 & M_2 & M_3 \end{pmatrix} D^{J_3*}_{M_3' M_3}(R) \qquad (3.116)
$$

$$
D^{J_3*}_{M_3' M_3}(R) = \sum_{M_1 M_1'} (2J_3 + 1) \begin{pmatrix} J_1 & J_2 & J_3 \\ M_1' & M_2' & M_3' \end{pmatrix} \begin{pmatrix} J_1 & J_2 & J_3 \\ M_1 & M_2 & M_3 \end{pmatrix}
$$

$$
\times D^{J_1}_{M_1' M_1}(R)\, D^{J_2}_{M_2' M_2}(R) \qquad (3.117)
$$

$$
\int D^{J_3}_{M_3' M_3}(R)\, D^{J_2}_{M_2' M_2}(R)\, D^{J_1}_{M_1' M_1}(R)\, d\Omega
$$

$$
= 8\pi^2 \begin{pmatrix} J_1 & J_2 & J_3 \\ M_1' & M_2' & M_3' \end{pmatrix} \begin{pmatrix} J_1 & J_2 & J_3 \\ M_1 & M_2 & M_3 \end{pmatrix} \qquad (3.118)
$$

and that

$$
\int_0^{2\pi} \int_0^{\pi} Y_{J_3 M_3}(\theta, \phi)\, Y_{J_2 M_2}(\theta, \phi)\, Y_{J_1 M_1}(\theta, \phi)\, \sin\theta\, d\theta\, d\phi
$$

$$
= \left[\frac{(2J_1 + 1)(2J_2 + 1)(2J_3 + 1)}{4\pi} \right]^{\frac{1}{2}} \begin{pmatrix} J_1 & J_2 & J_3 \\ 0 & 0 & 0 \end{pmatrix} \begin{pmatrix} J_1 & J_2 & J_3 \\ M_1 & M_2 & M_3 \end{pmatrix}
$$

$$
(3.119)
$$

3.10 THE ROTATION MATRICES AS RIGID BODY ROTATIONAL WAVE FUNCTIONS

Let $\psi_{JM}(\phi,\theta,\chi)$ be the rotational wave function of a rigid body so that ψ_{JM} is an eigenfunction of the Hamiltonian $\mathbf{H}$ with the eigenvalue E_{JM}, the square of the total angular momentum operator $\mathbf{J}^2$ with an eigenvalue $J(J+1)$, and the Z component of the total angular momentum J_Z with an eigenvalue M. We imagine that a rotation $\mathbf{R}_1 = (\phi_1,\theta_1,\chi_1)$ is applied to the system. Then ψ_{JM} will be transformed according to

$$\mathbf{R}_1\psi_{JM}(\phi,\theta,\chi) = \sum_{M'} D^J_{M'M}(\phi_1,\theta_1,\chi_1)\,\psi_{JM'}(\phi,\theta,\chi) \qquad (3.120)$$

where the value of the rotated wave function $\psi'_{JM} = \mathbf{R}_1\psi_{JM}$ is the same as the original wave function ψ_{JM} at point P' that is carried into P by the rotation $\mathbf{R}_1$ (see discussion on active vs. passive transformations):

$$\psi_{JM}(\phi',\theta',\chi') = \sum_{M'} D^J_{M'M}(\phi_1,\theta_1,\chi_1)\,\psi_{JM'}(\phi,\theta,\chi) \qquad (3.121)$$

Here we are using the Euler angles in two senses: (1) as coordinates to describe the rigid-body wave functions ψ_{JM} and $\psi_{JM'}$ relative to the space-fixed frame, and (2) as the arguments of the rotation $\mathbf{R}_1 = \mathbf{R}_1(\phi_1,\theta_1,\chi_1)$ in the rotation matrix $D^J_{M'M}$.

We wish to set $\mathbf{R}_1 = \mathbf{R}$, that is, $\phi_1 = \phi$, $\theta_1 = \theta$, and $\chi_1 = \chi$ in Eq. (3.121). The effect of this is to bring point P' to the new $Z' = z$ axis so that $\phi' = 0$, $\theta' = 0$, and $\chi' = 0$. Then Eq. (3.121) reads

$$\psi_{JM}(0,0,0) = \sum_{M'} D^J_{M'M}(\phi,\theta,\chi)\,\psi_{JM'}(\phi,\theta,\chi) \qquad (3.122)$$

We multiply both sides of this equation by $D^{J*}_{M''M}(\phi,\theta,\chi)$ and sum over the index M. From the unitarity of the rotation matrices we find

$$\sum_M D^{J*}_{M''M}(\phi,\theta,\chi)\,\psi_{JM}(0,0,0)$$

$$= \sum_M \sum_{M'} D^{J*}_{M''M}(\phi,\theta,\chi)\,D^J_{M'M}(\phi,\theta,\chi)\,\psi_{JM'}(\phi,\theta,\chi)$$

$$= \sum_{M'} \delta_{M'M''}\psi_{JM'}(\phi,\theta,\chi)$$

$$= \psi_{JM''}(\phi,\theta,\chi) \qquad (3.123)$$

Writing M for M'' and K for M, we obtain

$$\psi_{JM}(\phi,\theta,\chi) = \sum_K D_{MK}^{J^*}(\phi,\theta,\chi)\,\psi_{JK}(0,0,0) \tag{3.124}$$

Thus the general wave function for an asymmetric top is a linear combination of rotation matrices.

Let us suppose that the rigid body has an axis of symmetry, which we may take to lie along the z axis. Then an arbitrary rotation about this axis leaves the wave function $\psi_{JM}(\phi,\theta,\chi)$ unaltered, except for a possible phase factor, that is, the probability $|\psi_{JM}|^2$ is unchanged[32]. Hence we require $\psi_{JM}(\phi,\theta,\chi)$ to be the same up to a phase factor for a rotation about z through the angle χ. Considering the form of the rotation matrices, $D_{MK}^{J^*}(\phi,\theta,\chi) = e^{iM\phi}d_{MK}^J(\theta)e^{iK\chi}$, this condition can be fulfilled only if $\psi_{JM}(\phi,\theta,\chi)$ is proportional to one particular rotation matrix, say, $D_{MK}^{J^*}(\phi,\theta,\chi)$ in the sum over K shown in Eq. (3.124). If we also insist that the wave function is normalized [see Eq. (3.113)], then the wave function of a symmetric top is given by

$$\boxed{|JKM\rangle = \left[\frac{2J+1}{8\pi^2}\right]^{\frac{1}{2}} D_{MK}^{J^*}(\phi,\theta,\chi) = (-1)^{M-K}\left[\frac{2J+1}{8\pi^2}\right]^{\frac{1}{2}} D_{-M-K}^{J}(\phi,\theta,\chi)} \tag{3.125}$$

For the special case that the rigid body is a rigid linear rotator, $K = 0$, that is, $\mathbf{J}$ is perpendicular to the z axis (the molecular axis), the symmetric top wave function $|J0M\rangle$ reduces to the spherical harmonic wave function $|JM\rangle$ [see Eq. (3.94)], provided we no longer integrate over the angle χ.

The $|JKM\rangle$ satisfy the differential equations [see Eqs. (3.41) and (3.45)]:

$$J_Z|JKM\rangle = -i\frac{\partial}{\partial\phi}|JKM\rangle = M|JKM\rangle$$

$$J_z|JKM\rangle = -i\frac{\partial}{\partial\chi}|JKM\rangle = K|JKM\rangle$$

$$\mathbf{J}^2|JKM\rangle = \left[-\frac{\partial^2}{\partial\theta^2} - \cot\theta\frac{\partial}{\partial\theta}\right.$$

$$\left. -\frac{1}{\sin^2\theta}\left(\frac{\partial^2}{\partial\phi^2} + \frac{\partial^2}{\partial\chi^2} - 2\cos\theta\frac{\partial^2}{\partial\phi\partial\chi}\right)\right]|JKM\rangle$$

$$= J(J+1)|JKM\rangle \tag{3.126}$$

Thus we identify M with the projection of $\mathbf{J}$ on the space-fixed Z axis and K with the projection of $\mathbf{J}$ on the body-fixed z axis (the figure axis). From the geometric interpretation of the rotation matrices, it is, of course, clear that they should be proportional to the symmetric top wave functions. Historically, the symmetric top wave functions were laboriously derived by solving the differential equations shown in Eq. (3.126) subject to suitable boundary conditions for the wave functions, rather than developing their identification with the rotation matrices[33–35]. It should be noted that Eq. (3.126) also applies to systems with half-integral angular momenta, although the $|JKM\rangle$ lose their meaning as wave functions of a classical symmetric top.

The symmetric top wave functions are simultaneous eigenfunctions of J_Z and J_z. By rewriting Eq. (3.125) as

$$|JKM\rangle = \left[\frac{2J+1}{8\pi^2}\right]^{\frac{1}{2}} \langle JK| \exp(i\chi J_z + i\theta J_N + i\phi J_Z) |JM\rangle \tag{3.127}$$

where the dependence on the variables ϕ, θ, and χ is explicit, it is easily seen that the simultaneous specifications of the projections K and M cause the corresponding eigenvectors of the states to transform as $\langle JK|$ and $|JM\rangle$, that is, reversed from each other (contragradiently). The transformation to a scheme of description in which $\langle JK|$ transforms as $|JK\rangle$ is attained by means of the so-called conjugation operator $\mathbf{K}$, which is equivalent to the quantum mechanical operation of time reversal[19]. Under time reversal, as we have mentioned, the angular momentum $\mathbf{J}$ also changes sign:

$$\mathbf{KJK}^{-1} = -\mathbf{J} \tag{3.128}$$

that is, J_X, J_Y, and J_Z are transformed into $-J_X$, $-J_Y$, and $-J_Z$, respectively, so that the components of the angular momentum operator obey anomalous commutation rules. Thus the anomalous commutation rules of $\mathbf{J}$ referred to the rotating frame is a consequence of the simultaneous specification of the eigenvalues of J_Z and J_z. The internal angular momenta $\mathbf{L}$ and $\mathbf{S}$ have normal commutation rules in the molecular rotating frame essentially because these momenta involve the motion and spins of the electrons for fixed nuclei, so that no question concerning molecular rotation can enter.

NOTES AND REFERENCES

1. H. Goldstein, *Classical Mechanics* (Addison-Wesley, Cambridge, MA, 1953).

2. L. D. Landau and E. M. Lifshitz, *Mechanics* (Pergamon Press, Oxford, 1969).

3. E. Merzbacher, *Quantum Mechanics* (Wiley, New York, 1961).

4. A. A. Wolf, *Am. J. Phys.*, **37**, 531 (1969).

5. M. E. Rose, *Elementary Theory of Angular Momentum* (Wiley, New York, 1957).

6. A. R. Edmonds, *Angular Momentum in Quantum Mechanics* (Princeton University Press, 1957), 3rd printing with corrections, 1974.

7. D. M. Brink and G. R. Satchler, *Angular Momentum* (Oxford University Press, 1962), 2nd edition with corrections 1979.

8. M. Bouten, *Physica*, **42**, 572 (1969) points out a major inconsistency in the way Edmonds[6] treats active and passive rotations. Unfortunately, Bouten also finds that the treatment of Rose[5] has this confusion, but two errors are made that cancel each other. We treat rotations here in the same manner as Brink and Satchler[7].

9. O. Klein, *Z. Physik*, **58**, 730 (1929).

10. H. B. G. Casimir, *Rotation of a Rigid Body in Quantum Mechanics* (J. H. Woltjers, The Hague, 1931).

11. J. H. Van Vleck, *Rev. Mod. Phys.*, **23**, 213 (1951).

12. R. F. Curl, Jr. and J. L. Kinsey, *J. Chem. Phys.*, **35**, 1758 (1961).

13. K. F. Freed, *J. Chem. Phys.*, **45**, 4214 (1966).

14. J. T. Hougen, "The Calculation of Rotational Energy Levels and Rotational Line Intensities in Diatomic Molecules," National Bureau of Standards Monograph 115, U.S. Government Printing Office, Washington DC, June 1970; *J. Chem. Phys.*, **36**, 519 (1962).

15. I. Kovács, *Rotational Structure in the Spectra of Diatomic Molecules* (Akadémiai Kiadó, Budapest, 1969).

16. A. Carrington, D. H. Levy, and T. A. Miller, *Adv. Chem. Phys.*, **18**, 149 (1970).

17. I. C. Bowater, J. M. Brown, and A. Carrington, *Proc. Roy. Soc.*, **A333**, 265 (1973).

18. J. M. Brown and B. J. Howard, *Molec. Phys.*, **31**, 1517 (1976).

19. See E. P. Wigner, *Group Theory* (Academic Press, New York, 1959), Chapter 26, "Time Inversion." Actually only under special circumstances is the time reversal operator θ simply the same as the conjugation operator K. For example, one must disregard spin and work in the coordinate (not momentum) representation. See also A. S. Davydov, *Quantum Mechanics* (Addison-Wesley, Reading, MA, 1965), pp. 431–437.

20. This method is not the only means of remedying the inconsistency of the commutation rules of $\mathbf{J}$ and $\mathbf{P}$ in the molecule-fixed frame, where $\overline{\mathbf{P}}$ is an internal angular momentum operator. An alternative procedure, first suggested by Van Vleck[11] is to "reverse" all internal angular momentum operators, that is, to replace $\overline{\mathbf{P}}$ by $-\mathbf{P}$, where the superscript bar is omitted to denote the reversed operator. One must also relabel the basis set functions $|Pp\rangle$ that diagonalize $\overline{P}_z$ by $|P - p\rangle$. It is then readily verified that the reversed internal angular momenta also satisfy anomalous commutation rules, for example, $\overline{P}_x\overline{P}_y - \overline{P}_y\overline{P}_x = P_xP_y - P_yP_x = i\overline{P}_z = -iP_z$. If we also agree to change the sign of i everywhere it explicitly occurs, such as in the use of raising and lowering operators, the behavior of the angular momentum operators in the molecule-fixed frame is the same as in the space-fixed frame. To quote Van Vleck[11]: "The sign of i is never important as long as one is consistent in usage, and a consistent anomaly is effectively no anomaly at all." This is the procedure also advocated by Freed[13], who introduced the symbol $\xrightarrow{\text{RAM}}$ to describe this procedure. It is apparent that Eqs. (3.47), (3.48), and (3.49) result. Moreover, we see that, for example,

$$\langle S'\Sigma' | S^2 | S\Sigma \rangle \xrightarrow{\text{RAM}} \langle S' - \Sigma' | S^2 | S - \Sigma \rangle = S(S + 1)\delta_{S'S}\delta_{\Sigma'\Sigma}$$

and

$$\langle S'\Sigma' | S_z | S\Sigma \rangle \xrightarrow{\text{RAM}} \langle S' - \Sigma' | - S_z | S - \Sigma \rangle = \Sigma \delta_{S'S} \delta_{\Sigma'\Sigma}$$

but that

$$\langle S'\Sigma' | S_\pm | S\Sigma \rangle \xrightarrow{\text{RAM}} \langle S' - \Sigma' | S_\mp | S - \Sigma \rangle$$

$$= -[S(S+1) - (-\Sigma)(-\Sigma \mp 1)]^{\frac{1}{2}} \delta_{S'S} \delta_{-\Sigma', -\Sigma \mp 1}$$

$$= -[S(S+1) - \Sigma(\Sigma \pm 1)]^{\frac{1}{2}} \delta_{S'S} \delta_{\Sigma', \Sigma \pm 1}$$

This causes off-diagonal matrix elements of $\mathbf{J} \cdot \overline{\mathbf{P}}$ to differ in sign. While both procedures give the same energies when the molecular Hamiltonian is diagonalized, the reversed angular momentum method implies a different phase convention for the resulting molecular eigenfunctions.

21. E. P. Wigner, *Group Theory* (Academic Press, New York, 1959); U. Fano and G. Racah, *Irreducible Tensorial Sets* (Academic Press, New York, 1959). In the language of group theory, each D forms an irreducible representation (Darstellung) of the three-dimensional rotation group of dimension ($2J + 1$).

22. H. A. Buckmaster, *Can. J. Phys.*, **42**, 386 (1964); **44**, 2525 (1966).

23. P. J. Brussaard and H. A. Tolhoek, *Physica*, **23**, 955 (1957).

24. N. F. Ramsey, *Molecular Beams* (Oxford University Press, 1963).

25. E. Majorana, *Nuovo Cimento*, **9**, 43 (1932).

26. F. Bloch and I. I. Rabi, *Rev. Mod. Phys.*, **17**, 237 (1945).

27. For the electric field analog, see J. C. Zorn and T. C. English, *Adv. Atomic Molec. Phys.*, **9**, 244 (1973).

28. Not everyone can set up a triple Stern–Gerlach experiment, but the results can be readily appreciated by the analogy with linear polarizers for a light beam. When two polarizers are crossed, no light is transmitted, but if a third linear polarizer is inserted between the pair of crossed polarizers, light is transmitted provided the middle linear polarizer is parallel to neither of its neighboring linear polarizers. See T. S. Carlton, *Am. J. Phys.*, **42**, 408 (1974); *J. Chem. Ed.*, **52**, 322 (1975).

29. The explicit form of this unitary transformation is given in reference [5], pp. 62–67.

30. Equation (3.113) is the great orthogonality theorem for the continuous irreducible representations of the three-dimensional rotation group; see reference [21]. It serves as a jumping off point to look at finite point groups—those symmetry operations acting on an object of finite dimensions in which at least one point of the object remains fixed. Each symmetry operation $\mathbf{R}$ leaves the object unchanged, and hence each symmetry operation corresponds to a unitary transformation. In some suitable orthonormal basis set the symmetry operation $\mathbf{R}$ will be represented by a unitary matrix whose elements satisfy the relation $\Gamma^i(R)_{\mu\nu}^\dagger = \Gamma^i(R)_{\nu\mu}^*$. The set of square matrices of all the symmetry operations in the group is called a *representation* of the group. The set of orthonormal functions $\psi_1, \psi_2, \ldots, \psi_k$ with respect to which these matrices are defined is called the *basis of the representation*. Here the number k gives the dimensionality of the representation. A representation is said to be irreducible if it cannot be decomposed

into a sum of representations of smaller dimensions each of which satisfy the symmetry properties of the group. The analog of Eq. (3.113) for a finite group of order h (i.e., with h symmetry elements) is

$$\sum_R \Gamma^i(R)_{\mu\nu}^* \Gamma^j(R)_{\alpha\beta} = \frac{h}{\ell_i}\delta_{ij}\delta_{\mu\alpha}\delta_{\nu\beta} \qquad (3.129)$$

where the summation index R runs over all h symmetry operations in the group and ℓ_i is the dimensionality of the i th irreducible representation Γ_i.

For many purposes what is more important than the group representations is the corresponding group representation characters, defined as the sum of the diagonal elements (trace) of the matrix representation:

$$\chi(R) = \sum_\mu \Gamma(R)_{\mu\mu} \qquad (3.130)$$

The reason for this is that equivalent representations, which are related to each other by a unitary transformation, have the same character since the trace of a matrix is unaffected by a unitary transformation. By summing both sides of Eq. (3.129) over the indices μ, ν, α, and β, we obtain the result

$$\sum_R \chi^{(i)}(R)^* \chi^{(j)}(R) = h\delta_{ij} \qquad (3.131)$$

where we denote the character of the irreducible representation $\Gamma^i(R)$ of the symmetry operation R by $\chi^{(i)}(R)$. Equation (3.131) states that the sum of the squared moduli of the characters of an irreducible representation equals the order of the group.

Let $\chi(R)$ be the characters of some reducible representation of dimension d, and let $n^{(i)}$ be the number of times the irreducible representation Γ^i with dimension $d^{(i)}$ appears in the reducible representation. Then

$$\sum_{i=1}^p n^{(i)}d^{(i)} = d \qquad (3.132)$$

and it follows that

$$\chi(R) = \sum_{i=1}^p n^{(i)}\chi^{(i)}(R) \qquad (3.133)$$

We can solve for $n^{(i)}$ by multiplying both sides of Eq. (3.133) by $\chi^{(i')}(R)^*$ and summing over all symmetry operations in the group:

$$n^{(i)} = \frac{1}{h}\sum_R \chi(R)\chi^{(i)}(R)^* \qquad (3.134)$$

Recall that representations corresponding to physically equivalent symmetry operations have the same character. These are said to belong to the same class. Thus, instead of summing over all symmetry operations in a group, we need only sum over the different classes weighted by the number of symmetry operations belonging to that class. Hence Eq. (3.134) may be rewritten as

$$n^{(i)} = \frac{1}{h} \sum_c h_c \chi(R_c) \chi^{(i)}(R_c)^*$$

(3.135)

where h_c is the number of elements in the cth class and the sum runs over all classes of the group. Equation (3.134) or its equivalent form, Eq. (3.135), serves as the common starting point in applications of the theory of finite groups to many problems of interest, such as the crystal field splittings of atomic ions [see H. Bethe, *Ann. Physik*, **3**, 133 (1929); J. S. Griffith, *The Theory of Transition-Metal Ions* (Cambridge University Press, 1964); S. B. Piepho and P. N. Schatz, *Group Theory in Spectroscopy* (Wiley-Interscience, New York, 1983)] and the symmetry classification of vibrational normal modes [see E. B. Wilson, Jr., J. C. Decius, and P. C. Cross, *Molecular Vibrations* (McGraw-Hill, New York, 1955); and S. Califano, *Vibrational States* (Wiley, New York, 1976)]. Tables of the characters of irreducible representations of finite point groups may be found elsewhere; see F. A. Cotton, *Chemical Applications of Group Theory* (Wiley-Interscience, New York, 1963, 1971), M. Tinkham, *Group Theory and Quantum Mechanics* (McGraw-Hill, New York, 1964), and P. W. Atkins, M. S. Child, and C. S. G. Phillips, *Tables for Group Theory* (Oxford University Press, 1970, 1982 with corrections).

31. E. U. Condon and G. H. Shortley, *Theory of Atomic Spectra* (Cambridge University Press, 1935).

32. Prototypical examples of molecular symmetric tops are the monohalogenated and trihalogenated methanes, CH_3X and CHX_3. These molecules have a three-fold axis of rotational symmetry. It might be wondered why their rotational wave functions are unaltered by an arbitrary rotation about the symmetry axis. The answer to this question can be found in terms of the rotational symmetry of the moment of inertia ellipsoid of the molecule, as discussed in Section 6.1.

33. F. Reiche and H. Rademacher, *Z. Phys.*, **39**, 444 (1926); H. Rademacher and F. Reiche, *Z. Phys.*, **41**, 453 (1927).

34. R. L. Kronig and I. I. Rabi, *Phys. Rev.*, **29**, 262 (1927).

35. D. M. Dennison, *Rev. Mod. Phys.*, **3**, 310 (1931).

APPLICATION 4
ENERGY LEVELS OF ATOMS WITH TWO VALENCE ELECTRONS

The Central Field Approximation

The nonrelativistic Hamiltonian for an n-electron atom is

$$\mathcal{H} = -\frac{\hbar^2}{2m} \sum_{i=1}^{n} \nabla_i^2 - \sum_{i=1}^{n} \frac{Ze^2}{r_i} + \sum_{i=2}^{n} \sum_{j=1}^{i-1} \frac{e^2}{r_{ij}} \tag{1}$$

where Ze is the charge on the nucleus, e is the charge on the electron, r_i is the distance between the i th electron and the nucleus (taken as the origin), and $r_{ij} = |\mathbf{r}_i - \mathbf{r}_j|$ is the distance between the i th and j th electrons. In the central field approximation the preceding Hamiltonian is replaced by an effective Hamiltonian of the form

$$\mathcal{H}_0 = -\frac{\hbar^2}{2m} \sum_{i=1}^{n} \nabla_i^2 + \sum_{i=1}^{n} V_i(r_i) \tag{2}$$

This allows each electron to be expressed as a spin–orbital

$$\phi_{n\ell m_\ell m_s} = R_{n\ell}(r) Y_{\ell m}(\theta, \phi) |sm_s\rangle \tag{3}$$

where $R_{n\ell}$ is the radial wave function, $Y_{\ell m}$ is a spherical harmonic, and $|sm_s\rangle$ is the spin function, which may be written α for spin up ($|\frac{1}{2}\frac{1}{2}\rangle$) or β for spin down ($|\frac{1}{2} - \frac{1}{2}\rangle$). Because of the Pauli exclusion principle, the total wave function of the central field Hamiltonian is the antisymmetrized product of the spin–orbitals

$$|\psi\rangle = \mathcal{A} \left| \phi_{n_1 \ell_1 m_{\ell_1} m_{s_1}}(1) \phi_{n_2 \ell_2 m_{\ell_2} m_{s_2}}(2) \cdots \phi_{n_n \ell_n m_{\ell_n} m_{s_n}}(n) \right\rangle \tag{4}$$

which is conveniently written as a Slater determinant

$$|\psi\rangle = \sqrt{\frac{1}{n!}} \begin{vmatrix} \phi_{n_1 \ell_1 m_{\ell_1} m_{s_1}}(1) & \phi_{n_2 \ell_2 m_{\ell_2} m_{s_2}}(1) & \cdots & \phi_{n_n \ell_n m_{\ell_n} m_{s_n}}(1) \\ \phi_{n_1 \ell_1 m_{\ell_1} m_{s_1}}(2) & \phi_{n_2 \ell_2 m_{\ell_2} m_{s_2}}(2) & \cdots & \phi_{n_n \ell_n m_{\ell_n} m_{s_n}}(2) \\ \vdots & \vdots & & \vdots \\ \phi_{n_1 \ell_1 m_{\ell_1} m_{s_1}}(n) & \phi_{n_2 \ell_2 m_{\ell_2} m_{s_2}}(n) & \cdots & \phi_{n_n \ell_n m_{\ell_n} m_{s_n}}(n) \end{vmatrix} \tag{5}$$

We see that if two electrons have the same spatial and spin coordinates, that is, occupy the same orbital, the determinant has two identical columns and hence vanishes. Moreover, if any two electrons are interchanged, this corresponds to the interchange of any two rows of the Slater determinant and hence $|\psi\rangle$ changes sign.

A specification of an electronic configuration gives the $n\ell$ values of all the orbitals but it does not specify $|\psi\rangle$ completely since there are m_ℓ and m_s quantum numbers as well. In the central field approximation all the possible ^{2S+1}L terms arising from the same configuration are degenerate, that is, have the same energy. The form of $|\psi\rangle$ suggests that the eigenvalues of $\boldsymbol{\ell}_i$ and $\mathbf{s}_i$ are all observables. This implies that the total angular momentum of the atom, expressed as $\mathbf{L}$ and $\mathbf{S}$, may also be observable. This conclusion is certainly valid for the central field Hamiltonian, but for it to hold for the nonrelativistic Hamiltonian, it is necessary to show that the noncentral part, namely, the Coulomb repulsion terms, commute with $\mathbf{L}^2$, L_Z, $\mathbf{S}^2$, and S_Z

Consider the i th and j th electrons.

A. Show that ℓ_{iz} and ℓ_{jz} do *not* commute with the nonrelativistic Hamiltonian by finding explicitly the nonzero values of the commutators

$$\left[\ell_{iz}, \frac{e^2}{r_{ij}}\right]$$

and

$$\left[\ell_{jz}, \frac{e^2}{r_{ij}}\right]$$

B. Use the preceding results to show that the sum $\ell_{iz} + \ell_{jz}$ does commute with e^2/r_{ij}, that is, that

$$\left[\ell_{iz} + \ell_{jz}, \frac{e^2}{r_{ij}}\right] = 0 \tag{6}$$

Hence it follows that L_Z commutes with $\mathcal{H}$. By symmetry, L_X and L_Y also commute with $\mathcal{H}$, from which it is concluded that $\mathbf{L}^2 = L_X^2 + L_Y^2 + L_Z^2$ commutes with $\mathcal{H}$. Since the Coulomb repulsion terms contain no spin coordinates, $\mathbf{S}^2$ and S_Z also commute with $\mathcal{H}$. Finally, the inversion operator leaves terms such as e^2/r_{ij} unchanged. It follows that *the Coulomb repulsion terms have nonzero matrix elements only between states of the same L, S, and parity.*

Construction of Antisymmetrized Two-Electron Wave Functions in a Central Field

Let us restrict our attention to atoms having two valence electrons; we regard the closed shells as simply effectively screening the nuclear charge. They play no role in determining the relative energies of terms, although they do, of course, make the dominant contribution of the absolute energy of a configuration and its terms.

Consider two particles 1 and 2 having orbital angular momenta ℓ_1 and ℓ_2 with projections m_{ℓ_1} and m_{ℓ_2}. We may write the coupled representation in either of two ways:

$$|\ell_1\ell_2 L M_L\rangle = \sum_{m_{\ell_1} m_{\ell_2}} \langle \ell_1 m_{\ell_1}, \ell_2 m_{\ell_2} | L M_L\rangle |\ell_1 m_{\ell_1}\rangle |\ell_2 m_{\ell_2}\rangle \tag{7}$$

or

$$|\ell_2\ell_1 L M_L\rangle = \sum_{m_{\ell_1} m_{\ell_2}} \langle \ell_2 m_{\ell_2}, \ell_1 m_{\ell_1} | L M_L\rangle |\ell_1 m_{\ell_1}\rangle |\ell_2 m_{\ell_2}\rangle \tag{8}$$

C. Show that

$$|\ell_1\ell_2 L M_L\rangle = (-1)^{\ell_1+\ell_2-L} |\ell_2\ell_1 L M_L\rangle \tag{9}$$

If the two particles are electrons, we must insist that the Pauli exclusion principle is satisfied, specifically, that the total wave function (both space and spin part) be antisymmetric with respect to the interchange of the two indistinguishable electrons. If the two electrons are *equivalent*, then $n_1\ell_1 \equiv n_2\ell_2$, so that

$$|\ell_1(1)\ell_2(2) L M_L\rangle = (-1)^{\ell_1+\ell_2-L} |\ell_2(1)\ell_1(2) L M_L\rangle$$

$$= (-1)^L |\ell_2(1)\ell_1(2) L M_L\rangle \tag{10}$$

where $(-1)^{\ell_1+\ell_2} = 1$ when $\ell_1 = \ell_2$. Similarly

$$|s_1(1)s_2(2) S M_S\rangle = (-1)^{s_1+s_2-S} |s_2(1)s_1(2) S M_S\rangle$$

$$= (-1)^{S-1} |s_2(1)s_1(2) S M_S\rangle \tag{11}$$

For a two-electron wave function, we may always write it as a product of a space and a spin part. For two equivalent electrons

$$\psi(1,2) = |s_1\ell_1\, s_1\ell_1\, S L M_S M_L\rangle = |\ell_1\ell_1 L M_L\rangle |s_1 s_1 S M_S\rangle \tag{12}$$

Therefore, for the total wave function to be antisymmetric, it is necessary that $(-1)^{L+S-1} = -1$, that is, that $L + S$ be even.

D. Determine the terms that arise from the $(ns)^2$ configuration, the $(np)^2$ configuration, and the $(nd)^2$ configuration.

If the two electrons are *inequivalent,* then this implies that $n_1 \ell_1$ is not identical to $n_2 \ell_2$. The spatial part of the wave function may be written as

$$\frac{1}{\sqrt{2}} \left[|\ell_1(1)\ell_2(2)LM_L\rangle \pm (-1)^{\ell_1+\ell_2-L} |\ell_2(1)\ell_1(2)LM_L\rangle \right]$$

This is symmetric in the coordinates 1 and 2 for the plus sign and antisymmetric for the minus sign. Thus the entire wave function will be antisymmetric if we multiply by an antisymmetric spin function when we take the plus sign and a symmetric spin function when we take the minus sign. The antisymmetric spin combination is

$$\frac{1}{\sqrt{2}} \left[\alpha(1)\beta(2) - \alpha(2)\beta(1) \right]$$

and the symmetric combinations are

$$\alpha(1)\alpha(2)$$

$$\frac{1}{\sqrt{2}} \left[\alpha(1)\beta(2) + \alpha(2)\beta(1) \right]$$

$$\beta(1)\beta(2)$$

E. Show that $\mathbf{S}^2$ acts on the antisymmetric spin combination to give $S(S+1) = 0$ and the symmetric combination to give $S(S+1) = 2$. Combining these results, we conclude that the plus sign is for a singlet state (^1L) and the minus sign is for a triplet state (^3L).

Calculation of the Term Separations Arising from a Single Configuration

To find the relative term energies, we need only consider the Coulomb repulsion term e^2/r_{12} and calculate its nonzero matrix elements between the states of the same L and S (and parity, but of course they all have the same parity) of a configuration. We use the well-known expansion

$$\frac{e^2}{r_{12}} = e^2 \sum_{k=0}^{\infty} \frac{r_<^k}{r_>^{k+1}} P_k(\cos \Theta) \tag{13}$$

where Θ is the angle between $\mathbf{r}_1$ and $\mathbf{r}_2$ and $r_<$ is the lesser and $r_>$ the greater of r_1 and r_2. The Legendre polynomial $P_k(\cos \Theta) = [4\pi/(2k+1)]^{\frac{1}{2}} Y_{k0}(\theta, \phi)$ may be expressed as a sum over the products of spherical harmonics by using the spherical harmonic addition theorem. Hence we obtain

$$\frac{e^2}{r_{12}} = e^2 \sum_{k=0}^{\infty} \frac{r_<^k}{r_>^{k+1}} \left(\frac{4\pi}{2k+1} \right) \sum_{q=-k}^{k} Y_{kq}^*(\theta_1, \phi_1) Y_{kq}(\theta_2, \phi_2)$$

$$= e^2 \sum_{k=0}^{\infty} \frac{r_<^k}{r_>^{k+1}} \left(\frac{4\pi}{2k+1} \right) \sum_{q=-k}^{k} (-1)^q Y_{k-q}(\theta_1, \phi_1) Y_{kq}(\theta_2, \phi_2) \qquad (14)$$

A general matrix element in the coupled representation has the form

$$\langle LM | e^2 / r_{12} | LM \rangle$$

because different L or different M cannot be connected by the Coulomb repulsion term. If we expand these states in an uncoupled representation, we obtain one or more terms of the form

$$\langle n_a \ell_a m_{\ell_a} m_{s_a}(1) | \langle n_b \ell_b m_{\ell_b} m_{s_b}(2) | \frac{e^2}{r_{12}} | n_c \ell_c m_{\ell_c} m_{s_c}(1) \rangle | n_d \ell_d m_{\ell_d} m_{s_d}(2) \rangle$$

$$= \sum_k R^k(n_a \ell_a, n_b \ell_b; n_c \ell_c, n_d \ell_d)$$

$$\times \frac{4\pi}{2k+1} \sum_{q=-k}^{k} (-1)^q \left\langle Y_{\ell_a m_{\ell_a}} \middle| Y_{k-q} \middle| Y_{\ell_c m_{\ell_c}} \right\rangle \left\langle Y_{\ell_b m_{\ell_b}} \middle| Y_{kq} \middle| Y_{\ell_d m_{\ell_d}} \right\rangle \delta_{m_{s_a} m_{s_c}} \delta_{m_{s_b} m_{s_d}}$$

$$(15)$$

where

$$R^k = e^2 \int_0^{\infty} r_1^2 \, dr_1 \int_0^{\infty} r_2^2 \, dr_2 \, R_{n_a \ell_a}(r_1) R_{n_b \ell_b}(r_2) \frac{r_<^k}{r_>^{k+1}} R_{n_c \ell_c}(r_1) R_{n_d \ell_d}(r_2) \qquad (16)$$

Here the labels a, b, c, and d refer to electrons in different spin–orbitals (most general case).

It is customary to define the Slater–Condon parameters as

$$F^k(n_a \ell_a; n_b \ell_b) = e^2 \left\langle R_{n_a \ell_a}(1) R_{n_b \ell_b}(2) \middle| \frac{r_<^k}{r_>^{k+1}} \middle| R_{n_a \ell_a}(1) R_{n_b \ell_b}(2) \right\rangle \qquad (17)$$

and

$$G^k(n_a \ell_a; n_b \ell_b) = e^2 \left\langle R_{n_a \ell_a}(1) R_{n_b \ell_b}(2) \middle| \frac{r_<^k}{r_>^{k+1}} \middle| R_{n_b \ell_b}(1) R_{n_a \ell_a}(2) \right\rangle \qquad (18)$$

Provided we consider matrix elements of e^2 / r_{12} within the same configuration, the radial integrals reduce to only those of the type F^k and G^k.

F. Work out the relative term energies for the terms arising from the p^2 configuration. Show that

$$\frac{E(^1S) - E(^1D)}{E(^1D) - E(^3P)} = \frac{3}{2} \qquad (19)$$

Actual observations do not agree well with this result, showing that the theory is approximate. Much better agreement is obtained by considering the interaction between terms of the same L, S, and parity arising from different configurations. This effect is called *configuration interaction*.

The angular momentum machinery mastered so far allows this application to be worked out in a rather straightforward manner but with much algebra. In Application 11 we revisit this topic to show how spherical tensor techniques allow this problem to be solved almost effortlessly.

APPLICATION 5
ANGULAR DISTRIBUTION OF RIGID ROTOR AXES
FOLLOWING ABSORPTION OF PLANE POLARIZED LIGHT

The wave function of a rigid rotor is characterized by the total angular momentum $\mathbf{J}$ and its projection M on the axis of quantization, that is, by $|JM\rangle$. It has the explicit form

$$|JM\rangle = Y_{JM}(\theta, \phi) \tag{1}$$

where θ and ϕ are the polar and azimuthal angles, repectively. Thus the probability $P_{JM}(\theta, \phi)$ of finding the rotor axis pointing into the solid angle element $d\Omega = \sin\theta\, d\theta\, d\phi$ when the rotor is in the state $|JM\rangle$ is given by

$$P_{JM}(\theta) = |Y_{JM}(\theta, \phi)|^2 \sin\theta d\theta d\phi \tag{2}$$

which is seen to be independent of the azimuthal angle ϕ. Note that this is normalized:

$$\int d\Omega\, |Y_{JM}(\theta, \phi)|^2 = 1 \tag{3}$$

The probability *amplitude* of finding the rigid rotor in the state $|JM\rangle$ following the dipole absorption of plane polarized radiation ($j_{ph} = 1$, $m_{ph} = 0$) is proportional to the Wigner coefficient (Clebsch–Gordan coefficient) $\langle J''M'', 10|JM\rangle$, where $|J''M''\rangle$ is the initial state of the rotor. Thus the distribution of rigid rotor axes for this particular M state is

$$P_{JM}(\theta) = \langle J''M'', 10|JM\rangle^2 \, |Y_{JM}(\theta, \phi)|^2$$

(which has not been normalized). The M'' states are assumed to be equally populated with no phase relations (meaning random phase relations) among them. Thus the total probability is obtained by simply summing over all initial M'' states, giving

$$P_J(\theta) = \sum_{M''} \langle J''M'', 10|JM\rangle^2 \, |Y_{JM}(\theta, \phi)|^2 \tag{4}$$

The remainder of this application concerns the evaluation of Eq. (4).

A. Use the Clebsch–Gordan series or its inverse to show that

$$\langle j_1 m_1, j_2 m_2 | j_3 m_3 \rangle D_{k_3 m_3}^{j_3} = \sum_{k_1} \langle j_1 k_1, j_2 k_3 - k_1 | j_3 k_3 \rangle D_{k_1 m_1}^{j_1} D_{k_3 - k_1, m_2}^{j_2}$$

$$\tag{5}$$

B. Make the identification

$$j_1 = J'' \quad m_1 = M'' \quad k_1 = K''$$
$$j_2 = 1 \quad m_2 = 0 \quad k_2 = K - K''$$
$$j_3 = J \quad m_3 = M \quad k_3 = K$$

Then Eq. (5) becomes

$$\langle J''M'', 10|JM\rangle D_{KM}^{J}$$

$$= \sum_{K''} \langle J''K'', 1\,K - K''|JK\rangle D_{K''M''}^{J''} D_{K-K''\,0}^{1} \tag{6}$$

Square both sides of Eq. (6) and sum over M''. Use this to prove that

$$\sum_{M''} \langle J''M'', 10|JM\rangle^2 \, |D_{KM}^{J}|^2$$

$$= \sum_{K''} \langle J''K'', 1\,K - K''|JK\rangle^2 \, |D_{K-K''\,0}^{1}|^2 \tag{7}$$

where the sum over K'' is from $K - 1$ to $K + 1$. Thus Eq. (7) has the explicit value

$$\sum_{M''} \langle J''M'', 10|JM\rangle^2 \, |D_{KM}^{J}|^2$$

$$= \langle J''K - 1, 11|JK\rangle^2 \, |D_{10}^{1}|^2$$

$$+ \langle J''K, 10|JK\rangle^2 \, |D_{00}^{1}|^2 + \langle J''K + 1, 1 - 1|JK\rangle^2 \, |D_{-1\,0}^{1}|^2 \tag{8}$$

C. Use the fact that $|D_{KM}^{J}|^2$ for $K = 0$ is equal to $[4\pi/(2J+1)]|Y_{JM}|^2$ to express $P_J(\theta)$ in the form

$$P_J(\theta) = a \sin^2\theta + b \cos^2\theta \tag{9}$$

Find the coefficients a and b as an explicit function of J'' for the two types of transitions, a P branch in which $\Delta J = J - J'' = -1$ and an R branch in which $\Delta J = J - J'' = +1$.

D. $P_J(\theta)$ can be recast into the familiar form

$$P_J(\theta) = \frac{1 + \mathcal{A}_0(J)P_2(\cos\theta)}{4\pi} \tag{10}$$

where

$$A_0 = \frac{2(b-a)}{2a+b} \tag{11}$$

and

$$P_2(\cos\theta) = \frac{3\cos^2\theta - 1}{2} \tag{12}$$

Show that $A_0 \rightarrow \frac{1}{2}$ for either branch as $J \rightarrow \infty$.

A_0 is called the *alignment parameter*. It ranges in value from +2 for a pure $\cos^2\theta$ distribution to -1 for a pure $\sin^2\theta$ distribution. For $A_0 = 0$, the distribution is isotropic. This problem illustrates the use of plane polarized light to prepare aligned molecules that may then be used as projectiles or targets in scattering experiments[1]. It is a special case of the production of anisotropically distributed targets by beam excitation[2].

NOTES AND REFERENCES

1. R. N. Zare, *Berichte der Bunsen-Gesellschaft*, **86**, 422 (1982).

2. R. Bersohn and S. H. Lin, *Adv. Chem. Phys.*, **41**, 67 (1969).

APPLICATION 6
PHOTOFRAGMENT ANGULAR DISTRIBUTION
(CLASSICAL TREATMENT)

Here we consider the form of the spatial anisotropy of the photodissociation fragments when a molecule is dissociated by a beam of linearly polarized light. We restrict ourselves to a one-photon electric dipole transition. Let (ϕ, θ, χ) denote the Euler angles that describe the molecule-fixed axes relative to the space-fixed axes. We choose the Z axis of the space-fixed frame to lie along the electric vector $\mathbf{E}$ of the light beam and the z axis of the molecule-fixed frame to be along the direction of the transition dipole moment $\boldsymbol{\mu}$ in the molecule. Then the probability P_{diss} of making a dissociative transition is proportional to $|\boldsymbol{\mu} \cdot \mathbf{E}|^2$. For a given molecular orientation (ϕ, θ, χ) we write

$$
P_{\text{diss}}(\phi, \theta, \chi) = \frac{3 \cos^2 \theta}{8 \pi^2}
$$

$$
= \frac{1 + 2 P_2(\cos \theta)}{8 \pi^2}
$$

$$
= \frac{D_{00}^0(\phi, \theta, \chi) + 2 D_{00}^2(\phi, \theta, \chi)}{8 \pi^2} \tag{1}
$$

where P_{diss} has been normalized to unity:

$$
\int_0^{2\pi} d\phi \int_0^{\pi} \sin\theta d\theta \int_0^{2\pi} d\chi \, P_{\text{diss}}(\phi, \theta, \chi) = 1 \tag{2}
$$

Let $f(\theta_m, \phi_m)$ denote the final recoil distribution of the fragments of interest *in the molecular frame*. Here (θ_m, ϕ_m) are the polar and azimuthal angles about the z axis (the direction of $\boldsymbol{\mu}$). For some cases $f(\theta_m, \phi_m)$ may take a particularly simple form, namely,

$$
f(\theta_m, \phi_m) = \frac{\delta(\theta_m - \theta_m^0)}{2 \pi \sin \theta} \tag{3}
$$

An example is that of a diatomic molecule when the dissociation occurs sufficiently promptly that the fragments have the same final direction as their initial orientation when the dissociative transition occurred (the axial recoil approximation). Then $\theta_m^0 = 0$ if $\boldsymbol{\mu}$ lies along the internuclear axis (a parallel-type transition) and $\theta_m^0 = \pi/2$ if $\boldsymbol{\mu}$ lies in a plane at right angles to the internuclear axis (a perpendicular-type transition). In the general case, $f(\theta_m, \phi_m)$ is a more complicated distribution whose exact form depends on (1) the relative orientation of the bond that cleaves with respect to $\boldsymbol{\mu}$, (2) the internal motions of the fragment (molecular rotation, bending, torsion, etc.), and (3) the time elapsed between absorption and fragmentation.

Whatever its form, $f(\theta_m, \phi_m)$ can always be expanded in the complete set of spherical harmonics

$$f(\theta_m, \phi_m) = \sum_{k,q} b_{kq} Y_{kq}(\theta_m, \phi_m) \tag{4}$$

where the expansion coefficients are given by

$$b_{kq} = \int_0^{2\pi} d\phi_m \int_0^{\pi} \sin\theta_m d\theta_m \, Y_{kq}^*(\theta_m, \phi_m) f(\theta_m, \phi_m) \tag{5}$$

Let $I(\theta_s, \phi_s)$ denote the angular distribution of the fragment of interest in the lab frame where (θ_s, ϕ_s) are the polar and azimuthal angles about the Z axis (the direction of $\mathbf{E}$). A particular molecular orientation (ϕ, θ, χ) makes the contribution $P_{\text{diss}}(\phi, \theta, \chi) f(\theta_m, \phi_m)$ to $I(\theta_s, \phi_s)$, and the complete fragment angular distribution is obtained by integrating over all molecular orientations (assumed to be randomly distributed):

$$I(\theta_s, \phi_s) = \int_0^{2\pi} d\phi \int_0^{\pi} \sin\theta d\theta \int_0^{2\pi} d\chi \, P_{\text{diss}}(\phi, \theta, \chi) f(\theta_m, \phi_m) \tag{6}$$

A. Show that the fragment angular distribution has the form

$$I(\theta_s, \phi_s) = \frac{\sigma}{4\pi}[1 + \beta P_2(\cos\theta_s)] \tag{7}$$

where

$$\sigma = (4\pi)^{\frac{1}{2}} b_{00}$$

$$= \int_0^{2\pi} d\phi_m \int_0^{\pi} \sin\theta_m d\theta_m \, f(\theta_m, \phi_m) \tag{8}$$

is the total cross section and

$$\beta = \frac{2 b_{20}}{5^{\frac{1}{2}} b_{00}}$$

$$= \frac{2 \int_0^{2\pi} d\phi_m \int_0^{\pi} \sin\theta_m d\theta_m \, P_2(\cos\theta_m) f(\theta_m, \phi_m)}{\int_0^{2\pi} d\phi_m \int_0^{\pi} \sin\theta_m d\theta_m \, f(\theta_m, \phi_m)}$$

$$= 2 \langle P_2(\cos\theta_m) \rangle \tag{9}$$

is the asymmetry parameter. Here $\langle \cdots \rangle$ denotes an average over the recoil distribution $f(\theta_m, \phi_m)$. Note that β ranges from 2 to -1.

B. Consider once more the limiting case of diatomic photodissociation in the axial recoil approximation. Show that $\beta = 2$, that is, that $I(\theta_s, \phi_s)$ is proportional to $\cos^2 \theta_s$, for a parallel-type transition (μ parallel to internuclear axis), while $\beta = -1$, that is, $I(\theta_s, \phi_s)$ is proportional to $\sin^2 \theta_s$, for a perpendicular-type transition (μ perpendicular to internuclear axis).

APPLICATION 7
INTRODUCTION TO SYMMETRIC TOPS

Consider a symmetric top with total angular momentum $\mathbf{J}$ that makes a projection M on the space-fixed Z axis and a projection K on the body-fixed z axis. Let ψ_{JKM} denote its wave function. By symmetry, its permanent dipole moment (if it has one) must lie along z. Then the probability of the dipole moment pointing between θ and $\theta + d\theta$ is given by

$$
P_{JKM}(\theta)\sin\theta\, d\theta = \left[\int_0^{2\pi} d\phi \int_0^{2\pi} d\chi\, \psi^*_{JKM}(\phi,\theta,\chi)\, \psi_{JKM}(\phi,\theta,\chi) \right] \sin\theta\, d\theta
$$

(1)

where θ is the angle between the z and Z axes.

A. Show that

$$
P_{JKM}(\theta) = (-1)^{M-K}\, \frac{2J+1}{2}
$$

$$
\times \sum_{n=0}^{2J} (2n+1) \begin{pmatrix} J & J & n \\ M & -M & 0 \end{pmatrix} \begin{pmatrix} J & J & n \\ K & -K & 0 \end{pmatrix} P_n(\cos\theta) \quad (2)
$$

where $P_n(\cos\theta)$ is the nth-order Legendre polynomial. This result has been obtained by S. E. Choi and R. B. Bernstein, *J. Chem. Phys.*, **85**, 150 (1986).

B. Find the orientation of this system by evaluating the expectation value of $\cos\theta$. Show that

$$
\langle \cos\theta \rangle = \frac{MK}{J(J+1)}
$$

(3)

The effect of a static electric field on the energy level structure of a symmetric top can be calculated by perturbation theory provided the external electric field is much smaller than the internal electric field of the electron–nuclei interactions (as is usually the case). Then the perturbation term is

$$
H' = -\boldsymbol{\mu} \cdot \mathbf{E} = -\mu_z E_Z \cos\theta
$$

(4)

where the electric field $\mathbf{E}$ has been chosen to lie along the Z axis in the space-fixed frame. Thus, according to first-order perturbation theory,

$$
\Delta E^{(1)} = \langle JKM | H' | JKM \rangle
$$

$$= -\mu_z E_Z \langle \cos \theta \rangle$$

$$= -\mu_z E_Z \frac{MK}{J(J+1)} \tag{5}$$

Hence symmetric top states with $K = 0$ do not show a first-order Stark effect (nor do linear molecules in their vibrational ground states with $K = 0$ or diatomic molecules in $^1\Sigma^+$ or $^1\Sigma^-$ states). The factor $\langle \cos \theta \rangle = MK/[J(J+1)]$ controls the rotational state selection of polar symmetric top molecules as achieved by the electric hexapole focusing technique; see S. R. Gandhi, T. J. Curtiss, Q.-X. Xu, S. E. Choi, and R. B. Bernstein, *Chem. Phys. Lett.*, **132**, 6 (1986).

C. In second-order perturbation theory, the molecular wave function is changed by the presence of the static electric field and the resulting energy change is given by

$$\Delta E^{(2)} = \sum_{J'K'M'} \frac{\langle JKM | H' | J'K'M' \rangle \langle J'K'M' | H' | JKM \rangle}{E_{JKM} - E_{J'K'M'}} \tag{6}$$

where the summation is over all $J'K'M'$ not equal to JKM. As will be shown in Chapter 5,

$$E_{JKM} = BJ(J+1) + (C - B)K^2 \tag{7}$$

where B and C are rotational constants. Show that the second-order energy is affected only by the two neighboring states $J' = J + 1$ and $J' = J - 1$ with $K' = K$ and $M' = M$, and that the sum of their effects is

$$\Delta E^{(2)} = \frac{\mu_z^2 E_Z^2}{2B} \left[\frac{-\left[(J+1)^2 - K^2\right]\left[(J+1)^2 - M^2\right]}{(J+1)^3(2J+1)(2J+3)} \right.$$

$$\left. + \frac{\left[J^2 - K^2\right]\left[J^2 - M^2\right]}{J^3(2J+1)(2J-1)} \right] \tag{8}$$

For $K = 0$ we obtain the second-order Stark shift for a linear molecule in the ground vibrational level or for a $^1\Sigma^+$ or $^1\Sigma^-$ diatomic molecule:

$$\Delta E^{(2)} = \frac{\mu_z^2 E_Z^2}{2B} \left[\frac{J(J+1) - 3M^2}{J(J+1)(2J-1)(2J+3)} \right] \tag{9}$$

For the special case of the $J = 0$ level, only the $J = 1$ level can interact with it, and we have

$$\Delta E^{(2)}(J = 0) = \frac{-\mu_z^2 E_Z^2}{6B} \tag{10}$$

APPLICATION 8
POLARIZED RESONANCE FLUORESCENCE AND
POLARIZED RAMAN SCATTERING: CLASSICAL EXPRESSIONS

Because molecular systems are associated with states of large angular momentum $\mathbf{J}$, it is useful in many cases to seek an understanding of their radiative properties in the limit where $\mathbf{J}$ approaches infinity. Here we assume that we may describe the absorption and emission process classically where we replace the electric dipole transition moment μ of the molecule by a Hertzian dipole oscillator vibrating at the transition frequency and pointing in the same direction as μ and attached rigidly to the molecular framework. Let the incident light be linearly polarized with its electric vector $\mathbf{E}$ pointing along $F = X, Y, Z$ in the laboratory-fixed frame. Let it excite an absorption oscillator μ pointing along $\hat{\mathbf{g}}$ referred to the molecule-fixed axes. We choose $\hat{\mathbf{g}}$ to be a unit vector along x, y, or z and denote the unit vectors perpendicular to $\hat{\mathbf{g}}$ by $\hat{\mathbf{g}}'$ and $\hat{\mathbf{g}}''$. The probability for excitation is then proportional to $|\mathbf{E}_F \cdot \mu_g|^2$, that is, to Φ_{Fg}^2 where Φ_{Fg} is the direction cosine matrix element that relates the two unit vector $\hat{\mathbf{F}}$ and $\hat{\mathbf{g}}$ having as their common origin the center of mass of the molecule. Let the emission oscillator be along $\hat{\mathbf{h}}$ where $\hat{\mathbf{h}}$ does not necessarily coincide with $\hat{\mathbf{g}}$. The probability for emission with light plane polarized along $\hat{\mathbf{H}}$ is proportional to $|\mathbf{E}_H \cdot \mu_h|^2$, that is, to Φ_{Hh}^2. It is traditional to call $I_\parallel$ the intensity of the light emitted plane polarized in the same direction as the electric vector of the incident light and $I_\perp$ the intensity of the light emitted plane polarized perpendicular to the direction of the electric vector of the incident light. Thus we have

$$I_\parallel = A \, \overline{\Phi_{Fg}^2 \, \Phi_{Fh}^2} \tag{1}$$

and

$$I_\perp = A \, \overline{\Phi_{Fg}^2 \, \Phi_{F'h}^2} \tag{2}$$

where $\hat{\mathbf{F}}'$ is a unit vector perpendicular to $\hat{\mathbf{F}}$, the bars represent an ensemble average over all initial orientations of the molecules, and A is a proportionality constant.

It is traditional to define the degree of polarization P by

$$P = \frac{I_\parallel - I_\perp}{I_\parallel + I_\perp} \tag{3}$$

and the polarization anisotropy R by

$$R = \frac{I_\parallel - I_\perp}{I_\parallel + 2\,I_\perp} \tag{4}$$

We wish to develop expressions for P and R in terms of $\cos^2 \gamma$, the average cosine squared of the angle γ between the absorption and emission oscillators

$$\cos^2 \gamma = (\hat{\mathbf{g}} \cdot \hat{\mathbf{h}})^2 = \Phi_{gh}^2 \tag{5}$$

We expand $\hat{\mathbf{h}}$ in terms of the components of the g frame

$$\hat{\mathbf{h}} = \sum_g (\hat{\mathbf{h}} \cdot \hat{\mathbf{g}})\hat{\mathbf{g}} = \sum_g \Phi_{gh}\hat{\mathbf{g}} \tag{6}$$

Hence

$$\hat{\mathbf{H}} \cdot \hat{\mathbf{h}} = \Phi_{Hh} = \sum_g (\hat{\mathbf{H}} \cdot \hat{\mathbf{g}})(\hat{\mathbf{g}} \cdot \hat{\mathbf{h}}) = \sum_g \Phi_{Hg}\Phi_{gh} \tag{7}$$

and thus

$$(\hat{\mathbf{H}} \cdot \hat{\mathbf{h}})^2 = \sum_{g_1} \sum_{g_2} \Phi_{Hg_1}\Phi_{g_1 h}\Phi_{Hg_2}\Phi_{g_2 h} \tag{8}$$

Substitution of Eq. (8) into Eqs. (1) and (2) yields

$$I_{\parallel} = A \sum_{g_1} \sum_{g_2} \overline{\Phi_{Fg}^2 \Phi_{Fg_1} \Phi_{Fg_2}} \; \Phi_{g_1 h}\Phi_{g_2 h} \tag{9}$$

and

$$I_{\perp} = A \sum_{g_1} \sum_{g_2} \overline{\Phi_{Fg}^2 \Phi_{F'g_1} \Phi_{F'g_2}} \; \Phi_{g_1 h}\Phi_{g_2 h} \tag{10}$$

We list below the only nonvanishing products of four direction cosine matrix elements averaged over all molecular orientations:

$$\overline{\Phi_{Ai}^4} = \frac{1}{5}, \qquad\qquad \overline{\Phi_{Ai}^2 \Phi_{Aj}^2} = \frac{1}{15}$$

$$\overline{\Phi_{Ai}^2 \Phi_{Bi}^2} = \frac{1}{15}, \qquad \overline{\Phi_{Ai}^2 \Phi_{Bj}^2} = \frac{2}{15}$$

$$\overline{\Phi_{Ai}\Phi_{Aj}\Phi_{Bi}\Phi_{Bj}} = -\frac{1}{30} \tag{11}$$

A. Using Eq. (11), show that

$$I_{\parallel} = \frac{A}{15}(2\cos^2 \gamma + 1), \qquad I_{\perp} = \frac{A}{15}(2 - \cos^2 \gamma) \tag{12}$$

from which it follows that

$$P = \frac{3 \cos^2 \gamma - 1}{\cos^2 \gamma + 3} \tag{13}$$

and

$$R = \frac{2}{5} P_2 (\cos \gamma) \tag{14}$$

Hint: Make use of the identity

$$\Phi_{gh}^2 + \Phi_{g'h}^2 + \Phi_{g''h}^2 = 1$$

B. Derive Eq. (11). *Hint*: It is readily verified that

$$\overline{\Phi_{Fg}^4} = \frac{1}{4\pi} \int_0^\pi \sin\theta\, d\theta \int_0^{2\pi} d\phi\, \cos^4\theta = \frac{1}{5}$$

where θ is the angle between $\hat{\mathbf{F}}$ and $\hat{\mathbf{g}}$. Consider the two equations

$$\sum_g \Phi_{Fg}^2 = 1, \qquad \sum_g \Phi_{Fg} \Phi_{F'g} = 0$$

Square each equation and carry out the average over all molecular orientations. Complete this problem by considering the identity $(\sum_F \Phi_{Fg}^2)(\sum_{F'} \Phi_{F'g'}^2) = 1$.

For a Q branch transition ($\Delta J = 0$), group theory arguments show that the transition moment lies along $\mathbf{J}$ while for P and R branch transitions ($\Delta J = -1$ and $+1$, respectively), the transition moment lies in a plane perpendicular to $\mathbf{J}$. Let $\uparrow$ denote the absorption process and $\downarrow$ the emission process. Then for a $(Q\uparrow, Q\downarrow)$ resonance fluorescence transition, the absorption and emission oscillators coincide so that $\gamma = 0$, $\cos^2 \gamma = 1$, and $P = \frac{1}{2}$, while for $(Q\uparrow, P\downarrow)$, $(Q\uparrow, R\downarrow)$, $(P\uparrow, Q\downarrow)$, and $(R\uparrow, Q\downarrow)$ resonance fluorescence transitions, the absorption and emission oscillators are at right angles so that $\gamma = \pi/2$, $\cos^2 \gamma = 0$, and $P = -\frac{1}{3}$. Finally for a resonance fluorescence transition of the type $(P\uparrow, P\downarrow)$, $(P\uparrow, R\downarrow)$, $(R\uparrow, P\downarrow)$, or $(R\uparrow, R\downarrow)$, the absorption and emission oscillators are both in a plane perpendicular to $\mathbf{J}$. Because radiative lifetimes are typically thousands of rotational periods or longer, the absorption and emission oscillators are distributed uniformly about a circle (uncorrelated) and make an average (acute) angle $\gamma = \pi/4$ so that $\cos^2 \gamma = \frac{1}{2}$ and $P = \frac{1}{7}$. Hence, although P can span the range -1 to 1, P can assume only three values, $\frac{1}{2}$, $\frac{1}{7}$, and $-\frac{1}{3}$, in the classical limit[1–5].

C. Often molecular resonance fluorescence is not resolved. Find an expression for the apparent polarization if the resonance fluorescence process consists of two types of emission, one of total intensity I_1 of polarization P_1 and the other of total intensity I_2 of polarization P_2. *Hint*: R is linear in $\cos^2 \gamma$ while P is not.

Off resonance we may observe what is called the *Raman effect*. Here the electric field of the incident light $\mathbf{E}$ induces a dipole moment μ according to the relation

$$\mu = \underset{\sim}{\alpha}\mathbf{E} \tag{15}$$

where $\underset{\sim}{\alpha}$ is a second-rank Cartesian tensor called the *molecular polarizability tensor*. The induced Hertzian dipole then radiates. Its frequency of oscillation may be just that of the electric field impressed on the molecule (Rayleigh scattering), or it may have other frequency components caused by other internal periodic motions of the molecule that modulate the frequency of the Hertzian dipole oscillator (rotational and vibrational Raman scattering).

When Eq. (15) is written out in full in the laboratory frame, it looks like

$$\begin{pmatrix} \mu_X \\ \mu_Y \\ \mu_Z \end{pmatrix} = \begin{pmatrix} \alpha_{XX} & \alpha_{XY} & \alpha_{XZ} \\ \alpha_{YX} & \alpha_{YY} & \alpha_{YZ} \\ \alpha_{ZX} & \alpha_{ZY} & \alpha_{ZZ} \end{pmatrix} \begin{pmatrix} E_X \\ E_Y \\ E_Z \end{pmatrix} \tag{16}$$

As in all tensors relating to physical quantities $\underset{\sim}{\alpha}$ is symmetric, that is, $\alpha_{AB} = \alpha_{BA}$.

Let us consider an excitation–detection geometry in which the molecular sample is located at the origin, the incident light propagates along the X axis with its electric vector pointing along the Z axis and the scattered light is detected along the Y axis for polarization along Z (giving the signal $I_{\parallel}$) or for polarization along X (giving the signal $I_{\perp}$). In this arrangement

$$I_{\parallel} = C\,\overline{\mu_Z^2} = C'\,\overline{\alpha_{ZZ}^2} \tag{17}$$

and

$$I_{\perp} = C\,\overline{\mu_X^2} = C'\,\overline{\alpha_{XZ}^2} \tag{18}$$

where the bars again denote an average over all molecular orientations.

The direction cosine matrix $\boldsymbol{\Phi}$ expresses the transformation from the space-fixed to the molecule-fixed frames. We have

$$\mu^{\text{mol}} = \underset{\sim}{\Phi}\mu^{\text{space}} \qquad\qquad \mathbf{E}^{\text{mol}} = \underset{\sim}{\Phi}\mathbf{E}^{\text{space}} \tag{19}$$

Then the polarizability tensor in the molecule-fixed frame is given by

$$\mu^{\text{mol}} = \underset{\sim}{\alpha}^{\text{mol}}\,\mathbf{E}^{\text{mol}} \tag{20}$$

D. Use Eqs. (15)–(20) to show that

$$\underset{\sim}{\alpha}^{\text{space}} = \underset{\sim}{\Phi}^{-1}\underset{\sim}{\alpha}^{\text{mol}}\underset{\sim}{\Phi}, \qquad \underset{\sim}{\alpha}^{\text{mol}} = \underset{\sim}{\Phi}\underset{\sim}{\alpha}^{\text{space}}\underset{\sim}{\Phi}^{-1} \tag{21}$$

Among the possible molecule-fixed coordinate frames we choose x, y, z so that $\underset{\sim}{\alpha}^{\text{mol}}$ is diagonal

$$\underset{\sim}{\alpha}^{\text{mol}} = \begin{pmatrix} \alpha_{xx} & 0 & 0 \\ 0 & \alpha_{yy} & 0 \\ 0 & 0 & \alpha_{zz} \end{pmatrix} \tag{22}$$

where x, y, and z are the principal axes of the polarizability tensor.

Let us introduce the quantities

$$a = \frac{1}{3}(\alpha_{xx} + \alpha_{yy} + \alpha_{zz}) \tag{23}$$

and

$$\gamma^2 = \frac{1}{2}\left[(\alpha_{xx} - \alpha_{yy})^2 + (\alpha_{yy} - \alpha_{zz})^2 + (\alpha_{zz} - \alpha_{xx})^2\right] \tag{24}$$

where a is called the mean value or *isotropic* part of the polarizability and γ^2 is called the *anisotropic* part of the polarizability.

E. Show that the depolarization ratio

$$\rho = \frac{I_\perp}{I_\parallel} \tag{25}$$

is given by[6–8]

$$\rho = \frac{3\gamma^2}{45\,a^2 + 4\gamma^2} \tag{26}$$

Hint: Use Eqs. (21) and (22) to find expressions for α_{ZZ} and α_{XZ}. Square these expressions and use Eq. (11) to carry out the averages over all molecular orientations.

The depolarization ratio ρ can range from 0 when γ^2 vanishes to $\frac{3}{4}$ when a vanishes. The latter limit cannot be reached in Rayleigh scattering since this would imply $\alpha_{xx} = \alpha_{yy} = \alpha_{zz} = 0$ and thus there would be no scattered light. However, it can occur in Raman scattering for bands in which the symmetry of the normal vibration *differs* from that of the ground state.

F. Calculate ρ for Rayleigh scattering from CH_4.

Resonance fluorescence and Raman scattering are only two of several phenomena involving the interaction of two photons with a molecule[9, 10]. Some other examples are two-photon absorption and two-photon emission. In all of these processes (unlike one-photon spectroscopies with randomly oriented molecules), the intensities are dependent on the polarization of the photons involved. In fact, the form of the polarization dependence is the same in all cases. If the molecules are isotropically distributed in space and do not move appreciably during the two-photon process (as would be the case in a viscous fluid or a glassy matrix, but not in a crystalline sample), the dependence is given by

$$I = K_2 \left\langle |\hat{\mathbf{E}}_A \hat{\mathbf{E}}_B \mathbf{T}_{AB}|^2 \right\rangle \tag{27}$$

where I is the observed intensity, $\hat{\mathbf{E}}_A$, $\hat{\mathbf{E}}_B$ are complex unit vectors representing the polarizations of the two photons, and $\mathbf{T}_{AB}$ is a second-rank Cartesian tensor describing the response of the molecule to the radiation fields. Here K_2 is a proportionality constant that in the dilute absorber approximation[11] contains the number of absorbers as well as the intensities of the two applied fields. The subscripts A and B are taken to run over the space-fixed coordinates X, Y, Z, and repeated indices imply a summation over all Cartesian components. The brackets $\langle \cdots \rangle$ indicate an average over all the possible orientations of a single molecule. The Cartesian tensor $\mathbf{T}$ is most conveniently expressed (and evaluated) in terms of the molecular coordinates x, y, z[12]. The coordinate transformation is

$$\mathbf{T}_{AB} = \Phi_{Aa} \Phi_{Bb} \mathbf{T}_{ab} \tag{28}$$

where Φ_{Aa} is the direction cosine between $\hat{\mathbf{A}}$ and $\hat{\mathbf{a}}$ and so on. Thus we want to average

$$|\hat{\mathbf{E}}_A \hat{\mathbf{E}}_B \mathbf{T}_{AB}|^2 = (\hat{\mathbf{E}}_A \hat{\mathbf{E}}_B \mathbf{T}_{AB})(\hat{\mathbf{E}}_C \hat{\mathbf{E}}_D \mathbf{T}_{CD})^*$$

$$= (\hat{\mathbf{E}}_A \hat{\mathbf{E}}_B \Phi_{Aa} \Phi_{Bb} \mathbf{T}_{ab})(\hat{\mathbf{E}}_C \hat{\mathbf{E}}_D \Phi_{Cc} \Phi_{Dd} \mathbf{T}_{cd})^*$$

$$= (\hat{\mathbf{E}}_A \hat{\mathbf{E}}_B \hat{\mathbf{E}}_C^* \hat{\mathbf{E}}_D^*)(\Phi_{Aa} \Phi_{Bb} \Phi_{Cc} \Phi_{Dd})(\mathbf{T}_{ab} \mathbf{T}_{cd}^*) \tag{29}$$

where we make use of the fact that the direction cosines are real quantities. The expression for the intensity becomes

$$I = K_2 [\hat{\mathbf{E}}_A \hat{\mathbf{E}}_B \hat{\mathbf{E}}_C^* \hat{\mathbf{E}}_D^*][\langle \Phi_{Aa} \Phi_{Bb} \Phi_{Cc} \Phi_{Dd} \rangle][\mathbf{T}_{ab} \mathbf{T}_{cd}^*] \tag{30}$$

All polarization information appears in the first factor in square brackets, all molecular information in the last factor, and all orientation information in the middle factor. Thus the general polarization dependence of a two-photon process is contained in the rotationally averaged product of four direction cosines[13]. There are formally $3^8 = 6561$ such products. However, as Eq. (11) shows, only a small number of these products (of five different types) are nonvanishing.

Similar considerations apply to higher-order multiphoton processes[14]. For example, the general polarization dependence of a three-photon process is contained in

$$\langle \Phi_{Aa}\Phi_{Bb}\Phi_{Cc}\Phi_{Dd}\Phi_{Ee}\Phi_{Ff} \rangle$$

and of a four-photon process in

$$\langle \Phi_{Aa}\Phi_{Bb}\Phi_{Cc}\Phi_{Dd}\Phi_{Ee}\Phi_{Ff}\Phi_{Gg}\Phi_{Hh} \rangle$$

The explicit integrations are straightforward but so numerous (3^{16} different cases for a four-photon process!) as to be overwhelmingly tedious[15]. Fortunately, a systematic method for evaluating direction cosine averages of this type has been found by Andrews and Thirunamachandran[16], and the results for three-photon and four-photon processes are available in the literature[16–20].

The polarization dependence is often used to determine the symmetry of the excited state, particularly for molecules in a rigid medium[21]. This serves to introduce us to the topic of photoselection[22–24], which is a sequential process involving polarized absorption followed by a polarized probing step in a rigid isotropic medium. In principle, either or both processes might be multiphoton in nature, although single-photon absorption followed by a single-photon probe step is by far the most common application. Note that the state that is ultimately probed need not be the same state initially prepared since a rigid medium preserves the spatial anisotropy created in the polarized absorption step.

NOTES AND REFERENCES

1. P. P. Feofilov, *The Physical Basis of Polarized Emission*, Consultants Bureau Enterprises, New York, 1961.

2. V. L. Levshin, *Z. Phys.*, **32**, 307 (1925).

3. F. Perrin, *Ann. Phys.* (Paris), **12**, 169 (1929).

4. A. Jablonski, *Acta Phys. Polonica*, **4**, 371 (1935); **5**, 271 (1936); **10**, 33 (1950).

5. M. McClintock, W. Demtröder, and R. N. Zare, *J. Chem. Phys.*, **51**, 5509 (1969).

6. A. Anderson, ed., *The Raman Effect*, Vols. I and II, Marcel Dekker, New York, 1967.

7. E. B. Wilson, Jr., J. C. Decius, and P. C. Cross, *Molecular Vibrations*, McGraw-Hill, New York, 1955.

8. G. Placzek, in *Handbuch der Radiologie*, Vol. VI, E. Marx, ed., Akademische Verlagsgesellschaft, Leipzig, 1934.

9. W. M. McClain, *Acc. Chem. Res.*, **7**, 129 (1974); W. M. McClain and R. A. Harris, in *Excited States*, Vol. 3, E. C. Lim, ed., Academic Press, New York, 1977, pp. 1–56.

10. D. M. Friedrich and W. M. McClain, *Ann. Rev. Phys. Chem.*, **31**, 559 (1980).

11. In this approximation each molecule contributes to the overall transition matrix element a term with a different phase factor. Consequently, for a large number of molecules the interference between matrix elements of different molecules provides a vanishingly small contribution to the magnitude squared of the overall transition matrix element. Thus the N molecules contribute incoherently, and the contribution is equivalent to N times the contribution of a single molecule.

12. For example, in single-color two-photon (electric dipole) absorption, the second-rank Cartesian tensor $\underset{\sim}{T}_{ab}$ has the form

$$\underset{\sim}{T}_{ab} = \sum_{e} \frac{\langle f \mid \mu_a \mid e \rangle \langle e \mid \mu_b \mid i \rangle}{E_{ei} - h\nu + i\Gamma_e/2}$$

for a transition from the initial state i to the final state f via the intermediate states e with lifetimes $1/\Gamma_e$. Here μ is the electric dipole transition operator and E_{ei} is the energy difference $E_e - E_i$.

13. P. R. Monson and W. M. McClain, *J. Chem. Phys.*, **53**, 29 (1970).

14. D. L. Andrews and W. A. Ghoul, *J. Chem. Phys.*, **75**, 530 (1981).

15. An appealing alternative is to reexpress the Cartesian tensors in terms of spherical irreducible tensors (see Chapter 5), which transform under rotation as the Wigner rotation matrices. Then the integration over all orientations is readily accomplished. See K. S. Haber, Ph.D. thesis, Cornell University, Ithaca, New York, 1984. See also B. Dick, *Chem. Phys.*, **96**, 199 (1985).

16. D. L. Andrews and T. Thirunamachandran, *J. Chem. Phys.*, **67**, 5026 (1977).

17. S. J. Cyvin, J. E. Rauch, and J. C. Decius, *J. Chem. Phys.*, **43**, 4083 (1965).

18. W. M. McClain, *J. Chem. Phys.*, **57**, 2264 (1972).

19. D. L. Andrews and W. A. Ghoul, *J. Phys. A.*, **14**, 1281 (1981).

20. J. R. Cable and A. C. Albrecht, *J. Chem. Phys.*, **85**, 3145 (1986).

21. F. Dörr, "Polarized Light in Spectroscopy and Photochemistry," in *Creation and Detection of the Excited State*, Vol. I, A. A. Lamola, ed., Marcel Dekker, New York, 1971, pp. 53–122.

22. A. C. Albrecht, *J. Molec. Spectrosc.*, **6**, 84 (1961).

23. A. C. Albrecht, *Progr. Reaction Kinetics*, **5**, 301 (1970).

24. T. W. Scott and A. C. Albrecht, *J. Chem. Phys.*, **78**, 150 (1983).

APPLICATION 9
MAGNETIC DEPOLARIZATION OF RESONANCE FLUORESCENCE:
ZEEMAN QUANTUM BEATS AND THE HANLE EFFECT

Let us consider the resonance fluorescence process when an external magnetic field $\mathbf{H}$ is applied. This causes the excited state of the molecule to undergo Larmor precession before radiating. During a time t the total angular momentum vector $\mathbf{J}$ precesses about $\mathbf{H}$ by the angle $\omega_L t$, where the Larmor precession frequency is given by

$$\omega_L = \mu_0 g H / \hbar \tag{1}$$

Here μ_0 is the electronic Bohr magneton and g is the electronic Landé factor expressing the quotient of the magnitude of the average magnetic moment along $\mathbf{J}$ divided by the magnitude of the total angular momentum of the system.

Let the direction of $\mathbf{H}$ define the Z axis in the space-fixed frame. Then from the viewpoint of the molecule, the presence of the magnetic field causes the X and Y axes to precess at the frequency ω_L:

$$\begin{pmatrix} X^P \\ Y^P \\ Z^P \end{pmatrix} = \begin{pmatrix} \cos \omega_L t & \sin \omega_L t & 0 \\ -\sin \omega_L t & \cos \omega_L t & 0 \\ 0 & 0 & 1 \end{pmatrix} \begin{pmatrix} X \\ Y \\ Z \end{pmatrix} \tag{2}$$

We specialize our considerations to a mutually orthogonal geometry in which the molecular sample is located at the origin, the magnetic field is directly along the Z axis, the incident plane polarized light beam propagates along the X axis toward the origin, with its electric vector pointing along the Y axis, and the resulting resonance fluorescence is detected along the Y axis and analyzed for its polarization along the X axis. Suppose that at $t = t_0$ a light pulse of short duration is incident on the sample. Then at time t the resonance fluorescence signal is given by

$$I(H, t - t_0) = A \overline{\Phi_{Yg}^2 \Phi_{Xh}^2 (H, t - t_0)} \exp[-(t - t_0)/\tau] \tag{3}$$

where A is a proportionality constant, Φ_{Yg}^2 expresses the probability of exciting the Hertzian dipole absorption oscillator along $\hat{\mathbf{g}}$, $\Phi_{Xh}^2(H, t - t_0)$ expresses the probability that the emission oscillator along $\hat{\mathbf{h}}$ emits light plane polarized along $\hat{\mathbf{X}}$ at the field strength H at time t, the bar means that an average must be taken over all initial molecular orientations, and the factor $\exp[-(t - t_0)/\tau]$ expresses the fact that the excited-state population decays radiatively with a lifetime τ.

A. Show that

$$I(H, t - t_0) = A'[1 - P \cos 2\omega_L(t - t_0)] \exp[-(t - t_0)/\tau] \tag{4}$$

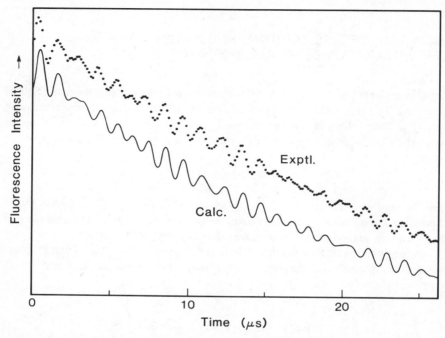

FIGURE 1   Time-resolved fluorescence decay following pulsed excitation of NO_2 on the $F_2 9_{09}-10_{0,10}$ transition (16,840.90 cm^{-1}) of the 5933 Å band in an applied magnetic field of 2.6 G. The calculated spectrum is displaced for clarity. The NO_2 excited state level has three hyperfine components ($I = 1$ for ^{14}N), each with slightly different g values and hence slightly different Larmor precession frequencies, ω_L. This causes the appearance of the super-beat structure as the contribution from each excited state hyperfine level comes in and out of phase. Reproduced with permission from P. J. Brucat and R. N. Zare, *J. Chem. Phys.*, **78**, 100 (1983).

where $P = (I_{\parallel} - I_{\perp})/(I_{\parallel} + I_{\perp})$ as defined previously. Hence the resonance fluorescence decays exponentially but is modulated at twice the Larmor precession frequency. This modulation is called *Zeeman quantum beats*[1]. The degree of modulation equals the degree of fluorescence polarization P. An example of this effect is shown in Figure 1.

B. Suppose that either the incident light is plane polarized along **H** (the Z axis) or the resonance fluorescence is analyzed for polarization along **H** (the Z axis). Discuss how the resonance fluorescence signal depends on the magnetic field strength for these excitation–detection geometries.

C. Instead of a light pulse of short duration, consider a continuous light beam. Show that $I(H)$ has a Lorentzian shape as a function of magnetic field strength with a half-width at half-maximum of

$$H_{\frac{1}{2}} = \frac{\hbar}{2\mu_0 g\tau} \qquad (5)$$

This type of magnetic depolarization of the resonance fluorescence is called the *Hanle effect* [2, 3]. Determination of the value of $H_{\frac{1}{2}}$ permits the measurement of the product $g\tau$. Classically, at small field strengths the Larmor precession rate is much slower than the rate of radiative decay so that the emission is polarized with a value close to that for $H = 0$. As the field strength increases, the Larmor precession frequency increases. For large field strengths, the molecule precesses many times on the average before radiating and the fluorescence becomes completely depolarized.

D. Suppose that the intensity of the incident light beam can be modulated at the frequency ω according to

$$I(t_0) = I_0 \left(\tfrac{1}{2} + \tfrac{1}{2} \cos \omega t_0 \right) \qquad (6)$$

Use Eq. (4), the response of the resonance fluorescence signal to a pulse of light at time t_0, to determine the steady-state response of the system. Consider what happens to the resonance fluorescence signal if the modulation frequency is swept or if the magnetic field strength is swept[4]. Explain how to design an experiment to determine g and τ independently.

NOTES AND REFERENCES

1. For examples, see E. B. Alexandrov, *Opt. Spectrosc.*, **17**, 522 (1964); J. N. Dodd, D. M. Warrington, and R. D. Kaul, *Proc. Phys. Soc.*, (London), **84**, 176 (1964); P. J. Brucat and R. N. Zare, *J. Chem. Phys.*, **78**, 100 (1983).

2. W. Hanle, *Z. Phys.*, **30**, 93 (1924).

3. R. N. Zare, *Acc. Chem. Res.*, **4**, 361 (1971).

4. J. N. Dodd and G. W. Series, "Time-Resolved Fluorescence Spectroscopy," in *Progress in Atomic Spectroscopy*, W. Hanle and H. Kleinpoppen, eds., Plenum, New York, 1978, Part A, Chapter 14, pp. 639–677.

APPLICATION 10
CORRELATION FUNCTIONS IN MOLECULAR SPECTROSCOPY

Spectroscopy concerns the frequency response of a system subjected to radiation. The most common picture of this process is a static one in which radiation with frequency ω is absorbed or emitted by the system provided the Bohr condition $\omega = \omega_{fi}$ is satisfied where

$$\omega_{fi} = \frac{(E_f - E_i)}{\hbar} \tag{1}$$

This time-independent view stresses finding the possible energy levels of the system and the selection rules that govern their radiative connections. In this picture it is useful to define a lineshape function (for absorption or emission) by

$$I(\omega) = 3 \sum_i \sum_f \rho_i \, |\langle f| \, \hat{\mathbf{e}} \cdot \boldsymbol{\mu} \, |i\rangle|^2 \, \delta(\omega_{fi} - \omega) \tag{2}$$

Here the initial state $|i\rangle$ has energy E_i, the final state $|f\rangle$ has energy E_f, ρ_i is the probability of the system being in state $|i\rangle$, and $|\langle f| \, \hat{\mathbf{e}} \cdot \boldsymbol{\mu} \, |i\rangle|^2$ is the square of the transition dipole matrix element connecting i to f, where $\hat{\mathbf{e}}$ is a unit vector pointing along the electric field of the light wave, and $\boldsymbol{\mu}$ is the transition dipole moment vector of the molecule. By switching to a time-dependent picture, we can extract the dynamics of the system undergoing the i to f transition[1, 2].

We introduce the Fourier expansion of the Dirac delta function

$$\delta(\omega) = \frac{1}{2\pi} \int_{-\infty}^{\infty} \exp(i\omega t) \, dt \tag{3}$$

and recall that

$$\boldsymbol{\mu}(t) = \exp(i\mathcal{H}t/\hbar) \, \boldsymbol{\mu}(0) \, \exp(-i\mathcal{H}t/\hbar) \tag{4}$$

and

$$|\langle i| \, \hat{\mathbf{e}} \cdot \boldsymbol{\mu} \, |f\rangle|^2 = \langle i| \, \hat{\mathbf{e}} \cdot \boldsymbol{\mu} \, |f\rangle \, \langle f| \, \hat{\mathbf{e}} \cdot \boldsymbol{\mu} \, |i\rangle \tag{5}$$

where we take $\hat{\mathbf{e}}$ and $\boldsymbol{\mu}$ as real functions.

A. Show that the lineshape factor may be written as

$$I(\omega) = \frac{3}{2\pi} \int_{-\infty}^{\infty} dt \, \exp(-i\omega t) \sum_i \rho_i \, \langle i| \, \hat{\mathbf{e}} \cdot \boldsymbol{\mu}(0) \, \hat{\mathbf{e}} \cdot \boldsymbol{\mu}(t) \, |i\rangle \tag{6}$$

The sum over initial states i weighted by ρ_i is simply the ensemble average, which we denote by $\langle \cdots \rangle$:

$$I(\omega) = \frac{3}{2\pi} \int_{-\infty}^{\infty} dt \exp(-i\omega t) \langle \hat{\mathbf{e}} \cdot \boldsymbol{\mu}(0)\, \hat{\mathbf{e}} \cdot \boldsymbol{\mu}(t) \rangle \tag{7}$$

For an isotropic system, the same result is obtained independent of the direction of polarization $\hat{\mathbf{e}}$. Hence

$$I(\omega) = \frac{1}{2\pi} \int_{-\infty}^{\infty} dt \exp(-i\omega t) \langle \boldsymbol{\mu}(0) \cdot \boldsymbol{\mu}(t) \rangle \tag{8}$$

We call

$$G(t) = \langle \boldsymbol{\mu}(0) \cdot \boldsymbol{\mu}(t) \rangle \tag{9}$$

the *dipole correlation function*. Equation (8) expresses the general result that the lineshape or frequency response of a system interacting with radiation is the Fourier transform of its dipole correlation function.

The dipole correlation function may be obtained directly from experiment by inverting the Fourier transform given in Eq. (8):

$$\langle \boldsymbol{\mu}(0) \cdot \boldsymbol{\mu}(t) \rangle = \int_{\text{band}} d\omega \exp(i\omega t) I(\omega) \tag{10}$$

where the integration is over the spectral band of interest. It is convenient to normalize the correlation function to its initial value:

$$\langle \boldsymbol{\mu}^2(0) \rangle = \int_{\text{band}} d\omega\, I(\omega) \tag{11}$$

Thus we define a correlation function of a unit vector $\hat{\mathbf{u}}(t)$ along $\boldsymbol{\mu}(t)$ by

$$\langle \hat{\mathbf{u}}(0) \cdot \hat{\mathbf{u}}(t) \rangle = \frac{\langle \boldsymbol{\mu}(0) \cdot \boldsymbol{\mu}(t) \rangle}{\langle \boldsymbol{\mu}^2(0) \rangle} \tag{12}$$

and a normalized intensity distribution by

$$\hat{I}(\omega) = \frac{I(\omega)}{\int_{\text{band}} d\omega\, I(\omega)} \tag{13}$$

Then

$$\langle \hat{\mathbf{u}}(0) \cdot \hat{\mathbf{u}}(t) \rangle = \int_{\text{band}} d\omega \exp(i\omega t) \hat{I}(\omega) \tag{14}$$

Some experimentally derived correlation functions (real part) obtained from the shape of the near-infrared rotation–vibration band of CO in various environments are

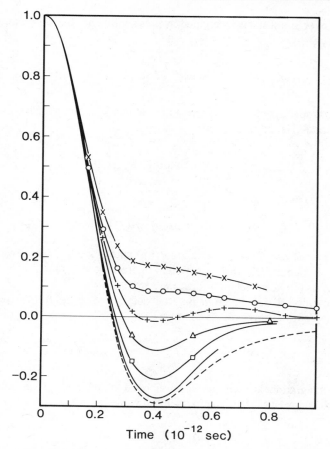

FIGURE 1   Carbon monoxide dipole correlation functions in various environments. Spectra were obtained from the following references: CHCl₃ and n-C₇H₁₆. M. O. Bulanin and N. D. Orlova, *Opt. i Spektroskopiya*, **15**, 112 (1963) [*Opt. Spectr. USSR (Eng. transl.)*, **15**, 208 (1963)]: CCl₄, J. Lascombe, P. V. Huong, and M. Josien, *Bull. Soc. Chim. France*, 1175 (1959); A. R. Coulon, L. Galalry, B. Oksengorn, S. Robin, and B. Vodar, *J. Phys. Radium*, **15**, 58, 641 (1954). [×, CO in CHCl₃ (liquid); ⊙, CO in CCl₄ (liquid); +, CO in n-C₇H₁₆ (liquid); △, CO in argon (gas, 510 amagat); ▢, CO in argon (gas, 270 amagat); —, CO in argon (gas, 66 amagat); - - -, CO free (calculated)]. Reprinted with permission from R. G. Gordon, *Adv. Magn. Res.*, **3**, 1 (1968).

shown in Figure 1[1]. The negative correlations in this gas-phase system mean that after a time t, it is more probable that the CO molecule has swung around to a point more than $90°$ from its direction at $t = 0$. For CO in liquids we anticipate that the dipole correlation function is always positive.

B. Because $I(\omega)$ is a real quantity, that is, $I(\omega) = I^*(\omega)$, show that

$$G(t) = G^*(-t) \tag{15}$$

from which it follows that

$$\text{Re } G(t) = \text{Re } G(-t) \tag{16}$$

and

$$\text{Im } G(t) = -\text{Im } G(-t) \tag{17}$$

Thus the real part of the correlation function is an even function of the time variable t, while the imaginary part of the correlation function is an odd function of t. This permits us to calculate the correlation function for positive time only:

$$
\begin{aligned}
I(\omega) &= \frac{1}{2\pi} \left[\int_{-\infty}^{0} dt \, \exp(-i\omega t) G(t) + \int_{0}^{\infty} dt \, \exp(-i\omega t) G(t) \right] \\
&= \frac{1}{2\pi} \left[\int_{0}^{\infty} dt \, \exp(i\omega t) G^*(t) + \int_{0}^{\infty} dt \, \exp(-i\omega t) G(t) \right] \\
&= \frac{1}{\pi} \text{Re} \int_{0}^{\infty} dt \, \exp(-i\omega t) G(t)
\end{aligned}
\tag{18}
$$

C. Suppose that the dipole correlation function varies sinusoidally in time with a frequency ω_0, that is, $G(t) = \exp(i\omega_0 t)$. Find $I(\omega)$ and discuss its form.

D. Suppose that $G(t)$ has a sinusoidal variation in time with frequency ω_0 but also decreases exponentially with a half-life τ, that is,

$$G(t) = \tau^{-1} \exp(-|t|/\tau) \exp(i\omega_0 t)$$

Find $I(\omega)$ and discuss its form.

E. As another example, consider the classical infrared spectrum[3, 4] of an equilibrium distribution of linear molecules of moment of inertia I at a temperature T. Here the transition dipole moment vector lies along the molecular axis, and if the latter rotates at an angular frequency Ω, then

$$\hat{\mathbf{u}}(0) \cdot \hat{\mathbf{u}}(t) = \cos \Omega t \tag{19}$$

The rotational energy of a linear molecule is given to a first approximation by

$$W_J = \frac{J(J+1)h^2}{8\pi^2 I} \tag{20}$$

where $J = 0, 1, 2$, and so on. The rotational partition function[5] is

$$Q_{\text{rot}} = \sum_J (2J + 1) \exp(-W_J/kT) \tag{21}$$

where $2J + 1$ is the rotational degeneracy factor. For sufficiently large T or large I, J may be regarded as a continuous variable, and this sum can be approximated by the integral

$$Q_{\text{rot}} = \int_0^\infty (2J + 1)\exp\left[\frac{-J(J+1)h^2}{8\pi^2 IkT}\right] dJ = \frac{8\pi^2 IkT}{h^2} \tag{22}$$

Then the fraction of molecules between J and $J + dJ$ is given by

$$f(J)\, dJ = \frac{(2J+1)}{Q_{\text{rot}}} \exp\left[\frac{-J(J+1)h^2}{8\pi^2 IkT}\right] dJ$$

$$= \frac{(2J+1)h^2}{8\pi^2 IkT} \exp\left[\frac{-J(J+1)h^2}{8\pi^2 IkT}\right] dJ \tag{23}$$

To obtain the probability $f(\Omega)\, d\Omega$ of finding a linear molecule rotating with an angular frequency magnitude between Ω and $\Omega + d\Omega$, we identify $J(J+1) \times h^2/8\pi^2 I$ with the classical rotational energy $\frac{1}{2}I\Omega^2$, from which it follows that $(J + \frac{1}{2})h$ equals approximately $2\pi I|\Omega|$, where the absolute value sign reminds us that classically the angular frequency can be positive or negative. Show then that

$$f(\Omega)\, d\Omega = \left(\frac{I}{kT}\right) \Omega \exp\left(\frac{-\frac{1}{2}I\Omega^2}{kT}\right) d\Omega \tag{24}$$

for $0 \leq \Omega \leq \infty$.

Thus the normalized dipole correlation function is found by taking the ensemble average over Eq. (19):

$$\langle \hat{\mathbf{u}}(0) \cdot \hat{\mathbf{u}}(t) \rangle = \left(\frac{I}{kT}\right) \int_0^\infty \Omega \exp\left(\frac{-\frac{1}{2}I\Omega^2}{kT}\right) \cos \Omega t \, d\Omega$$

$$= \left(\frac{I}{kT}\right) \text{Re} \int_0^\infty \Omega \exp\left(\frac{-\frac{1}{2}I\Omega^2}{kT}\right) \exp(i\Omega t) \, d\Omega \tag{25}$$

Show that

$$I(\omega) = \left(\frac{I}{kT}\right) |\omega| \exp\left(\frac{-\frac{1}{2}I\omega^2}{kT}\right) \tag{26}$$

Hint: Do not explicitly evaluate Eq. (25); instead, think Fourier transform pairs.

Equation (26) has a simple interpretation: the classical infrared spectrum of linear molecules at a temperature T has the form of a broad symmetric doublet of P and R branch envelopes, wider at the outer wings and narrower at the inner wings than that of the corresponding Gaussian curve. This example illustrates how dynamical information can be extracted from a spectrum without resolving individual lines!

The dipole correlation function can be used not only to determine the spectral density $I(\omega)$ but also the degree of polarization of scattered or emitted radiation. For example, the polarization anisotropy of resonance fluorescence at time t may be written as

$$R(t) = \tfrac{2}{5}\langle P_2[\,\hat{\mathbf{u}}(0) \cdot \hat{\mathbf{u}}(t)\,]\rangle \tag{27}$$

[see Eq. (14) of Application 8], from which R may be determined by taking a suitable average over the time history of the system[6–8]. Once it was understood by spectroscopists that the spectral density $I(\omega)$ and its depolarization features are related to the dipole correlation function $G(t)$ and its various moments, much work has been done to explore these connections[2, 9, 10] and even in some cases to calculate infrared or Raman spectra from first principles[11, 12].

NOTES AND REFERENCES

1. R. G. Gordon, *Adv. Magn. Res.*, **3**, 1 (1968).

2. B. J. Berne and R. Pecora, *Dynamic Light Scattering*, Wiley, New York, 1976.

3. H. Shimizu, *J. Chem. Phys.*, **43**, 2453 (1965).

4. A. G. St. Pierre and W. A. Steele, *Phys. Rev.*, **184**, 172 (1969).

5. G. Herzberg, *Spectra of Diatomic Molecules*, Van Nostrand, New York, 1950.

6. R. G. Gordon, *J. Chem. Phys.*, **45**, 1643 (1966).

7. J. Husain, J. R. Wiesenfeld, and R. N. Zare, *J. Chem. Phys.*, **72**, 2479 (1980).

8. T. Nagata, T. Kondow, K. Kuchitsu, G. W. Loge, and R. N. Zare, *Molec. Phys.*, **50**, 49 (1983).

9. R. E. D. McClung, *Adv. Molec. Relaxation Interaction Processes*, **10**, 83 (1977).

10. J.-Cl. Leicknam and Y. Guissani, *J. Molec. Struct.*, **80**, 377 (1982).

11. E. J. Heller, *Acc. Chem. Res.*, **14**, 368 (1981).

12. D. R. Fredkin, A. Komornicki, S. R. White, and K. R. Wilson, *J. Chem. Phys.*, **78**, 7077 (1983).

Chapter 4

COUPLING OF MORE THAN TWO ANGULAR MOMENTUM VECTORS

4.1 THE $6\text{-}j$ AND $9\text{-}j$ SYMBOLS

When we couple two angular momentum vectors $\mathbf{j}_1$ and $\mathbf{j}_2$, the resultant angular momentum states are completely characterized by $|jm\rangle$. However, if we couple together three angular momentum vectors, $\mathbf{j}_1$, $\mathbf{j}_2$, and $\mathbf{j}_3$, the resulting state $|jm\rangle$ does not completely specify the system because there is more than one way in which these vectors can be added to form the same resultant. Let us first couple $\mathbf{j}_1$ and $\mathbf{j}_2$ to give $\mathbf{j}_{12}$, then add $\mathbf{j}_{12}$ to $\mathbf{j}_3$ to give the resultant $\mathbf{j}$. The eigenfunctions are

$$|j_{12}j_3jm\rangle = \sum_{m_{12},m_3} \langle j_{12}\,m_{12}, j_3\,m_3|jm\rangle\,|j_{12}\,m_{12}, j_3\,m_3\rangle$$

$$= \sum_{\substack{m_1,m_2,\\m_3,m_{12}}} \langle j_1\,m_1, j_2\,m_2|j_{12}\,m_{12}\rangle\,\langle j_{12}\,m_{12}, j_3\,m_3|jm\rangle$$

$$\times |j_1\,m_1, j_2\,m_2, j_3\,m_3\rangle \tag{4.1}$$

where $|j_1\,m_1, j_2\,m_2, j_3\,m_3\rangle \equiv |j_1\,m_1\rangle\,|j_2\,m_2\rangle\,|j_3\,m_3\rangle$ are eigenfunctions of the completely uncoupled representations. Alternatively, we could add $\mathbf{j}_1$ to the resultant $\mathbf{j}_{23}$ of having coupled $\mathbf{j}_2$ with $\mathbf{j}_3$. The eigenfunctions are then

$$|j_1j_{23}jm\rangle = \sum_{m_1,m_{23}} \langle j_1\,m_1, j_{23}\,m_{23}|jm\rangle\,|j_1\,m_1, j_{23}\,m_{23}\rangle$$

$$= \sum_{\substack{m_1, m_2, \\ m_3, m_{23}}} \langle j_1 m_1, j_{23} m_{23} | j m \rangle \langle j_2 m_2, j_3 m_3 | j_{23} m_{23} \rangle$$

$$\times | j_1 m_1, j_2 m_2, j_3 m_3 \rangle \tag{4.2}$$

These two representations of the same set of states must be physically equivalent. Hence they are connected by a unitary transformation [just as the coupled and uncoupled representations of the addition of two angular momenta are related; see Eqs. (2.5)–(2.9)]. We may write

$$| j_1 j_{23} j m \rangle = \sum_{j_{12}} \langle j_{12} j_3 j | j_1 j_{23} j' \rangle | j_{12} j_3 j' m' \rangle \, \delta_{jj'} \delta_{mm'} \tag{4.3}$$

where each expansion coefficient $\langle j_{12} j_3 j | j_1 j_{23} j' \rangle \equiv \langle j_{12} j_3 j m | j_1 j_{23} j' m' \rangle$ is the scalar product between the eigenfunctions taken from the two coupling schemes. Because of orthogonality, this scalar product vanishes unless $j = j'$ and $m = m'$. Moreover, its value is independent of m because a scalar product does not depend on the orientation of the coordinate system. The expansion coefficients are also called *recoupling coefficients* because they describe the transformation between two different coupling schemes, in this case the $| j_1 j_{23} j m \rangle$ scheme in which the operators $\mathbf{j}^2$, j_z, $\mathbf{j}_1^2$, and $\mathbf{j}_{23}^2$ are diagonal, and the $| j_{12} j_3 j m \rangle$ scheme, in which the operators $\mathbf{j}^2$, j_z, $\mathbf{j}_{12}^2$, and $\mathbf{j}_3^2$ are diagonal.

In what follows we develop an explicit expression for the recoupling coefficients. We begin by inserting Eqs. (4.1) and (4.2) into Eq. (4.3) and equating the coefficients of $| j_1 m_1, j_2 m_2, j_3 m_3 \rangle$ on both sides of this equation. We find

$$\sum_{j_{12}} \langle j_{12} j_3 j | j_1 j_{23} j \rangle \sum_{m_{12}} \langle j_1 m_1, j_2 m_2 | j_{12} m_{12} \rangle \langle j_{12} m_{12}, j_3 m_3 | j m \rangle$$

$$= \sum_{m_{23}} \langle j_1 m_1, j_{23} m_{23} | j m \rangle \langle j_2 m_2, j_3 m_3 | j_{23} m_{23} \rangle \tag{4.4}$$

Multiply both sides of Eq. (4.4) by $\langle j_1 m_1, j_2 m_2 | j_{12}' m_1 + m_2 \rangle$ and sum over m_1 with $m_1 + m_2 = m_{12}$ held constant. By the orthonormality of the Clebsch–Gordan coefficients, the sum over m_1 on the left-hand side gives $\delta_{j_{12} j_{12}'}$, reducing the sum over j_{12} to one term with j_{12} replaced by j_{12}'. We drop this prime and make the remaining summation explicit, using the fact that m_{12} and m_3 are fixed and $m_1 + m_2 + m_3 = m$. We obtain the result

$$\langle j_{12} j_3 j | j_1 j_{23} j \rangle \langle j_{12} m_{12}, j_3 m_3 | j m \rangle$$

$$= \sum_{m_1} \langle j_1 m_1, j_2 m_{12} - m_1 | j_{12} m_{12} \rangle$$

$$\times \langle j_1 m_1, j_{23} m_{12} + m_3 - m_1 | j m_{12} + m_3 \rangle$$

$$\times \langle j_2 m_{12} - m_1, j_3 m_3 | j_{23} m_{12} + m_3 - m_1 \rangle \qquad (4.5)$$

The same procedure may be used to "isolate" the recoupling coefficient. Multiply both sides of Eq. (4.5) by $\langle j_{12} m'_{12}, j_3 m_3 | j m \rangle$ and sum over m'_{12} with $m = m'_{12} + m_3$ held fixed. Dropping primes, we obtain

$$\langle j_{12} j_3 j | j_1 j_{23} j \rangle$$

$$= \sum_{m_1, m_{12}} \langle j_1 m_1, j_2 m_{12} - m_1 | j_{12} m_{12} \rangle \langle j_1 m_1, j_{23} m - m_1 | j m \rangle$$

$$\times \langle j_2 m_{12} - m_1, j_3 m_3 | j_{23} m - m_1 \rangle$$

$$\times \langle j_{12} m_{12}, j_3 m - m_{12} | j m \rangle \qquad (4.6)$$

Here we have found an expression for the recoupling coefficients as the contraction of the product of four Clebsch–Gordan coefficients. As the latter are known, we have in principle obtained all the recoupling coefficients for the addition of three angular momenta. In particular, Eq. (4.6) shows at once that the coefficients of this unitary transformation are real.

The first to characterize these recoupling coefficients was Racah[1], who defined his W coefficients (called today *Racah coefficients*) in terms of these recoupling coefficients by

$$W(j_1 j_2 j j_3; j_{12} j_{23}) \equiv [(2 j_{12} + 1)(2 j_{23} + 1)]^{-\frac{1}{2}} \langle j_{12} j_3 j | j_1 j_{23} j \rangle \qquad (4.7)$$

The Racah coefficients were introduced because of their symmetry properties compared to the recoupling coefficients. However, Wigner[2] has defined a closely related quantity, the 6-j symbol, which has even higher symmetry. It is related to the Racah coefficient by a simple phase factor

$$\begin{Bmatrix} j_1 & j_2 & j_{12} \\ j_3 & j & j_{23} \end{Bmatrix} = (-1)^{j_1 + j_2 + j_3 + j} \, W(j_1 j_2 j j_3; j_{12} j_{23})$$

$$= (-1)^{j_1 + j_2 + j_3 + j} [(2 j_{12} + 1)(2 j_{23} + 1)]^{-\frac{1}{2}} \langle j_{12} j_3 j | j_1 j_{23} j \rangle$$

$$(4.8)$$

In what follows we work exclusively with the 6-j symbols, which appear to be gaining favor over the Racah coefficients in literature usage. We list below the most commonly used properties of the 6-j symbols. All these relations can be verified by expressing the 6-j symbols in terms of 3-j symbols (or Clebsch–Gordan coefficients) and using their previously derived properties.

(a) Symmetries: The $6\text{-}j$ symbol is invariant to the interchange of any two columns; the $6\text{-}j$ symbol is also invariant to the interchange of the upper and lower arguments in each of any two columns. For example,

$$\begin{Bmatrix} j_1 & j_2 & j_3 \\ j_4 & j_5 & j_6 \end{Bmatrix} = \begin{Bmatrix} j_2 & j_1 & j_3 \\ j_5 & j_4 & j_6 \end{Bmatrix} = \begin{Bmatrix} j_1 & j_5 & j_6 \\ j_4 & j_2 & j_3 \end{Bmatrix} \tag{4.9}$$

and so on. For the $6\text{-}j$ symbol to be nonzero, the following four triangle conditions hold for the vector addition of its angular momentum arguments: $\Delta(j_1 j_2 j_3)$, $\Delta(j_1 j_5 j_6)$, $\Delta(j_4 j_2 j_6)$, and $\Delta(j_4 j_5 j_3)$. These conditions can be illustrated in the following way:

$$\begin{Bmatrix} \bullet\!-\!\bullet\!-\!\bullet \end{Bmatrix} \quad \begin{Bmatrix} \searrow_{\bullet} \end{Bmatrix} \quad \begin{Bmatrix} \wedge \end{Bmatrix} \quad \begin{Bmatrix} \bullet\!-\!\bullet^{\nearrow} \end{Bmatrix} \tag{4.10}$$

This has the consequence that, for example, $(-1)^{j_1+j_2+j_3} = (-1)^{-j_1-j_2-j_3}$ and that $(-1)^{2(j_1+j_2+j_3)} = 1$.

(b) Orthonormality:

$$\sum_{j_3} (2j_3 + 1)(2j_6 + 1) \begin{Bmatrix} j_1 & j_2 & j_3 \\ j_4 & j_5 & j_6 \end{Bmatrix} \begin{Bmatrix} j_1 & j_2 & j_3 \\ j_4 & j_5 & j_6' \end{Bmatrix} = \delta_{j_6 j_6'} \tag{4.11}$$

(c) Expressions relating $3\text{-}j$ and $6\text{-}j$ symbols: A $6\text{-}j$ symbol may be written as the contraction of four $3\text{-}j$ symbols:

$$\begin{Bmatrix} j_1 & j_2 & j_3 \\ j_4 & j_5 & j_6 \end{Bmatrix} = \sum_{\text{all } m} (-1)^{j_1-m_1+j_2-m_2+j_3-m_3+j_4-m_4+j_5-m_5+j_6-m_6}$$

$$\times \begin{pmatrix} j_1 & j_2 & j_3 \\ -m_1 & -m_2 & -m_3 \end{pmatrix} \begin{pmatrix} j_1 & j_5 & j_6 \\ m_1 & -m_5 & m_6 \end{pmatrix}$$

$$\times \begin{pmatrix} j_2 & j_6 & j_4 \\ m_2 & -m_6 & m_4 \end{pmatrix} \begin{pmatrix} j_3 & j_4 & j_5 \\ m_3 & -m_4 & m_5 \end{pmatrix}$$

$$\tag{4.12}$$

We may use the symmetry properties of the $3\text{-}j$ symbols along with the facts that $m_1 + m_2 + m_3$ sums to zero and that $j_1 + j_2 + j_3$ sums to an integer to rewrite Eq. (4.11) as

$$\begin{Bmatrix} j_1 & j_2 & j_3 \\ j_4 & j_5 & j_6 \end{Bmatrix} = \sum_{\text{all } m} (-1)^{j_4-m_4+j_5-m_5+j_6-m_6}$$

$$\times \begin{pmatrix} j_1 & j_2 & j_3 \\ m_1 & m_2 & m_3 \end{pmatrix} \begin{pmatrix} j_1 & j_5 & j_6 \\ m_1 & -m_5 & m_6 \end{pmatrix}$$

$$\times \begin{pmatrix} j_4 & j_2 & j_6 \\ m_4 & m_2 & -m_6 \end{pmatrix} \begin{pmatrix} j_4 & j_5 & j_3 \\ -m_4 & m_5 & m_3 \end{pmatrix}$$

$$(4.13)$$

Actually, only two of the six summation indices in Eqs. (4.12) and (4.13) are independent, as the magnetic quantum numbers in each of the four 3-j symbols must sum to zero. Hence if one index, call it κ, is omitted from the sum, the right-hand side of Eq. (4.12) or (4.13) must be multiplied by $(2j_\kappa + 1)$.

Equations (4.12) and (4.13) may take on a number of guises, which at first appear contradictory, if not confusing! For example, we can change m_3 to $-m_3$ and m_5 to $-m_5$ since we sum over all values of these magnetic quantum numbers in any case. If we also drop the sum over m_5, Eq. (4.13) becomes

$$\begin{Bmatrix} j_1 & j_2 & j_3 \\ j_4 & j_5 & j_6 \end{Bmatrix} = \sum_{\substack{m_1, m_2, m_3, \\ m_4, m_6}} (2j_5 + 1)(-1)^{j_1+j_2-j_3+j_4+j_5+j_6-m_1-m_4}$$

$$\times \begin{pmatrix} j_1 & j_2 & j_3 \\ m_1 & m_2 & -m_3 \end{pmatrix} \begin{pmatrix} j_4 & j_5 & j_3 \\ m_4 & m_5 & m_3 \end{pmatrix}$$

$$\times \begin{pmatrix} j_2 & j_4 & j_6 \\ m_2 & m_4 & -m_6 \end{pmatrix} \begin{pmatrix} j_5 & j_1 & j_6 \\ m_5 & m_1 & m_6 \end{pmatrix}$$

$$(4.14)$$

By using the orthogonality properties of the 3-j symbols, we may develop a number of other relations. For example, by multiplying both sides by the last 3-j symbol on the right-hand side of Eq. (4.14) we have

$$\begin{Bmatrix} j_1 & j_2 & j_3 \\ j_4 & j_5 & j_6 \end{Bmatrix} \begin{pmatrix} j_5 & j_1 & j_6 \\ m_5 & m_1 & m_6 \end{pmatrix}$$

$$= \sum_{m_2, m_3, m_4} (-1)^{j_1+j_2-j_3+j_4+j_5+j_6-m_1-m_4} \begin{pmatrix} j_1 & j_2 & j_3 \\ m_1 & m_2 & -m_3 \end{pmatrix}$$

$$\times \begin{pmatrix} j_4 & j_5 & j_3 \\ m_4 & m_5 & m_3 \end{pmatrix} \begin{pmatrix} j_2 & j_4 & j_6 \\ m_2 & m_4 & -m_6 \end{pmatrix} \qquad (4.15)$$

where we have used Eq. (2.32). Next we make use of Eq. (2.33) to rearrange Eq. (4.15) into the form

$$\sum_{j_6} (2j_6 + 1)(-1)^{j_1+j_2-j_3+j_4+j_5+j_6-m_1-m_4}$$

$$\times \begin{Bmatrix} j_1 & j_2 & j_3 \\ j_4 & j_5 & j_6 \end{Bmatrix} \begin{pmatrix} j_5 & j_1 & j_6 \\ m_5 & m_1 & m_6 \end{pmatrix} \begin{pmatrix} j_2 & j_4 & j_6 \\ m_2 & m_4 & -m_6 \end{pmatrix}$$

$$= \begin{pmatrix} j_1 & j_2 & j_3 \\ m_1 & m_2 & -m_3 \end{pmatrix} \begin{pmatrix} j_4 & j_5 & j_3 \\ m_4 & m_5 & m_3 \end{pmatrix} \tag{4.16}$$

where the formal sum over m_6 on the left-hand side and m_3 on the right-hand side of Eq. (4.16) is omitted since for fixed m_1, m_2, m_4, and m_5, these sums are only single terms. Continuing in the same manner, we obtain

$$\sum_{j_5, j_6} (2j_5 + 1)(2j_6 + 1)(-1)^{j_1+j_2-j_3+j_4+j_5+j_6-m_1-m_4} \begin{Bmatrix} j_1 & j_2 & j_3 \\ j_4 & j_5 & j_6 \end{Bmatrix}$$

$$\times \begin{pmatrix} j_5 & j_1 & j_6 \\ m_5 & m_1 & m_6 \end{pmatrix} \begin{pmatrix} j_2 & j_4 & j_6 \\ m_2 & m_4 & -m_6 \end{pmatrix} \begin{pmatrix} j_4 & j_5 & j_3 \\ m_4 & m_5 & m_3 \end{pmatrix}$$

$$= \begin{pmatrix} j_1 & j_2 & j_3 \\ m_1 & m_2 & -m_3 \end{pmatrix} \tag{4.17}$$

and

$$\sum_{j_3, j_5, j_6} (2j_3 + 1)(2j_5 + 1)(2j_6 + 1)(-1)^{j_1+j_2-j_3+j_4+j_5+j_6-m_1-m_4}$$

$$\times \begin{Bmatrix} j_1 & j_2 & j_3 \\ j_4 & j_5 & j_6 \end{Bmatrix} \begin{pmatrix} j_5 & j_1 & j_6 \\ m_5 & m_1 & m_6 \end{pmatrix} \begin{pmatrix} j_2 & j_4 & j_6 \\ m_2 & m_4 & -m_6 \end{pmatrix}$$

$$\times \begin{pmatrix} j_4 & j_5 & j_3 \\ m_4 & m_5 & m_3 \end{pmatrix} \begin{pmatrix} j_1 & j_2 & j_3 \\ m_1 & m_2 & -m_3 \end{pmatrix}$$

$$= 1 \tag{4.18}$$

(d) Special case when one argument vanishes:

$$\begin{Bmatrix} j_1 & j_2 & 0 \\ j_4 & j_5 & j_6 \end{Bmatrix} = (-1)^{j_1+j_4+j_6} [(2j_1 + 1)(2j_4 + 1)]^{-\frac{1}{2}} \delta_{j_1 j_2} \delta_{j_4 j_5} \tag{4.19}$$

By using the symmetry properties of the $6\text{-}j$ symbol, we can rearrange a $6\text{-}j$ symbol with a zero in another position for application of Eq. (4.19).

The most convenient tabulation of $6\text{-}j$ symbols is that of Rotenberg et al.[3], which gives numerical values of all symbols up to a maximum value of 8 for any argument. Algebraic expressions for the $6\text{-}j$ symbols are given by Edmonds[4] when one of the arguments is $\frac{1}{2}$, 1, $\frac{3}{2}$, or 2. In Table 4.1 we reproduce for convenience the algebraic expressions when one argument is less than or equal to 2. In the Appendix we present a simple computer program for the numerical evaluation of 6-j symbols.

The recoupling of four angular momenta leads to the 9-j symbol. Once again there is more than one (understatement!) possible coupling scheme, and all these coupling schemes are related by unitary transformations. For example, we might couple $\mathbf{j}_1, \mathbf{j}_2, \mathbf{j}_3$, and $\mathbf{j}_4$ together so that $\mathbf{j}_{14} = \mathbf{j}_1 + \mathbf{j}_4, \mathbf{j}_{23} = \mathbf{j}_2 + \mathbf{j}_3$, and $\mathbf{j} = \mathbf{j}_{14} + \mathbf{j}_{23}$. The eigenfunctions in this coupling scheme are denoted by $|(j_1 j_4)j_{14}(j_2 j_3)j_{23}jm\rangle$. Alternatively, we might couple $\mathbf{j}_{12} = \mathbf{j}_1 + \mathbf{j}_2, \mathbf{j}_{34} = \mathbf{j}_3 + \mathbf{j}_4$, and $\mathbf{j} = \mathbf{j}_{12} + \mathbf{j}_{34}$ with eigenfunctions $|(j_1 j_2)j_{12}(j_3 j_4)j_{34}jm\rangle$. In analogy to Eq. (4.3), we write

$$|(j_1 j_4)j_{14}(j_2 j_3)j_{23}jm\rangle$$

$$= \sum_{j_{12}} \sum_{j_{34}} \langle (j_1 j_2)j_{12}(j_3 j_4)j_{34}j|(j_1 j_4)j_{14}(j_2 j_3)j_{23}j\rangle$$

$$\times |(j_1 j_2)j_{12}(j_3 j_4)j_{34}jm\rangle \tag{4.20}$$

where the 9-j symbol is defined by

$$\langle (j_1 j_2)j_{12}(j_3 j_4)j_{34}j|(j_1 j_4)j_{14}(j_2 j_3)j_{23}j\rangle$$

$$= [(2j_{12} + 1)(2j_{34} + 1)(2j_{14} + 1)(2j_{23} + 1)]^{\frac{1}{2}} \begin{Bmatrix} j_1 & j_2 & j_{12} \\ j_3 & j_4 & j_{34} \\ j_{14} & j_{23} & j \end{Bmatrix} \tag{4.21}$$

In the same manner as before, the 9-j symbol may be rewritten as the product of 6-j symbols and ultimately as the product of 3-j symbols. We list below the most commonly used properties:

1. Symmetries: The 9-j symbol is unchanged by an even permutation of rows or columns but is multiplied by the phase factor

$$(-1)^{\sum (\text{all } j)} = (-1)^{j_1+j_2+j_{12}+j_3+j_4+j_{34}+j_{14}+j_{23}+j}$$

by an odd permutation of rows or columns.

2. Orthonormality:

$$\sum_{j_{12}, j_{34}} (2j_{12} + 1)(2j_{34} + 1)(2j_{14} + 1)(2j_{23} + 1)$$

$$\times \begin{Bmatrix} j_1 & j_2 & j_{12} \\ j_3 & j_4 & j_{34} \\ j_{14} & j_{23} & j \end{Bmatrix} \begin{Bmatrix} j_1 & j_2 & j_{12} \\ j_3 & j_4 & j_{34} \\ j'_{14} & j'_{23} & j \end{Bmatrix}$$

$$= \delta_{j_{14}j'_{14}} \delta_{j_{23}j'_{23}} \tag{4.22}$$

3. Contraction of 3-j symbols and 6-j symbols:

$$
\left\{
\begin{array}{ccc}
j_1 & j_2 & j_3 \\
j_4 & j_5 & j_6 \\
j_7 & j_8 & j_9
\end{array}
\right\}
$$

$$
= \sum_{\text{all } m}
\begin{pmatrix}
j_1 & j_2 & j_3 \\
m_1 & m_2 & m_3
\end{pmatrix}
\begin{pmatrix}
j_4 & j_5 & j_6 \\
m_4 & m_5 & m_6
\end{pmatrix}
\begin{pmatrix}
j_7 & j_8 & j_9 \\
m_7 & m_8 & m_9
\end{pmatrix}
$$

$$
\times
\begin{pmatrix}
j_1 & j_4 & j_7 \\
m_1 & m_4 & m_7
\end{pmatrix}
\begin{pmatrix}
j_2 & j_5 & j_8 \\
m_2 & m_5 & m_8
\end{pmatrix}
\begin{pmatrix}
j_3 & j_6 & j_9 \\
m_3 & m_6 & m_9
\end{pmatrix}
$$

$$(4.23)$$

and

$$
\left\{
\begin{array}{ccc}
j_1 & j_2 & j_3 \\
j_4 & j_5 & j_6 \\
j_7 & j_8 & j_9
\end{array}
\right\}
= \sum_k (-1)^{2k}(2k+1)
\left\{
\begin{array}{ccc}
j_1 & j_4 & j_7 \\
j_8 & j_9 & k
\end{array}
\right\}
$$

$$
\times
\left\{
\begin{array}{ccc}
j_2 & j_5 & j_8 \\
j_4 & k & j_6
\end{array}
\right\}
\left\{
\begin{array}{ccc}
j_3 & j_6 & j_9 \\
k & j_1 & j_2
\end{array}
\right\}
$$

$$(4.24)$$

4. Special case when one argument vanishes:

$$
\left\{
\begin{array}{ccc}
j_1 & j_2 & j_3 \\
j_4 & j_5 & j_6 \\
j_7 & j_8 & 0
\end{array}
\right\}
= (-1)^{j_2+j_3+j_4+j_7}[(2j_3+1)(2j_7+1)]^{-\frac{1}{2}}
$$

$$
\times
\left\{
\begin{array}{ccc}
j_1 & j_2 & j_3 \\
j_5 & j_4 & j_7
\end{array}
\right\}
\delta_{j_3 j_6}\delta_{j_7 j_8}
\qquad (4.25)
$$

By using the symmetry properties of the 9-j symbol, a 9-j symbol with a zero in another position can be rearranged for application of Eq. (4.25). Numerical evaluation of the 9-j symbols can often be conveniently carried out using Eq. (4.23) or (4.24) and the numerical values available for the 3-j and 6-j symbols. The Appendix presents a simple computer program that carries out this task.

Higher n-j symbols can be defined[5] by considering the recoupling of five or more angular momenta, but we shall have no occasion to use them.

4.2 GRAPHICAL METHODS

Many problems in angular momentum theory confront one with an angry sea awash with monsters, Clebsch–Gordan coefficients (Wigner coefficients), $3\text{-}j$ symbols, and the like, containing horrid sums over a variety of magnetic quantum numbers—and yet when these sums are evaluated, often the dependence on these magnetic quantum numbers disappears. Although such calculations can always be carried out algebraically, there exist graphical procedures that simplify these tedious manipulations. It is the purpose of this section to outline these diagrammatic methods in a manner that allows their facile application to many of the problems commonly encountered in angular momentum theory. We follow closely the graphical methods of Yutsis, Levinson, and Vanagas[5] as revised and extended by others[6–12].

A $3\text{-}j$ symbol is represented by three lines meeting at a common vertex or node. Each line is labeled by its value of j and its magnetic quantum number m. A plus sign or a minus sign is placed on the node indicating the order in which the arguments appear in the $3\text{-}j$ symbol, read left to right. A plus sign means that the arguments in the $3\text{-}j$ symbol appear in counterclockwise order about the node; a minus sign, clockwise order. Hence

$$
\begin{pmatrix} j_1 & j_2 & j_3 \\ m_1 & m_2 & m_3 \end{pmatrix} =
\overset{+}{\underset{j_1 m_1}{\diagup\!\!\diagdown}}\!\!\begin{matrix} j_3 m_3 \\ j_2 m_2 \end{matrix} =
\overset{-}{\underset{j_1 m_1}{\diagup\!\!\diagdown}}\!\!\begin{matrix} j_2 m_2 \\ j_3 m_3 \end{matrix} \tag{4.26}
$$

Note that the orientation of the graph in space does not matter, nor do the length of the lines or the angles between the lines. From Eq. (2.30) we see that changing the plus sign or minus sign on a node corresponds to multiplying by the phase factor $(-1)^{j_1+j_2+j_3}$. Thus the permutation symmetry of the $3\text{-}j$ symbol has been built into this graphical representation. In particular, a change in the order of the lines about the node accompanied by a change in the sign of the node is no change at all. An even permutation causes a simple rotation of the graph, and hence also results in no change.

The present notation does not handle the Clebsch–Gordan coefficient except by explicit use of Eqs. (2.27) and (2.28). The important difference is that in a $3\text{-}j$ symbol all the arguments have the same standing while in a Clebsch–Gordan coefficient one of the jm pairs is the resultant of coupling together the other two. To obtain a graphical method that incorporates both, we follow Brink and Satchler[6] by introducing an arrow notation whereby an undirected line means

$$
\underline{jm \qquad\qquad j'm'} = \delta_{jj'}\delta_{mm'} \tag{4.27}
$$

and a directed line means

$$
\underline{jm \qquad\!\!\longleftarrow\!\! \qquad j'm'} = (-1)^{j-m}\delta_{jj'}\delta_{m\,-m'} \tag{4.28}
$$

$$\underset{jm\qquad\qquad j'm'}{\longrightarrow} = (-1)^{j+m}\delta_{jj'}\delta_{m-m'} \qquad (4.29)$$

Equations (4.28) and (4.29) serve as definitions of ingoing and outgoing arrows from a jm pair. The m value changes its sign when passing through an arrow and a phase factor of $(-1)^{j-m}$ is introduced if the arrow points toward the jm pair, $(-1)^{j+m}$ if it points away[13]. It follows directly from (4.28) and (4.29) that reversing the direction of an arrow introduces a phase factor of $(-1)^{2m} = (-1)^{2j}$:

$$\underset{jm\qquad\qquad j'm'}{\longleftarrow} = (-1)^{2j}\ \underset{jm\qquad\qquad j'm'}{\longrightarrow} \qquad (4.30)$$

For a line with two arrows, we find

$$\underset{jm\qquad\quad j'm'}{\rightarrow\quad\leftarrow} = \underset{jm\qquad\quad j'm'}{\rule{2cm}{0.4pt}}$$

$$\underset{jm\qquad\quad j'm'}{\rightarrow\quad\rightarrow} = \underset{jm\qquad\quad j'm'}{\rule{2cm}{0.4pt}}$$

$$\underset{jm\qquad\quad j'm'}{\rightarrow\quad\rightarrow} = (-1)^{2j}\ \underset{jm\qquad\quad j'm'}{\rule{2cm}{0.4pt}}$$

$$\underset{jm\qquad\quad j'm'}{\leftarrow\quad\leftarrow} = (-1)^{2j}\ \underset{jm\qquad\quad j'm'}{\rule{2cm}{0.4pt}} \qquad (4.31)$$

It is often convenient to add or drop an arrow from a line in a diagram. From the above we can state the following rules:

1. To add or drop an arrow pointing toward a particular jm pair, multiply the diagram by $(-1)^{j+m}$ and change the sign of m in the diagram.

2. To add or drop an arrow pointing away from a particular jm pair, multiply the diagram by $(-1)^{j-m}$ and change the sign of m in the diagram.

Additional properties of the arrow notation are easily developed. In particular, we can add arrows on the lines coming from a vertex in a diagram provided they all point the same way, either into or out of the vertex:

$$(4.32)$$

The proof of this relies on Eq. (2.31) and the fact that $(-1)^{m_1+m_2+m_3} = 1$ since $m_1 + m_2 + m_3 = 0$ for any 3-j symbol that does not vanish.

With these conventions, and the notation that $[j] = 2j+1$, we represent Clebsch–Gordan coefficients as follows [see Eqs. (2.27) and (2.28)]:

$$\langle j_1 m_1, j_2 m_2 | j_3 m_3 \rangle = (-1)^{j_1-j_2+m_3}[j_3]^{\frac{1}{2}} \begin{pmatrix} j_1 & j_2 & j_3 \\ m_1 & m_2 & -m_3 \end{pmatrix}$$

$$= (-1)^{j_1-j_2+m_3}[j_3]^{\frac{1}{2}}$$

$$= (-1)^{j_1-j_2+m_3}[j_3]^{\frac{1}{2}}(-1)^{j_3+m_3}$$

$$= (-1)^{j_1-j_2-j_3}[j_3]^{\frac{1}{2}}$$

$$= (-1)^{j_1-j_2-j_3}[j_3]^{\frac{1}{2}}(-1)^{j_1+j_2+j_3}$$

$$= (-1)^{2j_1}[j_3]^{\frac{1}{2}}$$

$$= (-1)^{2j_1}[j_3]^{\frac{1}{2}}$$

$$= (-1)^{2j_1}[j_3]^{\frac{1}{2}}$$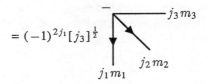

$$= (-1)^{2j_1}[j_3]^{\frac{1}{2}}$$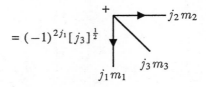

$$= [j_3]^{\frac{1}{2}}$$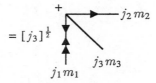

$$= [j_3]^{\frac{1}{2}}$$ (4.33)

In Eq. (4.33) we have made the operations explicit to illustrate graph manipulations. The reader is urged to check the logic of each step before continuing with this section.

Of particular interest is the case in which one of the angular momenta of a Clebsch–Gordan coefficient or a 3-j symbol is equal to zero. The results are already known (see the first entry in Tables 2.4 and 2.5). Hence in terms of diagrams we find that $\langle j_1 m_1, 00 | j_3 m_3 \rangle = \delta_{j_1 j_3} \delta_{m_1 m_3}$ becomes

$$[j_3]^{\frac{1}{2}}$$ $$= \frac{j_3 m_3 \qquad j_1 m_1}{}$$ (4.34)

and that $\begin{pmatrix} j_1 & j_2 & 0 \\ m_1 & m_2 & 0 \end{pmatrix} = (-1)^{j_1-m_1}[j_1]^{-\frac{1}{2}}\delta_{j_1j_2}\delta_{m_1-m_2}$ becomes

$$\text{(diagram)} = [j_1]^{-\frac{1}{2}} \xrightarrow{\hspace{0.3cm} j_2m_2 \hspace{1.2cm} j_1m_1 \hspace{0.3cm}} \hspace{1cm} (4.35)$$

Hence any node involving a zero line (a line with $j = 0$) may be removed by use of Eqs. (4.34) or (4.35).

More complicated diagrams are constructed by joining together simpler diagrams. Often we encounter calculations involving products of 3-j symbols in which the same angular momentum appears in two 3-j symbols with either the same or opposite magnetic quantum numbers. We represent this by joining together the free lines representing the same angular momenta. Thus, for example,

$$\begin{pmatrix} j_1 & j_2 & j_3 \\ m_1 & m_2 & m_3 \end{pmatrix}\begin{pmatrix} j_4 & j_5 & j_3 \\ m_4 & m_5 & m_3 \end{pmatrix}$$

$$= \text{(diagram)} \times \text{(diagram)}$$

$$= \text{(diagram)}$$

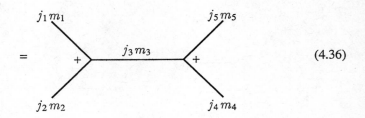

$$(4.36)$$

and similarly

$$(-1)^{j_3-m_3}\begin{pmatrix} j_1 & j_2 & j_3 \\ m_1 & m_2 & m_3 \end{pmatrix}\begin{pmatrix} j_4 & j_5 & j_3 \\ m_4 & m_5 & -m_3 \end{pmatrix}$$

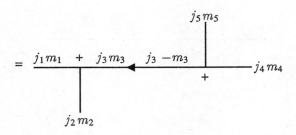

$$(4.37)$$

It is conventional to omit the m label on a line in representing a sum over magnetic quantum numbers. Hence

$$\sum_{m_3} \begin{pmatrix} j_1 & j_2 & j_3 \\ m_1 & m_2 & m_3 \end{pmatrix} \begin{pmatrix} j_4 & j_5 & j_3 \\ m_4 & m_5 & m_3 \end{pmatrix}$$

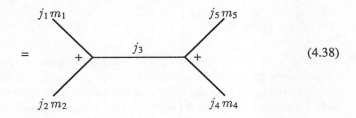 (4.38)

and

$$\sum_{m_3} (-1)^{j_3 - m_3} \begin{pmatrix} j_1 & j_2 & j_3 \\ m_1 & m_2 & m_3 \end{pmatrix} \begin{pmatrix} j_4 & j_5 & j_3 \\ m_4 & m_5 & -m_3 \end{pmatrix}$$

(4.39)

It is useful to introduce some additional definitions. An *internal line* is one that connnects two nodes. An *external line* has one end free and the other end connected to a node. *Closed diagrams* have no external lines. We illustrate the meaning of the above by constructing some simple diagrams.

The first orthogonality relation for the $3\text{-}j$ symbols [see Eq. (2.32)] becomes

(4.40)

while that for the second orthogonality relation [see Eq. (2.33)] becomes

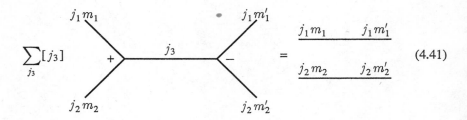

$$\sum_{j_3} [j_3] \qquad\qquad\qquad = \qquad \begin{array}{cc} j_1 m_1 & j_1 m_1' \\ \\ j_2 m_2 & j_2 m_2' \end{array} \qquad\qquad (4.41)$$

Equation (4.40) shows how a loop inserted into a line can be removed by a diagrammatic manipulation. Suppose we set $j_3 = j_3'$, $m_3 = m_3'$, and join $j_3 m_3$ to $j_3' m_3'$ in this figure and sum over m_3. Then, of course, we get unity provided $\mathbf{j}_1$, $\mathbf{j}_2$, and $\mathbf{j}_3$ form a triangle, denoted by $\Delta(j_1 j_2 j_3)$. This is represented graphically by

$$- \ \bigcirc \ + \ = \Delta(j_1 j_2 j_3) \qquad\qquad (4.42)$$

Equation (4.42) is our first encounter with a closed diagram. The symbol $\Delta(j_1 j_2 j_3)$ has the value 1 if $\mathbf{j}_1$, $\mathbf{j}_2$, and $\mathbf{j}_3$ satisfy the triangle condition [see Eq. (2.11)] and zero otherwise.

As another example of the connected line summation convention we give a graphical representation of the 6-j symbol [see Eq. (4.12)], which is a closed diagram with six angular momenta:

$$\left\{ \begin{array}{ccc} j_1 & j_2 & j_3 \\ j_4 & j_5 & j_6 \end{array} \right\} = \qquad\qquad\qquad\qquad (4.43)$$

It is noted that all vertices have plus signs; the arrows go around the perimeter in a clockwise manner; the 6-j symbol is nonvanishing only if the angular momenta meeting at each vertex satisfy the triangle condition [see Eq. (4.10)]; and the pairing of the angular momenta is such that (j_1, j_4), (j_2, j_5), and (j_3, j_6) have no vertices in common.

The symmetries of the 6-j symbol are readily verified by reference to the graph shown in (4.43) and with the help of the two identities: $(-1)^{4j} = 1$ for any integral or half-integral j; and $(-1)^{2j_a+2j_b+2j_c} = 1$ when j_a, j_b, j_c form a triangle (i.e., enter or leave a node), since $j_a + j_b + j_c$ must then be an integer. We illustrate some of these symmetries. For example, by rotating the diagram in Eq. (4.43) about an axis perpendicular to the plane of the figure and passing through the central vertex, we have

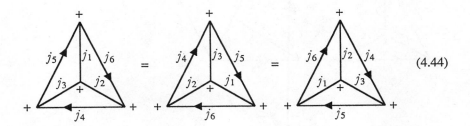

$$(4.44)$$

from which it is seen that the $6\text{-}j$ symbol is invariant to cyclic interchange of its columns. Consider next a rotation by $180°$ about one edge, for example, the j_4 line. We find that

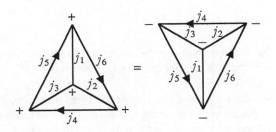

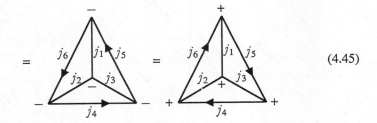

$$(4.45)$$

where the phase change from reversing the directions of all the arrows cancels that from reversing the signs on all nodes. Equations (4.44) and (4.45) establish that the interchange of any two columns of a $6\text{-}j$ symbol leaves the symbol unaltered. Finally, consider the operations of pulling the central vertex to the periphery:

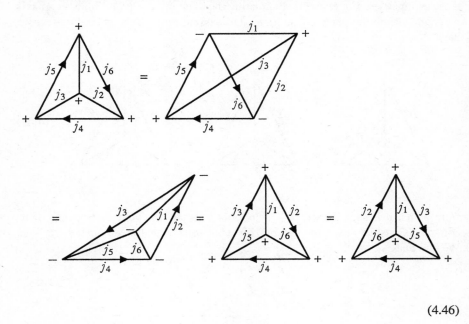

$$(4.46)$$

from which we find that the interchange of the upper and lower arguments in each of any two columns leaves the $6\text{-}j$ symbol unaltered. Equations (4.44)–(4.46) are the same as Eq. (4.9).

Often the $6\text{-}j$ graph does not have the symmetric form illustrated in Eq. (4.43) but is geometrically distorted with the wrong arrows and plus and minus signs. For example, by pulling the central vertex in Eq. (4.43) to the left, we obtain the following equivalent diagrammatic representations of the $6\text{-}j$ symbol:

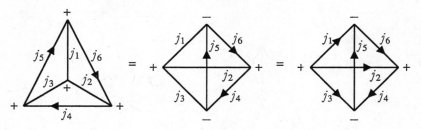

$$(4.47)$$

where we have added arrows to all the lines of the vertex on the left to obtain the diagram on the extreme right. A useful way for finding the appropriate $6\text{-}j$ symbol is to note that the angular momenta at each node, with a plus sign, taken in a clockwise manner, are in the same order as in the $6\text{-}j$ symbol read left to right [see Eq. (4.43)].

The next closed diagram in increasing order of complexity is that representing the $9\text{-}j$ symbol:

$$\begin{Bmatrix} j_1 & j_2 & j_3 \\ j_4 & j_5 & j_6 \\ j_7 & j_8 & j_9 \end{Bmatrix} = -$$ $$- \qquad (4.48)$$

Please note that in Eq. (4.48) the lines do not form a node at the center. Equation (4.48) has the properties that (1) all the vertices have minus signs (or equivalently all plus signs); (2) none of the lines have arrows, although the diagram can be put into so-called normal form by adding arrows to every other vertex so that each internal line has an arrow; and (3) the angular momentum arguments in each row of the $9\text{-}j$ symbol and in each column of the $9\text{-}j$ symbol meet at a vertex. Once more, the symmetries of the $9\text{-}j$ symbol are readily derived by graphical manipulations of Eq. (4.48).

The power of the graphical method is very much enhanced by the following techniques for the reduction of closed diagrams to products and sums of products of simpler diagrams. Consider a diagram that can be decomposed into two parts, called *blocks*, denoted by A and B, such that A has no external lines and is joined to B by only a single angular momentum line j. We assume that A is in or can be placed in so-called normal form with one and only one arrow on each line. In this situation rotational invariance of the diagram causes $j = 0$. The reason for this is the following argument. A closed block in normal form must represent a scalar because of the summation over all magnetic quantum numbers associated with each line in the closed block. For example, the $6\text{-}j$ and $9\text{-}j$ symbols are closed blocks. Hence the closed block A is proportional to some number times the zero line $j = 0$, $m = 0$. If a single line j joins to the closed block A, then by Eq. (4.27) this line must also be a zero line. We may apply Eq. (4.35) to erase the zero line and hence to factor the graph, as shown:

$$= [j_1]^{-\frac{1}{2}} [j_3]^{-\frac{1}{2}} \delta_{j_1 j_2} \delta_{j_3 j_4} \quad$$ $$\quad \qquad (4.49)$$

Here the dashed line on B indicates any number of external lines.

Equation (4.49) is of fundamental importance in reducing the complexity of graphs. Therefore, we pause to present a very simple instance where $j = 0$. Consider the expression

$$\sum_{m'} (-1)^{j'-m'} \begin{pmatrix} j' & j' & j \\ m' & -m' & m \end{pmatrix},$$

which has the value $\sum_{m'} \delta_{j0} \delta_{m0} (2j'+1)^{-\frac{1}{2}} = (2j'+1)^{\frac{1}{2}}$. This expression may be represented graphically by a closed diagram with one external angular momentum line:

$$(4.50)$$

Applying Eqs. (4.34) and (4.40) to Eq. (4.50), we find that

$$(4.51)$$

from which it is seen that the jm line is a zero line.

Next we examine the case where A and B are connected by two lines. We reduce this diagram by application of Eq. (4.41) followed by Eq. (4.49):

$$= [j_1]^{-1} \delta_{j_1 j_2} \quad \boxed{A} \!\!\!\rangle \times \langle\!\!\! \boxed{B} \text{ --- } \tag{4.52}$$

To ensure that the diagram to the left of the $j = 0$ line is normal, we incorporate any arrows on j_1 and j_2 in the B block. If the B block is completely empty, Eq. (4.52) becomes simply

$$\boxed{A} \begin{array}{l} \!\!\!\! -\!\!j_1 m_1 \\ \!\!\!\! -\!\!j_2 m_2 \end{array} = [j_1]^{-1} \delta_{j_1 j_2} \quad \boxed{A}\!\!\!\rangle_{j_1} \times \quad \begin{array}{l} j_1 m_1 \\ \uparrow \\ j_2 m_2 \end{array} \tag{4.53}$$

This result shows that if a normal diagram has two free lines, these must represent the same total angular momentum; moreover, the nonzero value of such a diagram is independent of the magnetic quantum numbers on the free lines.

For the case where A and B are connected by three and only three lines, the same procedure may be used to reduce the diagram into the product of two simpler diagrams:

$$\tag{4.54}$$

Here we can drop the two oppositely directed arrows in going from the second to last to the last line in Eq. (4.54) because the resulting product of phase factors cancels. Again we assume that A is normal and that any arrows on j_1, j_2, or j_3 are incorporated into B. Note that the plus and minus signs may be reversed with no change in the overall value of Eq. (4.54).

As a corollary, if the B block is completely empty, Eq. (4.54) becomes simply

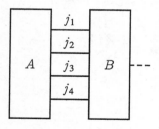

$$\hspace{10cm} (4.55)$$

This result is essentially the Wigner–Eckart theorem, which is discussed in more detail in the next chapter. It is seen that a normal diagram having three free lines may be replaced by a product of two factors: one factor is a closed diagram that can be put into normal form, and the other factor is a $3\text{-}j$ symbol made up of the three free lines. The first factor is independent of magnetic quantum number; the second factor contains completely the magnetic quantum number dependence.

This graphical reduction process can be continued, but if A and B are connected by four or more lines, the reduction must cause the appearance of sums of products of simpler graphs since the maximum number of lines meeting at a node is three. We illustrate this by reducing a diagram in which A and B are connected by four lines. We assume once more that A is normal and that all arrows on j_1, j_2, j_3, and j_4 are incorporated into B:

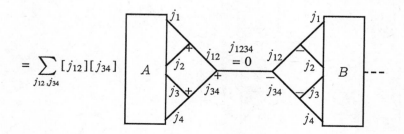

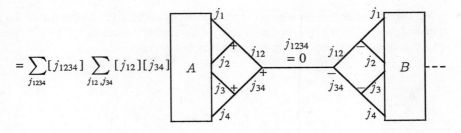

$$= \sum_{j_{1234}} [j_{1234}] \sum_{j_{12},j_{34}} [j_{12}][j_{34}]$$

$$= \sum_{j_{12},j_{34}} [j_{12}]^{-1} [j_{12}][j_{34}] \delta_{j_{12}j_{34}} \quad \times$$

$$= \sum_{j} [j] \qquad \times \qquad\qquad\qquad (4.56)$$

At this point the power of the graphical method may begin to become apparent. Of course, to every graphical reduction or manipulation there exists a corresponding algebraic operation, but the graphical method can have several advantages. First the notation is more compact because redundant magnetic quantum numbers have been suppressed. Second, reduction can be made by recognizing geometric patterns. Finally, the graphical method can serve as a convenient roadmap of what should be expected in carrying out some involved angular momentum recoupling operation, and this roadmap is available even if no care is taken to preserve phase factors, that is, even if no attention is given to the directions of lines containing arrows nor to the signs of nodes.

We conclude this section with some examples chosen to illustrate the use of the graphical method. We begin by presenting a diagrammatic proof of Eq. (4.19), the reduction of a $6\text{-}j$ symbol having one vanishing argument:

$$\begin{Bmatrix} j_1 & j_2 & 0 \\ j_4 & j_5 & j_6 \end{Bmatrix} =$$

$$= \delta_{j_4 j_5} \delta_{j_1 j_2} [j_1]^{-\frac{1}{2}} [j_4]^{-\frac{1}{2}} (-1)^{2j_4}$$

$$= \delta_{j_4 j_5} \delta_{j_1 j_2} [j_1]^{-\frac{1}{2}} [j_4]^{-\frac{1}{2}} (-1)^{j_1 + j_4 + j_6}$$

$$= \delta_{j_4 j_5} \delta_{j_1 j_2} [j_1]^{-\frac{1}{2}} [j_4]^{-\frac{1}{2}} (-1)^{j_1 + j_4 + j_6} \Delta(j_1 j_4 j_6) \quad (4.57)$$

Here $\Delta(j_1 j_4 j_6)$ reminds us that $\mathbf{j}_1, \mathbf{j}_4$, and $\mathbf{j}_6$ must form a triangle if this $6\text{-}j$ symbol is not to vanish. The corresponding situation of a $9\text{-}j$ symbol with one vanishing argument [see Eq. (4.24)] is reduced diagrammatically as follows:

$$\begin{Bmatrix} j_1 & j_2 & j_3 \\ j_4 & j_5 & j_6 \\ j_7 & j_8 & 0 \end{Bmatrix} =$$

$$= \delta_{j_3 j_6} \delta_{j_7 j_8} [j_3]^{-\frac{1}{2}} [j_7]^{-\frac{1}{2}} +$$

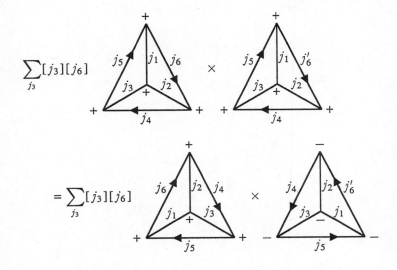

$$= \delta_{j_3 j_6} \delta_{j_7 j_8} [j_3]^{-\frac{1}{2}} [j_7]^{-\frac{1}{2}}$$

$$= \delta_{j_3 j_6} \delta_{j_7 j_8} [j_3]^{-\frac{1}{2}} [j_7]^{-\frac{1}{2}} (-1)^{j_2 + j_3 + j_4 + j_7} \begin{Bmatrix} j_1 & j_2 & j_3 \\ j_5 & j_4 & j_7 \end{Bmatrix}$$

$$(4.58)$$

where we have in the second to last step reversed the arrows on the j_4 and j_7 lines giving a factor of $(-1)^{2 j_4 + 2 j_7} = (-1)^{j_2 + j_3 + j_4 + j_7}$, in which we make use of the fact that $(j_4 j_7 j_5)$ and $(j_2 j_3 j_5)$ form triangles so that $(-1)^{2(j_2 + j_3 + j_5)} = (-1)^{2(j_4 + j_7 + j_5)}$.

As a final example we offer a diagrammatic proof of Eq. (4.11), the orthonormality property of the 6-j symbols:

$$\sum_{j_3} [j_3][j_6] \qquad \times$$

$$= \sum_{j_3} [j_3][j_6] \qquad \times$$

$$= \sum_{j_3} [j_3][j_6] \quad \cdots \quad = [j_6] \quad \cdots$$

$$= [j_6] \quad \cdots \quad = \delta_{j_6 j'_6} \quad \cdots \quad \times \quad \cdots$$

$$= \delta_{j_6 j'_6} \Delta(j_2 j_4 j_6) \Delta(j_1 j_5 j_6) \tag{4.59}$$

The first line of Eq. (4.59) uses (4.43), the second line uses Eqs. (4.44) and (4.45), the third line uses Eq. (4.54), the fourth line uses Eq. (4.41), the fifth line uses Eqs. (4.31) and (4.32), the sixth line uses Eq. (4.52), while the final line makes use of Eq. (4.42).

TABLE 4.1 Algebraic Expressions for Some Commonly Occurring $6\text{-}j$ Symbols[a]

$$\begin{Bmatrix} j_1 & j_2 & j_3 \\ 0 & j_3 & j_2 \end{Bmatrix} = (-1)^s [(2j_2 + 1)(2j_3 + 1)]^{-\frac{1}{2}}$$

$$\begin{Bmatrix} j_1 & j_2 & j_3 \\ \frac{1}{2} & j_3 - \frac{1}{2} & j_2 + \frac{1}{2} \end{Bmatrix}$$
$$= (-1)^s \left[\frac{(s - 2j_2)(s - 2j_3 + 1)}{(2j_2 + 1)(2j_2 + 2)(2j_3)(2j_3 + 1)} \right]^{\frac{1}{2}}$$

$$\begin{Bmatrix} j_1 & j_2 & j_3 \\ \frac{1}{2} & j_3 - \frac{1}{2} & j_2 - \frac{1}{2} \end{Bmatrix}$$
$$= (-1)^s \left[\frac{(s + 1)(s - 2j_1)}{(2j_2)(2j_2 + 1)(2j_3)(2j_3 + 1)} \right]^{\frac{1}{2}}$$

$$\begin{Bmatrix} j_1 & j_2 & j_3 \\ 1 & j_3 - 1 & j_2 - 1 \end{Bmatrix}$$
$$= (-1)^s \left[\frac{s(s + 1)(s - 2j_1 - 1)(s - 2j_1)}{(2j_2 - 1)(2j_2)(2j_2 + 1)(2j_3 - 1)(2j_3)(2j_3 + 1)} \right]^{\frac{1}{2}}$$

$$\begin{Bmatrix} j_1 & j_2 & j_3 \\ 1 & j_3 - 1 & j_2 \end{Bmatrix}$$
$$= (-1)^s \left[\frac{2(s + 1)(s - 2j_1)(s - 2j_2)(s - 2j_3 + 1)}{(2j_2)(2j_2 + 1)(2j_2 + 2)(2j_3 - 1)(2j_3)(2j_3 + 1)} \right]^{\frac{1}{2}}$$

$$\begin{Bmatrix} j_1 & j_2 & j_3 \\ 1 & j_3 - 1 & j_2 + 1 \end{Bmatrix}$$
$$= (-1)^s \left[\frac{(s - 2j_2 - 1)(s - 2j_2)(s - 2j_3 + 1)(s - 2j_3 + 2)}{(2j_2 + 1)(2j_2 + 2)(2j_2 + 3)(2j_3 - 1)(2j_3)(2j_3 + 1)} \right]^{\frac{1}{2}}$$

$$\begin{Bmatrix} j_1 & j_2 & j_3 \\ 1 & j_3 & j_2 \end{Bmatrix}$$
$$= (-1)^s 2 \frac{j_1(j_1 + 1) - j_2(j_2 + 1) - j_3(j_3 + 1)}{[(2j_2)(2j_2 + 1)(2j_2 + 2)(2j_3)(2j_3 + 1)(2j_3 + 2)]^{\frac{1}{2}}}$$

$$\begin{Bmatrix} j_1 & j_2 & j_3 \\ \frac{3}{2} & j_3 - \frac{3}{2} & j_2 - \frac{3}{2} \end{Bmatrix}$$
$$= (-1)^s A_{\frac{3}{2}}(2j_2 + 1) A_{\frac{3}{2}}(2j_3 + 1)$$
$$\times [(s - 1)s(s + 1)(s - 2j_1 - 2)(s - 2j_1 - 1)(s - 2j_1)]^{\frac{1}{2}}$$

$$\begin{Bmatrix} j_1 & j_2 & j_3 \\ \frac{3}{2} & j_3 - \frac{3}{2} & j_2 - \frac{1}{2} \end{Bmatrix}$$
$$= (-1)^s A_{\frac{3}{2}}(2j_2 + 2) A_{\frac{3}{2}}(2j_3 + 1)$$
$$\times [3s(s + 1)(s - 2j_1 - 1)(s - 2j_1)(s - 2j_2)(s - 2j_3 + 1)]^{\frac{1}{2}}$$

TABLE 4.1 (continued)

$$
\left\{
\begin{array}{ccc}
j_1 & j_2 & j_3 \\
\frac{3}{2} & j_3 - \frac{3}{2} & j_2 + \frac{1}{2}
\end{array}
\right\}
$$
$$
= (-1)^s A_{\frac{3}{2}}(2 j_2 + 3) A_{\frac{3}{2}}(2 j_3 + 1)
$$
$$
\times [3(s+1)(s - 2 j_1)(s - 2 j_2 - 1)(s - 2 j_2)
$$
$$
\times (s - 2 j_3 + 1)(s - 2 j_3 + 2)]^{\frac{1}{2}}
$$

$$
\left\{
\begin{array}{ccc}
j_1 & j_2 & j_3 \\
\frac{3}{2} & j_3 - \frac{3}{2} & j_2 + \frac{3}{2}
\end{array}
\right\}
$$
$$
= (-1)^s A_{\frac{3}{2}}(2 j_2 + 4) A_{\frac{3}{2}}(2 j_3 + 1)
$$
$$
\times [(s - 2 j_2 - 2)(s - 2 j_2 - 1)(s - 2 j_2)
$$
$$
\times (s - 2 j_3 + 1)(s - 2 j_3 + 2)(s - 2 j_3 + 3)]^{\frac{1}{2}}
$$

$$
\left\{
\begin{array}{ccc}
j_1 & j_2 & j_3 \\
\frac{3}{2} & j_3 - \frac{1}{2} & j_2 - \frac{1}{2}
\end{array}
\right\}
$$
$$
= (-1)^s A_{\frac{3}{2}}(2 j_2 + 2) A_{\frac{3}{2}}(2 j_3 + 2)
$$
$$
\times [2(s - 2 j_2)(s - 2 j_3) - (s + 2)(s - 2 j_1 - 1)]
$$
$$
\times [(s + 1)(s - 2 j_1)]^{\frac{1}{2}}
$$

$$
\left\{
\begin{array}{ccc}
j_1 & j_2 & j_3 \\
\frac{3}{2} & j_3 - \frac{1}{2} & j_2 + \frac{1}{2}
\end{array}
\right\}
$$
$$
= (-1)^s A_{\frac{3}{2}}(2 j_2 + 3) A_{\frac{3}{2}}(2 j_3 + 2)
$$
$$
\times [(s - 2 j_2 - 1)(s - 2 j_3) - 2(s + 2)(s - 2 j_1)]
$$
$$
\times [(s - 2 j_2)(s - 2 j_3 + 1)]^{\frac{1}{2}}
$$

$$
\left\{
\begin{array}{ccc}
j_1 & j_2 & j_3 \\
2 & j_3 - 2 & j_2 - 2
\end{array}
\right\}
$$
$$
= (-1)^s A_2(2 j_2 + 1) A_2(2 j_3 + 1)
$$
$$
\times [(s - 2)(s - 1)(s)(s + 1)(s - 2 j_1 - 3)
$$
$$
\times (s - 2 j_1 - 2)(s - 2 j_1 - 1)(s - 2 j_1)]^{\frac{1}{2}}
$$

$$
\left\{
\begin{array}{ccc}
j_1 & j_2 & j_3 \\
2 & j_3 - 2 & j_2 - 1
\end{array}
\right\}
$$
$$
= (-1)^s 2 A_2(2 j_2 + 2) A_2(2 j_3 + 1)
$$
$$
\times [(s - 1)(s)(s + 1)(s - 2 j_1 - 2)(s - 2 j_1 - 1)
$$
$$
\times (s - 2 j_1)(s - 2 j_2)(s - 2 j_3 + 1)]^{\frac{1}{2}}
$$

$$
\left\{
\begin{array}{ccc}
j_1 & j_2 & j_3 \\
2 & j_3 - 2 & j_2
\end{array}
\right\}
$$
$$
= (-1)^s A_2(2 j_2 + 3) A_2(2 j_3 + 1)
$$
$$
\times [6 s(s + 1)(s - 2 j_1 - 1)(s - 2 j_1)(s - 2 j_2 - 1)
$$
$$
\times (s - 2 j_2)(s - 2 j_3 + 1)(s - 2 j_3 + 2)]^{\frac{1}{2}}
$$

TABLE 4.1 (continued)

$$\begin{Bmatrix} j_1 & j_2 & j_3 \\ 2 & j_3 - 2 & j_2 + 1 \end{Bmatrix}$$
$$= (-1)^s 2\, A_2(2j_2 + 4)\, A_2(2j_3 + 1)$$
$$\times [(s+1)(s-2j_1)(s-2j_2-2)(s-2j_2-1)(s-2j_2)$$
$$\times (s-2j_3+1)(s-2j_3+2)(s-2j_3+3)]^{\frac{1}{2}}$$

$$\begin{Bmatrix} j_1 & j_2 & j_3 \\ 2 & j_3 - 2 & j_2 + 2 \end{Bmatrix}$$
$$= (-1)^s A_2(2j_2 + 5)\, A_2(2j_3 + 1)$$
$$\times [(s-2j_2-3)(s-2j_2-2)(s-2j_2-1)(s-2j_2)$$
$$\times (s-2j_3+1)(s-2j_3+2)(s-2j_3+3)(s-2j_3+4)]^{\frac{1}{2}}$$

$$\begin{Bmatrix} j_1 & j_2 & j_3 \\ 2 & j_3 - 1 & j_2 - 1 \end{Bmatrix}$$
$$= (-1)^s 4\, A_2(2j_2 + 2)\, A_2(2j_3 + 2)$$
$$\times [(j_1+j_2)(j_1-j_2+1) - (j_3-1)(j_3-j_2+1)]$$
$$\times [s(s+1)(s-2j_1-1)(s-2j_1)]^{\frac{1}{2}}$$

$$\begin{Bmatrix} j_1 & j_2 & j_3 \\ 2 & j_3 - 1 & j_2 \end{Bmatrix}$$
$$= (-1)^s 2\, A_2(2j_2 + 3)\, A_2(2j_3 + 2)$$
$$\times [(j_1+j_2+1)(j_1-j_2) - j_3^2 + 1]$$
$$\times [6(s+1)(s-2j_1)(s-2j_2)(s-2j_3+1)]^{\frac{1}{2}}$$

$$\begin{Bmatrix} j_1 & j_2 & j_3 \\ 2 & j_3 - 1 & j_2 + 1 \end{Bmatrix}$$
$$= (-1)^s 4\, A_2(2j_2 + 4)\, A_2(2j_3 + 2)$$
$$\times [(j_1+j_2+2)(j_1-j_2-1) - (j_3-1)(j_2+j_3+2)]$$
$$\times [(s-2j_2-1)(s-2j_2)(s-2j_3+1)(s-2j_3+2)]^{\frac{1}{2}}$$

$$\begin{Bmatrix} j_1 & j_2 & j_3 \\ 2 & j_3 & j_2 \end{Bmatrix}$$
$$= (-1)^s 2\, A_2(2j_2 + 3)\, A_2(2j_3 + 3)$$
$$\times [3C(C+1) - 4j_2(j_2+1)j_3(j_3+1)]$$

where
$$s = j_1 + j_2 + j_3$$
$$A_{\frac{3}{2}}(X) = [X(X-1)(X-2)(X-3)]^{-\frac{1}{2}}$$
$$A_2(X) = [X(X-1)(X-2)(X-3)(X-4)]^{-\frac{1}{2}}$$

and
$$C = j_1(j_1+1) - j_2(j_2+1) - j_3(j_3+1)$$

---

[a] Symmetry relations [see Eq. (4.9)] may be used in conjunction with this table to evaluate all nonvanishing 6-j symbols with one angular momentum argument equal to 0, $\frac{1}{2}$, 1, $\frac{3}{2}$, or 2.

NOTES AND REFERENCES

1. G. Racah, *Phys. Rev.*, **62**, 438 (1942); **63**, 367 (1943). See also L. C. Biedenharn, J. M. Blatt, and M. E. Rose, *Rev. Mod. Phys.*, **24**, 249 (1952).

2. E. P. Wigner, "On the Matrices which Reduce the Kronecker Products of Representations of S. R. Groups," in *Quantum Theory of Angular Momentum*, L. C. Biedenharn and H. Van Dam, eds. (Academic Press, New York, 1965), pp. 87–133.

3. M. Rotenberg, R. Bivins, N. Metropolis, and J. K. Wooten, Jr., *the 3-j and 6-j symbols* (The Technology Press, MIT, Cambridge, MA, 1959). See also "Numerical Tables for Angular Correlation Computations in α-, β-, and γ-Spectroscopy: 3-j, 6-j, 9-j Symbols, F- and Γ-Coefficients." in *Landolt–Börnstein Numerical Data and Functional Relationships in Science and Technology*, K.-H. Hellwege, ed. (Springer-Verlag, Berlin, 1968), Group I, Volume 3.

4. A. R. Edmonds, *Angular Momentum in Quantum Mechanics* (Princeton University Press, Princeton, NJ, 1974).

5. See A. P. Yutsis, I. B. Levinson, and V. V. Vanagas, *Mathematical Apparatus of the Theory of Angular Momentum* (Academy of Sciences of the Lithuanian SSR, 1960), translated into English by the Israel Program for Scientific Translations, Jerusalem, 1962.

6. D. M. Brink and G. R. Satchler, *Angular Momentum* (Clarendon Press, Oxford, 1979), Chapter VII.

7. J.-N. Massot, E. El-Baz, and J. LaFoucrière, *Rev. Mod. Phys.*, **39**, 288 (1967); E. El-Baz and B. Castel, *Graphical Methods of Spin Algebras* (Marcel Dekker, New York, 1972).

8. J. S. Briggs, *Rev. Mod. Phys.*, **43**, 189 (1971).

9. P. G. H. Sandars, "Graphical Methods in Angular Momentum Theory" in *Atomic Physics and Astrophysics*, Vol. 1, M. Chrétien and E. Lipworth, eds. (Gordon & Breach, New York, 1971) pp. 171–216.

10. I. Lindgren and J. Morrison, *Atomic Many-Body Theory* (Springer-Verlag, Berlin, 1982). See especially Chapters 3 and 4.

11. R. L. Chien, *J. Chem. Phys.*, **81**, 4023 (1984).

12. D. A. Varshalovich, A. N. Moskalev, and V. K. Khersonskii, *Quantum Theory of Angular Momentum* (Nauka, Leningrad, 1975), Chapters 11 and 12, pp. 349–404 (in Russian).

13. This convention agrees with that of Lindgren and Morrison[10] but is opposite to that of Brink and Satchler[6] and of Sandars[9].

PROBLEM SET 2

1. The vector model allows once more a simple geometric interpretation of the expression $\langle j_{12} j_3 j | j_1 j_{23} j \rangle^2$, the square of the recoupling coefficient for the two schemes $\mathbf{j} = \mathbf{j}_{12} + \mathbf{j}_3$ and $\mathbf{j} = \mathbf{j}_1 + \mathbf{j}_{23}$. According to Eq. (4.3), the probability

$$ P = \langle j_{12} j_3 j | j_1 j_{23} j \rangle^2 = (2 j_{12} + 1)(2 j_{23} + 1) \left\{ \begin{array}{ccc} j_1 & j_2 & j_{12} \\ j_3 & j & j_{23} \end{array} \right\}^2 \qquad (4.60) $$

represents the probability that a system prepared in a state of the coupling scheme $\mathbf{j} = \mathbf{j}_{12} + \mathbf{j}_3$ (having fixed magnitudes of $\mathbf{j}_1, \mathbf{j}_2, \mathbf{j}_3, \mathbf{j}_{12}$, and $\mathbf{j}$) will be found to be in a state of the coupling scheme $\mathbf{j} = \mathbf{j}_1 + \mathbf{j}_{23}$ (having fixed magnitudes of $\mathbf{j}_1, \mathbf{j}_2, \mathbf{j}_3, \mathbf{j}_{23}$, and $\mathbf{j}$) (see Figure 4.1). In Figure 4.1 it is seen at once that $\langle j_{12} j_3 j | j_1 j_{23} j \rangle$ is independent of the value of m (the direction of the z axis). It is also seen that the six angular momenta $\mathbf{j}_1, \mathbf{j}_2, \mathbf{j}_3, \mathbf{j}_{12}, \mathbf{j}_{23}$, and $\mathbf{j}$ may be regarded as forming the sides of an irregular tetrahedron whose volume V is given by one-third the area of a face times the slant height; that is,

$$ V = \tfrac{1}{3} \left[\tfrac{1}{2} (\mathbf{j}_{12} \times \mathbf{j}) \cdot \mathbf{j}_1 \right] \qquad (4.61) $$

If $\mathbf{j}_1, \mathbf{j}_2, \mathbf{j}_3, \mathbf{j}_{12}$, and $\mathbf{j}$ are fixed, the magnitude of $\mathbf{j}_{23}$ is free to vary in such a way that the locus of $\mathbf{j}_{23}$ describes a circle in space centered about $\mathbf{j}_{12}$. Then the dihedral angle ϕ between the two planes containing $(\mathbf{j}, \mathbf{j}_{12}, \mathbf{j}_3)$ and $(\mathbf{j}_1, \mathbf{j}_2, \mathbf{j}_{12})$

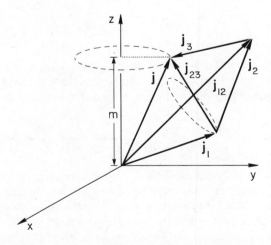

FIGURE 4.1   Vector model for the coupling of three angular momenta.

varies between 0 and 2π at a uniform rate. The unit vectors normal to these two planes are $(\mathbf{j}_{12} \times \mathbf{j})/|\mathbf{j}_{12} \times \mathbf{j}|$ and $(\mathbf{j}_{12} \times \mathbf{j}_1)/|\mathbf{j}_{12} \times \mathbf{j}_1|$, respectively. Hence

$$\cos \phi = \frac{(\mathbf{j}_{12} \times \mathbf{j}) \cdot (\mathbf{j}_{12} \times \mathbf{j}_1)}{|\mathbf{j}_{12} \times \mathbf{j}||\mathbf{j}_{12} \times \mathbf{j}_1|} \tag{4.62}$$

and

$$|\sin \phi| = \frac{|(\mathbf{j}_{12} \times \mathbf{j}) \times (\mathbf{j}_{12} \times \mathbf{j}_1)|}{|\mathbf{j}_{12} \times \mathbf{j}||\mathbf{j}_{12} \times \mathbf{j}_1|} \tag{4.63}$$

A. Differentiate Eq. (4.62) with respect to time to show that

$$\frac{d|\mathbf{j}_{23}|}{dt} = \frac{(\mathbf{j}_{12} \times \mathbf{j}) \cdot \mathbf{j}_1}{|\mathbf{j}_{12}||\mathbf{j}_{23}|} \frac{d\phi}{dt} \tag{4.64}$$

where $d\phi/dt$ may be replaced by 2π. *Hint*: Make use of the law of cosines

$$|\mathbf{j}_{23}|^2 = |\mathbf{j}_1|^2 + |\mathbf{j}_2|^2 - 2\mathbf{j}_1 \cdot \mathbf{j} \tag{4.65}$$

and

$$|\mathbf{j}_1|^2 = |\mathbf{j}_{23}|^2 + |\mathbf{j}|^2 - 2\mathbf{j}_{23} \cdot \mathbf{j} \tag{4.66}$$

to make the identification

$$\frac{d}{dt}(\mathbf{j}_1 \cdot \mathbf{j}) = \frac{d}{dt}(\mathbf{j}_{23} \cdot \mathbf{j}) = -|\mathbf{j}_{23}|\frac{d|\mathbf{j}_{23}|}{dt} \tag{4.67}$$

B. The probability density for a given value of $\mathbf{j}_{23}$ may then be found from

$$P(j_{23}) = 2 \left(\frac{d|\mathbf{j}_{23}|}{dt} \right)^{-1} \tag{4.68}$$

where the factor 2 arises from the fact that $\mathbf{j}_{23}$ takes on any of its values twice as ϕ goes from 0 to 2π. Hence show that in the classical limit for large angular momenta the square of the 6-j symbol has the limiting value

$$\left\{ \begin{matrix} j_1 & j_2 & j_{12} \\ j_3 & j & j_{23} \end{matrix} \right\}^2 = \frac{1}{4\pi(\mathbf{j}_1 \times \mathbf{j}) \cdot \mathbf{j}_1} = \frac{1}{24\pi V} \tag{4.69}$$

The behavior of the square of the 6-j symbol for large values of the six arguments is analogous to the semiclassical expression we found previously for the square of the 3-j symbol: in classically allowed regions where $V > 0$, the 6-j coefficient is rapidly oscillating with an envelope for the oscillations being given by Eq. (4.69), while in classically forbidden regions where $V < 0$ the square of the 6-j symbol is exponentially decreasing. At the transition regions corresponding to $V = 0$, Eq. (4.69) is singular while the quantum mechanical value for the square of the $6\text{-}j$ symbol is a finite fractional number.

2. Prove by diagrammatic means that

$$\sum_{m_4, m_5, m_6} (-1)^{j_4+m_4+j_5+m_5+j_6+m_6} \begin{pmatrix} j_1 & j_5 & j_6 \\ m_1 & m_5 & -m_6 \end{pmatrix}$$

$$\times \begin{pmatrix} j_4 & j_2 & j_6 \\ -m_4 & m_2 & m_6 \end{pmatrix} \begin{pmatrix} j_4 & j_5 & j_3 \\ m_4 & -m_5 & m_3 \end{pmatrix}$$

$$= \begin{Bmatrix} j_1 & j_2 & j_3 \\ j_4 & j_5 & j_6 \end{Bmatrix} \begin{pmatrix} j_1 & j_2 & j_3 \\ m_1 & m_2 & m_3 \end{pmatrix} \qquad (4.70)$$

which is a variant of Eq. (4.15). *Hint*: Start by showing that the left-hand side is given by

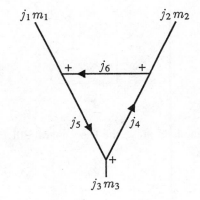

Next apply Eq. (4.55) to obtain the final result

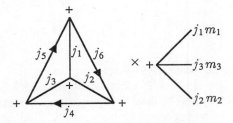

This identity allows us to remove a triangle from a diagram.

3. Use graphical methods to show that the sum

$$\sum_{\substack{M_1,M_2,\\M_3,M_4}} (-1)^{F_1-M_1+F_2-M_2+F_3-M_3+F_4-M_4} \begin{pmatrix} F_1 & 1 & F_2 \\ -M_1 & \mu & M_2 \end{pmatrix}$$

$$\times \begin{pmatrix} F_2 & 1 & F_3 \\ -M_2 & \mu' & M_3 \end{pmatrix} \begin{pmatrix} F_3 & 1 & F_4 \\ -M_3 & \nu & M_4 \end{pmatrix} \begin{pmatrix} F_4 & 1 & F_1 \\ -M_4 & \nu' & M_1 \end{pmatrix}$$

can be reexpressed as

$$\sum_{k,q} (-1)^{q+F_4-F_1+F_4-F_3} (2k+1) \begin{pmatrix} 1 & 1 & k \\ \mu & \mu' & q \end{pmatrix} \begin{pmatrix} 1 & 1 & k \\ \nu & \nu' & -q \end{pmatrix}$$

$$\times \begin{Bmatrix} 1 & 1 & k \\ F_1 & F_3 & F_2 \end{Bmatrix} \begin{Bmatrix} 1 & 1 & k \\ F_1 & F_3 & F_4 \end{Bmatrix}$$

Hint: Start by showing that the sum over the four 3-j symbols has the graph

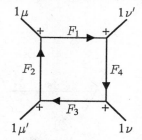

4. Use graphical methods to prove the Biedenharn–Elliott sum rule for 6-j symbols:

$$\begin{Bmatrix} j_1 & j_2 & j_3 \\ j_4 & j_5 & j_6 \end{Bmatrix} \begin{Bmatrix} j_1 & j_2 & j_3 \\ j_7 & j_8 & j_9 \end{Bmatrix}$$

$$= \sum_j (-1)^{j_1+j_2+j_3+j_4+j_5+j_6+j_7+j_8+j_9+j}(2j+1)$$

$$\times \begin{Bmatrix} j_2 & j_4 & j_6 \\ j & j_9 & j_7 \end{Bmatrix} \begin{Bmatrix} j_1 & j_5 & j_6 \\ j & j_9 & j_8 \end{Bmatrix} \begin{Bmatrix} j_3 & j_4 & j_5 \\ j & j_8 & j_9 \end{Bmatrix} \quad (4.71)$$

This was derived independently by algebraic means by L. C. Biedenharn, *J. Math. Phys.*, **31**, 287 (1953) and J. P. Elliott, *Proc. Roy. Soc.* (London), **A218**, 345 (1953). This identity may be used to obtain recursion relations for the 6-j symbols. For example, setting one of the arguments to $\frac{1}{2}$ on the left-hand side of the Biedenharn–Elliott sum ruleindexsum rule causes the sum on the right-hand side to reduce to two terms.

Chapter 5

SPHERICAL TENSOR OPERATORS

5.1 DEFINITION

So far we have concerned ourselves with the transformation properties of the angular momentum wave functions $|JM\rangle$ under rotation. We have found that when the rotation $\mathbf{R}$ acts on a particular $|JM\rangle$ it is transformed into a linear combination of a closed set of functions $|JM'\rangle$ $M' = -J, -J + 1, \ldots, J$. Thus the $2J + 1$ functions form a basis for an irreducible representation of the three-dimensional rotation group of dimension $2J + 1$. In the language of group theory, the elements of this irreducible representation are the Wigner rotation matrices $D^J_{M'M}(R)$. Viewed in this group theoretic manner, for example, the Clebsch–Gordan series merely states that the direct product $D^{J_1}_{M'_1 M_1} D^{J_2}_{M'_2 M_2}$, which spans a space $(2J_1 + 1) \times (2J_2 + 1)$, is a reducible representation that may be decomposed into irreducible representations $D^{J_3}_{M'_3 M_3}$ where J_3 runs from $J_1 + J_2$ to $|J_1 - J_2|$, and moreover, the coefficients of this reduction are just the Clebsch–Gordan coefficient products $\langle J_1 M'_1, J_2 M'_2 | J_3 M'_3 \rangle \langle J_1 M_1, J_2 M_2 | J_3 M_3 \rangle$.

The same idea can be applied to operators $\mathbf{Q}$, which we recall [see Eq. (3.6)] transform under a rotation of axes as $\mathbf{RQR}^{-1}$. We define a spherical irreducible tensor operator[1] of rank k to be a set of $2k + 1$ functions $T(k, q)$ with components $q = -k, -k + 1, \ldots, k$ that transform under rotation of the coordinate frame as

$$\mathbf{R} T(k, q) \mathbf{R}^{-1} = \sum_{q'} D^k_{q'q}(R) T(k, q') \tag{5.1}$$

that is, under the rotation $\mathbf{R}$ the operator $T(k, q)$ is transformed into a closed set of $(2k + 1)$ operators $T(k, q')$, and the coefficients of this expansion are the Wigner rotation matrix elements $D^k_{q'q}(R)$. Comparison of Eq. (5.1) with Eq. (3.87) shows that the tensor operators $T(k, q)$ are proportional to the spherical harmonics Y_{kq}.

Let us discover more about the nature of these tensor operators by looking at some special cases. Let us start with $k = 0$, for which there is only one spherical irreducible tensor operator $T(0,0)$. Under rotation it behaves as

$$\mathbf{R} T(0,0)\mathbf{R}^{-1} = D^0_{00}(R) T(0,0) = T(0,0) \tag{5.2}$$

in other words, $T(0,0)$ is *unchanged* by rotation. Hence $T(0,0)$ behaves just as a number and such a zero-rank tensor operator is called a *scalar operator*. It might at first be thought that scalar operators are of trivial importance, but this impression is not correct. Because the Hamiltonian of a physical system (in the absence of external fields) is invariant to rotation, this means that all the terms in the Hamiltonian are scalar operators, and our interest in scalar operators will be intense.

For $k = 1$ there are three operators $T(1,1)$, $T(1,0)$, and $T(1,-1)$ that constitute a first-rank tensor. Under rotation we have

$$\mathbf{R} T(1,q)\mathbf{R}^{-1} = D^1_{-1q}(R)T(1,-1) + D^1_{0q}(R)T(1,0) + D^1_{1q}(R)T(1,1) \tag{5.3}$$

We recognize that the $T(1,q)$ span the $2k + 1 = 3$ invariant subspace of the three-dimensional rotation group and hence behave as components of a vector or dipole. We call such a first-rank tensor a *vector* or *dipole* operator. A simple example is the position operator $\mathbf{r}$ with Cartesian components r_x, r_y, r_z. But recall that the Cartesian components of a vector transform under rotation according to the direction cosine matrix elements. To ensure (5.3), we need to rewrite these components in a spherical basis set, that is, in analogy with Eqs. (3.101)–(3.103):

$$T(1,1) = -2^{-\frac{1}{2}}(r_x + ir_y) = -2^{-\frac{1}{2}}(x + iy)$$

$$T(1,0) = r_z = z$$

$$T(1,-1) = 2^{-\frac{1}{2}}(r_x - ir_y) = 2^{-\frac{1}{2}}(x - iy) \tag{5.4}$$

As reference to Eqs. (3.99) and (3.100) shows, the operators defined in Eq. (5.4) are proportional to the spherical harmonics $Y_{lm}(\theta, \phi)$. Indeed, it is readily seen that the spherical harmonics themselves transform like spherical irreducible tensor operators:

$$\mathbf{R}Y_{\ell m}(\theta, \phi)\mathbf{R}^{-1} = \sum_{m'} D^\ell_{m'm}(R) Y_{\ell m'}(\theta, \phi) \tag{5.5}$$

At first, it may seem troubling that the spherical harmonics $Y_{\ell m}$ can be used as both functions and operators. However, this same duality applies to the functions x, y, z, which are also used as position operators. As eigenfunctions of ℓ^2 and ℓ_z, the $Y_{\ell m}(\theta, \phi)$ are clearly functions, while as part of the multipole expansion of an electrostatic potential (see Application 4), they are certainly operators.

For $k = 2$ there are five operators, $T(2,2)$, $T(2,1)$, $T(2,0)$, $T(2,-1)$, $T(2,-2)$, which constitute a second-rank tensor. Again they transform under rotation like the $Y_{2m}(\theta, \phi)$, that is, like the five d orbitals or the quadrupole moments of a charge distribution. Higher-rank tensors with $k > 2$ appear more seldom and represent higher-order multipole moments. The definition of the spherical irreducible tensor operators [Eq. (5.1)] does not exclude k being half-integral. However, if there were tensor operators of half-integral rank, they could have matrix elements that connect integral (boson) and half-integral (fermion) angular momentum states. As long as bosons and fermions are not interconvertible, we need not bother ourselves about the properties of such tensor operators. Hence we conclude that spherical irreducible tensor operators are operators that behave like the spherical harmonics under rotation.

Actually our definition is equivalent to defining the spherical tensor operators by the commutation relations

$$[J_z, T(k, q)] = q\, T(k, q) \tag{5.6}$$

and

$$[J_x \pm iJ_y, T(k, q)] = [k(k + 1) - q(q \pm 1)]^{\frac{1}{2}}\, T(k, q \pm 1) \tag{5.7}$$

This connection follows at once when it is realized that the $T(k, q)$ are proportional to the Y_{kq} and the proportionality constant is independent of rotation, that is, commutes with the components of $\mathbf{J}$.

We close this section by considering the always vexing problem of relating spherical tensor components defined in the space-fixed and molecule-fixed frames. Let us use the mnemonic that the component p is defined in the space-fixed frame and the component q in the molecule-fixed frame. Then Eq. (5.1) becomes

$$T(k, q) = \sum_p D_{pq}^k(R)\, T(k, p) \tag{5.8}$$

which expresses how the components of the tensor $T(k, q)$ in the molecule-fixed frame are related to those in the space-fixed frame $T(k, p)$ under the rotation $\mathbf{R}$. We derive the inverse of Eq. (5.9) by multiplying both sides by $D_{p'q}^{k*}(R)$, summing over q, and using the unitarity properties of the rotation matrices [Eqs. (3.77) and (3.78)]. By dropping the prime on p we find that

$$T(k, p) = \sum_q D_{pq}^{k*}(R)\, T(k, q) \tag{5.9}$$

Recall that the components of the angular momentum operator when referred to molecule-fixed axes have anomalous commutation relations, while those referred to the space-fixed frame have normal commutation relations (see Section 3.4). To apply the spherical tensor methods to be developed in this chapter it is necessary that all our tensor operators satisfy Eq. (5.6) in both frames. Some consistent procedure is mandatory. We choose to adopt a simple solution first introduced by Curl and Kinsey[2] and later strongly advocated by Brown and Howard[3], namely, to evaluate all spherical tensor matrix elements in the space-fixed frame, using Eq. (5.8) to refer the molecule-fixed components to the space-fixed axis system. In this way explicit consideration of the anomalous commutation relations is completely circumvented.

5.2   THE WIGNER–ECKART THEOREM

Probably you are asking at this point why bother to introduce spherical tensor operators. The reason is that it vastly simplifies the computation of matrix elements of these operators between states of sharp (definite) angular momentum, say, $\langle \alpha J M|$ and $|\alpha' J' M'\rangle$, for example. As we will show, such matrix elements factor into the product of two terms, one of which expresses the geometry, symmetry, and selection rules of the system, the other of which contains the dynamics. This disentanglement of geometrical from dynamical behavior is the power and the glory of the celebrated Wigner–Eckart theorem[4].

Consider the matrix element $\langle \alpha J M| T(k, q) |\alpha' J' M'\rangle$. The state vector $T(k, q) |\alpha' J' M'\rangle$ transforms under rotation according to the representation $\mathbf{D}^k \otimes \mathbf{D}^{J'}$, which spans a $(2k + 1)(2J' + 1)$ space of the rotation group. We decompose this into its irreducible representations. Specifically, under rotation the product $T(k, q) |\alpha' J' M'\rangle$ transforms as

$$\mathbf{R}[T(k, q) |\alpha' J' M'\rangle] = \mathbf{R} T(k, q) \mathbf{R}^{-1} \mathbf{R} |\alpha' J' M'\rangle$$

$$= \sum_{q'} D^k_{q'q}(R) \, T(k, q') \sum_{M''} D^{J'}_{M''M'}(R) \, |\alpha' J' M''\rangle$$

$$= \sum_{q', M''} D^k_{q'q}(R) D^{J'}_{M''M'}(R) \, \left[T(k, q') |\alpha' J' M''\rangle \right] \quad (5.10)$$

Equation (5.10) shows that the $(2k + 1)(2J' + 1)$ products $T(k, q) |\alpha' J' M'\rangle$ transform under rotation as the direct product representation $\mathbf{D}^k \otimes \mathbf{D}^{J'}$.

Recall the Clebsch–Gordan series in which the product of two rotation matrices is a sum over some Clebsch–Gordan coefficients times a rotation matrix $\mathbf{D}^K$ where K ranges from $k + J'$ to $|k - J'|$. We are led to ask what linear combination of the products $T(k, q) |\alpha' J' M'\rangle$ transforms as a particular state vector $|\beta K Q\rangle$ of the $2K + 1$ functions forming a basis for the representation $\mathbf{D}^K$. The answer to this question is that this problem is just what Clebsch–Gordan coefficients and angular momentum coupling were designed to solve

$$|\beta K Q\rangle = \sum_{q, M'} \langle kq', J' M''|K Q\rangle \, T(k, q') |\alpha' J' M''\rangle \quad (5.11)$$

We form the matrix element with $\langle \alpha J M|$:

$$\langle \alpha J M|\beta K Q\rangle = \sum_{q, M'} \langle kq', J' M''|K Q\rangle \, \langle \alpha J M| T(k, q') |\alpha' J' M''\rangle \quad (5.12)$$

By multiplying both sides by $\langle kq, J'M'|KQ\rangle$ and summing over K and Q, we find

$$\langle\alpha JM|T(k,q)|\alpha'J'M'\rangle = \sum_{K,Q}\langle\alpha JM|\beta KQ\rangle\langle kq, J'M'|KQ\rangle \qquad (5.13)$$

But the scalar product $\langle\alpha JM|\beta KQ\rangle$ vanishes unless $J = K$ and $M = Q$. When these conditions are satisfied, the value of the scalar product is *independent* of M, that is, unaltered by rotation. We therefore can write Eq. (5.13) as

$$\langle\alpha JM|T(k,q)|\alpha'J'M'\rangle = \langle kq, J'M'|JM\rangle\langle\alpha J\|T_k\|\alpha'J'\rangle$$

where K is replaced by J. Here $\langle\alpha J\|T_k\|\alpha'J'\rangle$, which equals $\langle\alpha JM|\beta JM\rangle$, is called the *reduced matrix element* of T_k. This result shows that the matrix elements of $T(k,q)$ separate into the physical part of the problem embodied in the reduced matrix element and the geometric part embodied in the Clebsch–Gordan coefficient.

If only angular momentum theory spoke to us with one tongue... but, alas, at this crucial juncture there is another of those damnable convention problems. Some authors, including Brink and Satchler[5], adopt the above, while others, such as, Edmonds[6], redefine the reduced matrix element to be a factor of $(2J+1)^{\frac{1}{2}}$ larger! We adopt the convention (Edmonds[6]) and state the Wigner–Eckart theorem as

$$\langle\alpha'j'm'|T(k,q)|\alpha jm\rangle = (-1)^{j'-m'}\begin{pmatrix} j' & k & j \\ -m' & q & m \end{pmatrix}\langle\alpha'j'\|T^k\|\alpha j\rangle \qquad (5.14)$$

where we remind ourselves of the extra $(2j+1)^{1/2}$ factor by writing $\langle\alpha'j'\|T^k\|\alpha j\rangle$ in place of $\langle\alpha'j'\|T_k\|\alpha j\rangle$. By knowing the value of the reduced matrix element $\langle\alpha'j'\|T^k\|\alpha j\rangle$, we can evaluate *all* the $(2j'+1)(2k+1)(2j+1)$ different possible matrix elements $\langle\alpha'j'm'|T(k,q)|\alpha jm\rangle$!

Some simple examples may serve as a guide to how to find reduced matrix elements. A particularly easy matrix element is calculated both directly and by the Wigner–Eckart theorem. Comparison of the two results yields the reduced matrix element being sought. As an example consider the orbital and spin angular momenta **L** and **S**. From their commutation rules it is readily verified that **L** and **S** are first-rank spherical tensor operators, that is, satisfy Eq. (5.6) with $k = 1$. The three components of **L** and **S** are

$$L_{+1} = -\frac{1}{\sqrt{2}}(L_x + iL_y) = -\frac{1}{\sqrt{2}}L_+$$

$$L_0 = L_z$$

$$L_{-1} = \frac{1}{\sqrt{2}}(L_x - iL_y) = \frac{1}{\sqrt{2}}L_- \qquad (5.15)$$

and

$$S_{+1} = -\frac{1}{\sqrt{2}}(S_x + iS_y) = -\frac{1}{\sqrt{2}}S_+$$

$$S_0 = S_z$$

$$S_{-1} = \frac{1}{\sqrt{2}}(S_x - iS_y) = \frac{1}{\sqrt{2}}S_- \tag{5.16}$$

The reduced matrix elements of **L** and **S** are easily found. For example, $\langle L'M'|L_z|LM\rangle$ is nonzero only for $L' = L$ and $M' = M$, in which case

$$\langle LM|L_z|LM\rangle = M \tag{5.17}$$

By the Wigner–Eckart theorem

$$\langle LM|L_z|LM\rangle = (-1)^{L-M} \begin{pmatrix} L & 1 & L \\ -M & 0 & M \end{pmatrix} \langle L\|L^{(1)}\|L\rangle \tag{5.18}$$

But

$$\begin{pmatrix} L & L & 1 \\ M & -M & 0 \end{pmatrix} = (-1)^{L-M} \frac{M}{[(2L+1)L(L+1)]^{\frac{1}{2}}} \tag{5.19}$$

so that Eq. (5.18) becomes

$$\langle LM|L_z|LM\rangle = \frac{M}{[(2L+1)L(L+1)]^{\frac{1}{2}}} \langle L\|L^{(1)}\|L\rangle \tag{5.20}$$

Comparison of Eqs. (5.17) and (5.20) yields the result

$$\langle L'\|L^{(1)}\|L\rangle = [(2L+1)L(L+1)]^{\frac{1}{2}}\delta_{L'L} \tag{5.21}$$

Similarly

$$\langle S'\|S^{(1)}\|S\rangle = [(2S+1)S(S+1)]^{\frac{1}{2}}\delta_{S'S} \tag{5.22}$$

which for the case $S = \frac{1}{2}$ becomes

$$\langle \tfrac{1}{2}\|S^{(1)}\|\tfrac{1}{2}\rangle = \left(\tfrac{3}{2}\right)^{\frac{1}{2}} \tag{5.23}$$

A trivial example of the use of these reduced matrix elements is the evaluation of

$$\langle 33|L_+|32\rangle = \langle 33|L_x + iL_y|32\rangle = [3(3+1) - (2)(3)]^{\frac{1}{2}} = 6^{\frac{1}{2}} \tag{5.24}$$

This may also be evaluated according to the Wigner–Eckart theorem by

$$\langle 33 | L_+ | 32 \rangle = -\sqrt{2} \, \langle 33 | L_{+1} | 32 \rangle$$

$$= -\sqrt{2}(-1)^{3-3} \begin{pmatrix} 3 & 1 & 3 \\ -3 & 1 & 2 \end{pmatrix} \langle 3 \| L \| 3 \rangle$$

$$= -\sqrt{2} \begin{pmatrix} 3 & 3 & 1 \\ 2 & -3 & 1 \end{pmatrix} [(7)(3)(4)]^{\frac{1}{2}}$$

$$= -\sqrt{2}(-1)^{3-2} \left[\frac{(2)(1)(6)}{(8)(7)(6)} \right]^{\frac{1}{2}} [(7)(3)(4)]^{\frac{1}{2}}$$

$$= 6^{\frac{1}{2}} \tag{5.25}$$

Of course, no one would normally use the Wigner–Eckart theorem to evaluate the matrix elements of L_+, L_-, or L_z, but it certainly is reassuring to know that all the pieces of this machinery do work!

Another example of practical importance is the reduced matrix element of a spherical harmonic. We have previously seen that for the integral of the triple product of spherical harmonics with $m_1 = m_2 = m_3 = 0$

$$\langle Y_{\ell_1 0} | Y_{\ell_2 0} | Y_{\ell_3 0} \rangle = \langle \ell_1 0 | Y_{\ell_2 0} | \ell_3 0 \rangle$$

$$= \left[\frac{(2\ell_1 + 1)(2\ell_2 + 1)(2\ell_3 + 1)}{4\pi} \right]^{\frac{1}{2}} \begin{pmatrix} \ell_1 & \ell_2 & \ell_3 \\ 0 & 0 & 0 \end{pmatrix}^2 \tag{5.26}$$

Alternatively, application of the Wigner–Eckart theorem yields

$$\langle \ell_1 0 | Y_{\ell_2 0} | \ell_3 0 \rangle = (-1)^{\ell_1} \begin{pmatrix} \ell_1 & \ell_2 & \ell_3 \\ 0 & 0 & 0 \end{pmatrix} \langle \ell_1 \| Y^{\ell_2} \| \ell_3 \rangle \tag{5.27}$$

Hence

$$\langle \ell_1 \| Y^{\ell_2} \| \ell_3 \rangle = (-1)^{\ell_1} \left[\frac{(2\ell_1 + 1)(2\ell_2 + 1)(2\ell_3 + 1)}{4\pi} \right]^{\frac{1}{2}} \begin{pmatrix} \ell_1 & \ell_2 & \ell_3 \\ 0 & 0 & 0 \end{pmatrix} \tag{5.28}$$

In practice it is often more convenient to work with a modified spherical harmonic $C(k, q)$, defined by

$$C(k, q) = C_{kq}(\theta, \phi) = \left(\frac{4\pi}{2k + 1} \right)^{\frac{1}{2}} Y_{kq}(\theta, \phi) \tag{5.29}$$

which differs from the spherical harmonic Y_{kq} by the use of a different normalization. Then we have

$$\langle \ell_1 \| C^k \| \ell_3 \rangle = (-1)^{\ell_1} [(2\ell_1 + 1)(2\ell_3 + 1)]^{\frac{1}{2}} \begin{pmatrix} \ell_1 & k & \ell_3 \\ 0 & 0 & 0 \end{pmatrix} \qquad (5.30)$$

We close this section by examining some consequences of the Wigner–Eckart theorem applied to electric and magnetic moments. We begin by looking for what conditions must be met for the system to possess a static moment. Let the system be in a state of definite angular momentum so that its state vector is $|\alpha j m\rangle$. We next show that such a system cannot have a permanent dipole moment unless $j \geq \frac{1}{2}$ or a permanent quadrupole moment unless $j \geq 1$. Let μ denote the dipole moment operator and $\mathbf{Q}$ the quadrupole moment operator. For μ and $\mathbf{Q}$ to be "permanent" means that the expectation value of μ and $\mathbf{Q}$ must be nonzero, that is, $\langle \alpha j m | \mu(1,q) | \alpha j m \rangle \neq 0$ and $\langle \alpha j m | Q(2,q) | \alpha j m \rangle \neq 0$ for at least one component q. Recognizing that μ and $\mathbf{Q}$ are spherical tensors of ranks $k = 1$ and $k = 2$, respectively, we apply the Wigner–Eckart theorem. For example,

$$\langle \alpha j m | \mu(1,q) | \alpha j m \rangle = (-1)^{j-m} \begin{pmatrix} j & 1 & j \\ -m & q & m \end{pmatrix} \langle \alpha j \| \mu \| \alpha j \rangle \qquad (5.31)$$

which vanishes if $j = 0$ or if $q \neq 0$. Hence only systems with $j \geq \frac{1}{2}$ have permanent dipole moments, and for such systems the dipole moment points along the axis of quantization. A similar argument pertains to the quadrupole moment operator, that is, only states with $j \geq 1$ can have a permanent quadrupole moment, and such systems possess only one such moment, the expectation value of $Q(2,0)$.

So far nothing has been said about the difference between electric and magnetic moments. However, these operators behave differently under inversion of the spatial coordinates. The electric dipole operator can be written as $\mu^{(e)} = \sum_i e_i \mathbf{r}_i$, where e_i is the charge of the ith particle with position vector $\mathbf{r}_i$. However, the magnetic dipole operator looks like $\mu^{(m)} = \mu_0 g_L \mathbf{L} + \mu_0 g_S \mathbf{S} + \mu_I g_I \mathbf{I} + \cdots$, where μ_0 is the electronic Bohr magneton, μ_I is the nuclear Bohr magneton, and g_L, g_S, g_I are the corresponding g factors for each type of angular momentum (current loop) possessed by the system. Both $\mu^{(e)}$ and $\mu^{(m)}$ are first-rank tensors (i.e., vectors), but $\mu^{(m)}$ is an axial vector (pseudovector) that does not change sign under inversion of all spatial coordinates, while $\mu^{(e)}$ is a polar vector and does. We say that $\mu^{(e)}$ has odd parity while $\mu^{(m)}$ has even parity. For the matrix element of $\mu^{(e)}$ or $\mu^{(m)}$ to be nonvanishing, its overall parity must be even. Therefore, in a state of definite angular momentum and definite parity there can be no permanent electric dipole moment, while there can be a permanent magnetic dipole moment provided $j \geq \frac{1}{2}$. Hence nondegenerate states (excluding the $(2j + 1)$-fold m degeneracy present in the absence of external fields) cannot have permanent electric dipole moments and cannot exhibit first-order Stark effects. In contrast, nondegenerate states generally have permanent magnetic moments and show linear Zeeman effects, unless $j = 0$. The parity of the 2^k multipole moments go as $(-1)^k$ for electric moments and $(-1)^{k-1}$

for magnetic moments so that there are in general only even electric multipole moments and odd magnetic multipole moments. Thus we understand why nuclei with $I \geq \frac{1}{2}$ have magnetic dipole moments while those with $I \geq 1$ have electric quadrupole moments. Moreover, ground-state nuclei do not have electric dipole moments (neglecting weak forces, parity violation, and all that jazz).

So far we have focused our attention on the expectation values (diagonal matrix elements) of the multipole moment operators. But the off-diagonal matrix elements are also of interest, of course, particularly with regard to radiative transitions. Here we see that electric dipole transitions can connect states of only opposite parity while magnetic dipole and electric quadrupole transitions can connect states of only the same parity. Moreover, the conditions for the nonvanishing of the 3-j symbol

$$\begin{pmatrix} J' & k & J \\ -M' & q & M \end{pmatrix}$$

give us at once the familiar selection rules: namely, for electric dipole transitions as well as magnetic dipole transitions $\Delta J = 0, \pm 1$ and $\Delta M = 0, \pm 1$, except that $J = 0$ cannot be connected to $J = 0$, and for electric quadrupole transitions (as well as magnetic quadrupole transitions) $\Delta J = 0, \pm 1, \pm 2$ and $\Delta M = 0, \pm 1, \pm 2$, except that transitions for which $J + J' < 2$ cannot occur. Whereas q must be zero for static moments of systems in which J_Z is diagonal, in general three transition dipole moment operators and five transition quadrupole moment operators must be considered in working out possible radiative transitions. This example illustrates how the Wigner–Eckart theorem allows at once a separation and identification of all the geometric factors, in this case the common selection rules for radiative transitions.

In many problems we encounter the square modulus of matrix elements. This motivates us to consider the Hermitian conjugate of the matrix elements of a spherical tensor operator. We begin by taking the Hermitian conjugate of both sides of Eq. (5.14):

$$\langle \alpha' j' m' | T(k, q) | \alpha j m \rangle^* = \langle \alpha j m | T(k, q)^\dagger | \alpha' j' m' \rangle$$

$$= (-1)^{j'-m'} \begin{pmatrix} j' & k & j \\ -m' & q & m \end{pmatrix} \langle \alpha' j' \| T^k \| \alpha j \rangle^*$$

$$(5.32)$$

Because we insist that $T(k, q)$ behave as a spherical harmonic of rank k and component q, we define the adjoint of a spherical tensor operator $T(k, q)$ as

$$T(k, q)^\dagger = (-1)^{-q} T(k, -q) \qquad (5.33)$$

in analogy with $Y_{kq}^\dagger(\theta, \phi) = Y_{kq}^*(\theta, \phi) = (-1)^{-q} Y_{k-q}(\theta, \phi)$. Then

$$\langle \alpha j m | T(k, q)^\dagger | \alpha' j' m' \rangle = (-1)^{-q} \langle \alpha j m | T(k, -q) | \alpha' j' m' \rangle$$

$$= (-1)^{j-m-q} \begin{pmatrix} j & k & j' \\ -m & -q & m' \end{pmatrix} \langle \alpha j || T^k || \alpha' j' \rangle$$

$$= (-1)^{j-m-q} \begin{pmatrix} j' & k & j \\ -m' & q & m \end{pmatrix} \langle \alpha j || T^k || \alpha' j' \rangle \quad (5.34)$$

Comparison of Eq. (5.32) with Eq. (5.34) yields the useful identity

$$\langle \alpha' j' || T^k || \alpha j \rangle^* = (-1)^{j-j'} \langle \alpha j || T^k || \alpha' j' \rangle \quad (5.35)$$

5.3 SPHERICAL TENSOR PRODUCTS

As we have seen, the spherical tensors behave like spherical harmonics. Recalling the transformation between coupled and uncoupled representations [Eq. (2.5)], it is not surprising to find that the multiplication (also called the *contraction*) of two spherical tensors $T(\ell_1, q_1)$ and $T(\ell_2, q_2)$ of ranks ℓ_1 and ℓ_2 to form a spherical tensor of rank ℓ is given by

$$T(\ell, q) = \sum_{q_1} \langle \ell_1 q_1, \ell_2 q_2 | \ell q \rangle T(\ell_1, q_1) T(\ell_2, q_2) \quad (5.36)$$

To establish that $T(\ell, q)$ is a tensor of rank ℓ we must show that the right-hand side of Eq. (5.36) transforms under rotation according to

$$\mathbf{R} T(\ell, q) \mathbf{R}^{-1} = \sum_{q'} D^{\ell}_{q'q}(R) T(\ell, q') \quad (5.37)$$

We apply the rotation $\mathbf{R}$ to both sides of (5.36):

$$\mathbf{R} T(\ell, q) \mathbf{R}^{-1} = \sum_{q_1} \langle \ell_1 q_1, \ell_2 q_2 | \ell q \rangle \, \mathbf{R} T(\ell_1, q_1) \mathbf{R}^{-1} \, \mathbf{R} T(\ell_2, q_2) \mathbf{R}^{-1}$$

$$= \sum_{q_1, q_1', q_2'} \langle \ell_1 q_1, \ell_2 q_2 | \ell q \rangle \, D^{\ell_1}_{q_1' q_1}(R) D^{\ell_2}_{q_2' q_2}(R) \, T(\ell_1, q_1') T(\ell_2, q_2')$$

$$= \sum_{\substack{\ell, q_1, \\ q_1', q_2'}} \langle \ell_1 q_1, \ell_2 q_2 | \ell q \rangle \langle \ell_1 q_1, \ell_2 q_2 | \ell' q \rangle$$

$$\times \langle \ell_1 q_1', \ell_2 q_2' | \ell' q' \rangle \, D^{\ell'}_{q_1' + q_2', q_1 + q_2}(R) \, T(\ell_1, q_1') T(\ell_2, q_2')$$

$$= \sum_{q_1', q_2'} D^{\ell}_{q_1' + q_2', q_1 + q_2}(R) \, \langle \ell_1 q_1', \ell_2 q_2' | \ell q' \rangle \, T(\ell_1, q_1') T(\ell_2, q_2')$$

$$= \sum_{q'} D^{\ell}_{q',q}(R) \, \langle \ell_1 q'_1, \ell_2 q'_2 | \ell q' \rangle \, T(\ell_1, q'_1) \, T(\ell_2, q'_2) \qquad (5.38)$$

where we have made use of the Clebsch–Gordan series [Eq. (3.105)] and the orthonormality of the Clebsch–Gordan coefficients [Eq. (2.8)]. Comparison of Eqs. (5.38) and (5.37) shows the validity of the identification made in Eq. (5.36). The tensor product of $T(\ell_1, q_1)$ and $T(\ell_2, q_2)$ may also be written as

$$T(\ell, q) = \left[T^{(\ell_1)} \otimes T^{(\ell_2)} \right]^{(\ell)}_q \qquad (5.39)$$

This notation reminds us that the direct product of the sets of operators $T(\ell_1, q_1)$ and $T(\ell_2, q_2)$ span the representation $\mathbf{D}^{\ell_1} \otimes \mathbf{D}^{\ell_2}$, which can be decomposed into the representation $\mathbf{D}^{\ell_1+\ell_2} + \mathbf{D}^{\ell_1+\ell_2-1} + \cdots + \mathbf{D}^{|\ell_1-\ell_2|}$. The particular linear combination of the products $T(\ell_1, q_1) T(\ell_2, q_2)$ that transform like $T(\ell, q)$ are given by the Clebsch–Gordan coefficients $\langle \ell_1 q_1, \ell_2 q_2 | \ell q \rangle$ according to Eq. (5.36). This result shows that the coupling of spherical tensors is mathematically the same as coupling angular momentum eigenvectors, for both use only the group theoretic properties of the operators or states under rotation.

To illustrate Eq. (5.36) consider, for example, the compound irreducible tensor operators formed by multiplying (contracting) two first-rank tensors (vectors) $\mathbf{A}$ and $\mathbf{B}$, which have the components of the form given in Eq. (5.4). By evaluating the expression

$$\left[A^{(1)} \otimes B^{(1)} \right]^{(k)}_q = \sum_m \langle 1m, 1q - m | kq \rangle A(1, m) B(1, q - m) \qquad (5.40)$$

we find explicitly that for $k = 0$

$$\left[A^{(1)} \otimes B^{(1)} \right]^{(0)}_0 = -3^{-\frac{1}{2}} \left[-A(1, 1) B(1, -1) \right.$$

$$+ A(1, 0) B(1, 0) - A(1, -1) B(1, 1) \right]$$

$$= -3^{-\frac{1}{2}} (A_x B_x + A_y B_y + A_z B_z) = -3^{-\frac{1}{2}} \mathbf{A} \cdot \mathbf{B} \qquad (5.41)$$

for $k = 1$

$$\left[A^{(1)} \otimes B^{(1)} \right]^{(1)}_0 = i2^{-\frac{1}{2}} (A_x B_y - A_y B_x)$$

$$= i2^{-\frac{1}{2}} (\mathbf{A} \times \mathbf{B}) \cdot \hat{\mathbf{Z}} \qquad (5.42)$$

$$\left[A^{(1)} \otimes B^{(1)} \right]^{(1)}_{\pm 1} = i2^{-\frac{1}{2}} \left[\frac{\mp (A_y B_z - A_z B_y) - i(A_z B_x - A_x B_z)}{\sqrt{2}} \right]$$

$$= \tfrac{1}{2} \left[(A_z B_x - A_x B_z) \pm i(A_z B_y - A_y B_z) \right] \qquad (5.43)$$

and for $k = 2$

$$\left[A^{(1)} \otimes B^{(1)}\right]_0^{(2)} = \left(3 A_z B_z - \mathbf{A} \cdot \mathbf{B}\right)/6^{\frac{1}{2}}$$

$$\left[A^{(1)} \otimes B^{(1)}\right]_{\pm 1}^{(2)} = \mp \tfrac{1}{2} \left[\left(A_z B_x + A_x B_z\right) \pm i\left(A_z B_y + A_y B_z\right)\right]$$

$$\left[A^{(1)} \otimes B^{(1)}\right]_{\pm 2}^{(2)} = \tfrac{1}{2} \left[\left(A_x B_x - A_y B_y\right) \pm i\left(A_x B_y + A_y B_x\right)\right] \qquad (5.44)$$

It is seen that the tensor product of rank $k = 0$ is related to the scalar product (dot product) of $\mathbf{A}$ and $\mathbf{B}$ while the tensor product of rank $k = 1$ is related to the cross product of $\mathbf{A}$ and $\mathbf{B}$. Indeed, the latter is just $(\mathbf{A} \times \mathbf{B}) \cdot \hat{\mathbf{e}}_q$, where $\hat{\mathbf{e}}_{+1} = -2^{-\frac{1}{2}}(\hat{\mathbf{X}} + i\hat{\mathbf{Y}})$, $\hat{\mathbf{e}}_0 = \hat{\mathbf{Z}}$, and $\hat{\mathbf{e}}_{-1} = 2^{-\frac{1}{2}}(\hat{\mathbf{X}} - i\hat{\mathbf{Y}})$ are unit vectors in the spherical basis that transform as $\mathbf{D}^1$ under rotation. Care must be taken here to use the definition for a scalar product of vectors having complex components, that is,

$$\hat{\mathbf{e}} \cdot \hat{\mathbf{e}}' = e_x^* e_x' + e_y^* e_y' + e_z^* e_z' \qquad (5.45)$$

from which it is seen that $\hat{\mathbf{e}}_q \cdot \hat{\mathbf{e}}_{q'} = \delta_{qq'}$.

A particularly important case of (5.36) is that of compound tensor scalar operators:

$$\left[T^{(k)} \otimes T^{(k')}\right]_0^{(0)} = \sum_q \langle kq, k'q' | 00 \rangle T(k,q) T(k',q')$$

$$= \sum_q (-1)^{k-q} (2k+1)^{-\frac{1}{2}} T(k,q) T(k',q') \delta_{q,-q'} \delta_{kk'}$$

$$= (2k+1)^{-\frac{1}{2}} \sum_q (-1)^{k-q} T(k,q) T(k,-q) \qquad (5.46)$$

Equation (5.46) shows that scalar operators can be built from tensor operators of rank k by contracting two tensor operators of the *same rank*. Equation (5.41), in which the scalar operator $[A^1 \otimes B^1]_0^{(0)}$ was constructed from the two vector operators $\mathbf{A}$ and $\mathbf{B}$, is a special example of Eq. (5.46). It is traditional to define the generalized dot product $A^{(k)} \cdot B^{(k)}$ by

$$A^{(k)} \cdot B^{(k)} = \sum_q (-1)^q A(k,q) B(k,-q) \qquad (5.47)$$

Then $A^{(k)} \cdot B^{(k)}$ is also a scalar operator and is related to $\left[A^{(k)} \otimes B^{(k)}\right]_0^{(0)}$ by

$$A^{(k)} \cdot B^{(k)} = (-1)^k (2k+1)^{\frac{1}{2}} \left[A^{(k)} \otimes B^{(k)}\right]_0^{(0)} \qquad (5.48)$$

Both $A^{(k)} \cdot B^{(k)}$ and $[A^{(k)} \otimes B^{(k)}]_0^{(0)}$ appear in the literature, so care is needed not to confuse these two forms of the scalar contraction.

With Eqs. (5.40)–(5.44) it is possible to rewrite products of vector operators in an alternative form by use of this angular momentum coupling algebra. As an example (which will prove useful later in discussing electric dipole absorption and/or emission of radiation) we consider the scalar operator $(\mathbf{a} \cdot \mathbf{b})(\mathbf{c} \cdot \mathbf{d})$. According to Eq. (5.41), this has the tensorial character

$$(\mathbf{a} \cdot \mathbf{b})(\mathbf{c} \cdot \mathbf{d}) = 3 \left[\left[a^{(1)} \otimes b^{(1)} \right]^{(0)} \otimes \left[c^{(1)} \otimes d^{(1)} \right]^{(0)} \right]_0^{(0)} \qquad (5.49)$$

In many applications it is necessary to regroup the operators in such an expression so that, for example, $a^{(1)}$ and $c^{(1)}$ are coupled into one resultant and $b^{(1)}$ and $d^{(1)}$ into another. This is accomplished by using the recoupling transformation of four angular momenta [Eq. (4.21)], with $j_1 = j_2 = j_3 = j_4 = 1$. Explicitly we have

$$(\mathbf{a} \cdot \mathbf{b})(\mathbf{c} \cdot \mathbf{d}) = 3 \sum_k \left[\left[a^{(1)} \otimes c^{(1)} \right]^{(k)} \otimes \left[b^{(1)} \otimes d^{(1)} \right]^{(k)} \right]_0^{(0)}$$

$$\times \left\langle (11)k(11)k0 \,|\, (11)0(11)00 \right\rangle \qquad (5.50)$$

Evaluation of the recoupling coefficient $\left\langle (11)k(11)k0 \,|\, (11)0(11)00 \right\rangle$ involves a $9\text{-}j$ coefficient, three of whose arguments are zero. Application of Eq. (4.25) followed by (4.19) yields the result that

$$\left\langle (11)k(11)k0 \,|\, (11)0(11)00 \right\rangle = (2k+1)^{\frac{1}{2}}/3 \qquad (5.51)$$

where k is restricted to have the values of $0, 1$, or 2. Hence

$$(\mathbf{a} \cdot \mathbf{b})(\mathbf{c} \cdot \mathbf{d}) = \sum_k (2k+1)^{\frac{1}{2}} \left[\left[a^{(1)} \otimes c^{(1)} \right]^{(k)} \otimes \left[b^{(1)} \otimes d^{(1)} \right]^{(k)} \right]_0^{(0)}$$

$$= \sum_{k,q} (-1)^{k-q} \left[a^{(1)} \otimes c^{(1)} \right]_q^{(k)} \left[b^{(1)} \otimes d^{(1)} \right]_{-q}^{(k)} \qquad (5.52)$$

where Eq. (5.46) has been used to obtain the last line of the above. Equation (5.52) has a very pleasing appearance in that it clearly displays the scalar contraction of the spherical operators $T(k, q)$ formed from the vector operators $\mathbf{a}$, $\mathbf{b}$, $\mathbf{c}$, and $\mathbf{d}$. This vector identity may also be written as

$$(\mathbf{a} \cdot \mathbf{b})(\mathbf{c} \cdot \mathbf{d}) = (\mathbf{a} \cdot \mathbf{c})(\mathbf{b} \cdot \mathbf{d})/3 + (\mathbf{a} \times \mathbf{c}) \cdot (\mathbf{b} \times \mathbf{d})/2$$

$$+ \sum_{q=-2}^{2} (-1)^{2-q} \left[a^{(1)} \otimes c^{(1)} \right]_q^{(2)} \left[b^{(1)} \otimes d^{(1)} \right]_{-q}^{(2)} \qquad (5.53)$$

This expression cannot be simplified further without writing out the actual vector components.

As remarked before, the Hamiltonian of the system must be invariant to coordinate rotation (the energy of the system must be independent of the choice of coordinates) and hence is a scalar operator. As such, it is expressible as a sum of scalars, each of which is a contraction of spherical tensors of the same rank. It is no wonder, then, that the interaction between a system of charged particles (atom, molecule) and an external electric field (from a radiation source, from a neighboring charge distribution, or from an applied external field) contains terms such as the dipole moment of the system times the electric field, the quadrupole moment of the system times the gradient of the electric field, the octupole moment of the charge distribution times the gradient of the electric field, and so on.

To conclude this section let us examine the scalar invariant that can be formed from the tensor product $C_{kq}(\theta, \phi)$ and $C_{kq}(\theta', \phi')$ of modified spherical harmonics. Using Eq. (5.47), we write

$$C^k(\theta, \phi) \cdot C^k(\theta', \phi') = \sum_q (-1)^q C_{kq}(\theta, \phi) C_{k-q}(\theta', \phi') \tag{5.54}$$

Because $C^k \cdot C^k$ must be unchanged by rotation, it follows that this scalar product must be some function of the angle Θ included between the directions (θ, ϕ) and (θ', ϕ'), as this angle is the only quantity that does not depend on the choice of axes. In particular, let us select our axes so that (θ', ϕ') becomes $(0, 0)$, that is, so that the (θ', ϕ') direction coincides with the new Z axis. Then $C_{k-q}(\theta', \phi')$ becomes $C_{k-q}(0, 0) = \delta_{q0}$ [see Eq. (1.55)] and the other factor in (5.54) $C_{kq}(\theta, \phi)$ becomes $C_{k0}(\Theta, \Phi) = P_k(\cos \Theta)$, where we recognize that (θ, ϕ) is rotated to (Θ, Φ) when (θ', ϕ') is rotated to $(0, 0)$. Hence

$$C^k(\theta, \phi) \cdot C^k(\theta', \phi') = P_k(\cos \Theta) \tag{5.55}$$

This result is the same as Eq. (3.85); that is, we have rediscovered the spherical harmonic addition theorem, which we now recognize as simply the scalar contraction of the spherical harmonics.

Spherical harmonics with different arguments may also be used to construct other tensors in addition to the scalar tensor contraction. These tensors, called *bipolar harmonics*, are defined by[3]

$$B_{KQ}(k, k') = \sum_q \langle kq, k'q' | KQ \rangle C_{kq}(\theta, \phi) C_{k'q'}(\theta', \phi')$$

$$= \sum_q (-1)^{k-k'+Q} (2K+1)^{\frac{1}{2}} \begin{pmatrix} k & k' & K \\ q & q' & -Q \end{pmatrix}$$

$$\times C_{kq}(\theta, \phi) C_{k'q'}(\theta', \phi') \tag{5.56}$$

They satisfy the orthogonality property

$$\int d\Omega \int d\Omega' \, B_{KQ}^*(k_1, k_2) B_{K'Q'}(k_1', k_2')$$

$$= \frac{16\pi^2 \delta(K, K')\delta(Q, Q')\delta(k_1, k_1')\delta(k_2, k_2')}{(2k_1 + 1)(2k_2 + 1)} \qquad (5.57)$$

and the closure condition

$$\sum_{K,Q} |B_{KQ}(k_1, k_2)|^2 = 1 \qquad (5.58)$$

For $K = Q = 0$, Equation (5.56) becomes simply another statement of the spherical harmonic addition theorem.

The bipolar harmonics often appear in problems involving two directions (correlation and coincidence measurements)[7–10]. A simple example is the ejection of two electrons from an atom or molecule, caused by some collision process or by the absorption of radiation. The two-electron angular distribution may be expressed as[11]

$$I(\theta_1, \phi_1, \theta_2, \phi_2) = \sum_{\substack{L,M, \\ \ell_1, \ell_2}} A_{LM}(\ell_1, \ell_2) B_{LM}(\ell_1, \ell_2) \qquad (5.59)$$

where the angular dependence is contained solely in the bipolar harmonics and the $A_{LM}(\ell_1, \ell_2)$ express the dynamics of the electron photoejection process. In Eq. (5.59) the sum is over the possible angular momenta ℓ_1 and ℓ_2 of electrons 1 and 2, which form the resultant total angular momentum $\mathbf{L}$ with projection M on the axis of quantization. The dynamical coefficients A_{LM} may depend on scalars (energy or time of emission) of the two electrons considered separately but not on the angle between their momenta. Other examples of two-vector correlations are readily formulated in reference to photofragment rotational and translational motions[12] or to the prototypical bimolecular reaction $A + BC \rightarrow AB + C$. In the latter case one may consider measuring the correlation between the initial relative velocity vector of the reagents and the recoil direction of one of the reaction products AB or C (the differential cross section), the correlation between the direction of the initial relative velocity of the reagents and the direction of the BC angular momentum on the reaction probability (the total cross-section anisotropy), the correlation between the initial relative velocity vector and the angular momentum direction of the AB product (the product polarization), or the correlation between the direction of the angular momentum of the BC reagent and the direction of the angular momentum of the AB product (the rotational tilt)[13]. Correlations between three or more vectors are also of interest and lead naturally to a generalization of Eq. (5.56) to n-polar harmonics[7, 8, 10, 13].

5.4 TENSOR PRODUCT MATRIX ELEMENTS

In this section we take full advantage of the entire angular momentum coupling machinery to derive general expressions for the matrix elements of spherical tensors that are themselves products of spherical tensor operators. These expressions form an important and powerful reference library in the application of angular momentum coupling techniques to problems of interest, particularly those involving composite systems. Such a composite system may consist of two or more separate systems or separate spaces of the same system. We consider the most general case and then specialize our results.

Let $|jm\rangle = |j_1 j_2 j\,m\rangle$ denote a coupled state built from two component states $|j_1 m_1\rangle$ and $|j_2 m_2\rangle$ in spaces 1 and 2 according to the tried-and-true recipe

$$|j_1 j_2 j\,m\rangle = \sum_{m_1,m_2} \langle j_1 m_1, j_2 m_2 | j m\rangle\, |j_1 m_1\rangle\, |j_2 m_2\rangle \qquad (5.60)$$

or lovingly recast into a form with $3\text{-}j$ symbols as

$$|j_1 j_2 j\,m\rangle = \sum_{m_1,m_2} (-1)^{j_1 - j_2 + m}(2j+1)^{\frac{1}{2}} \begin{pmatrix} j_1 & j_2 & j \\ m_1 & m_2 & -m \end{pmatrix} |j_1 m_1\rangle\,|j_2 m_2\rangle$$

$$(5.61)$$

We also introduce spherical tensor operators $T(k_1, q_1)$ and $U(k_2, q_2)$ acting on independent sets of variables in spaces 1 and 2 and hence commuting. We form the composite spherical tensor $X(k, q)$ built from $T(k_1, q_1)$ and $U(k_2, q_2)$ according to

$$X(k,q) = \sum_{q_1,q_2} \langle k_1 q_1, k_2 q_2 | kq\rangle\, T(k_1, q_1) U(k_2, q_2) \qquad (5.62)$$

which has as its $3\text{-}j$ counterpart

$$X(k,q) = \sum_{q_1,q_2} (-1)^{k_1 - k_2 + q}(2k+1)^{\frac{1}{2}} \begin{pmatrix} k_1 & k_2 & k \\ q_1 & q_2 & -q \end{pmatrix} T(k_1, q_1) U(k_2, q_2)$$

$$(5.63)$$

We desire to evaluate the general matrix element

$$\langle \gamma j_1 j_2 j m | X(k,q) | \gamma' j_1' j_2' j' m'\rangle$$

$$= (-1)^{j-m} \begin{pmatrix} j & k & j' \\ -m & q & m' \end{pmatrix} \langle \gamma j_1 j_2 j \| X^k \| \gamma' j_1' j_2' j'\rangle \qquad (5.64)$$

where we have applied the Wigner–Eckart theorem to obtain the right-hand side. Here γ and γ' represent all other quantum numbers of the states characterized by j and j'.

We use Eqs. (5.61) and (5.63) to rewrite Eq. (5.64) as

$$\langle \gamma j_1 j_2 j m | X(k,q) | \gamma' j_1' j_2' j' m' \rangle$$

$$= \sum_{q_1, q_2} \langle \gamma j_1 j_2 j m | T(k_1, q_1) U(k_2, q_2) | \gamma' j_1' j_2' j' m' \rangle$$

$$\times (-1)^{k_1 - k_2 + q} (2k+1)^{\frac{1}{2}} \begin{pmatrix} k_1 & k_2 & k \\ q_1 & q_2 & -q \end{pmatrix}$$

$$= \sum_{\gamma''} \sum_{m_1, m_2} \sum_{m_1', m_2'} \sum_{q_1, q_2} (-1)^{k_1 - k_2 + q + j_1 - j_2 + m + j_1' - j_2' + m'}$$

$$\times \left[(2j+1)(2j'+1)(2k+1) \right]^{\frac{1}{2}}$$

$$\times \begin{pmatrix} j_1 & j_2 & j \\ m_1 & m_2 & -m \end{pmatrix} \begin{pmatrix} j_1' & j_2' & j' \\ m_1' & m_2' & -m' \end{pmatrix} \begin{pmatrix} k_1 & k_2 & k \\ q_1 & q_2 & -q \end{pmatrix}$$

$$\times \langle \gamma j_1 m_1 | T(k_1, q_1) | \gamma'' j_1' m_1' \rangle \langle \gamma'' j_2 m_2 | U(k_2, q_2) | \gamma' j_2' m_2' \rangle$$

$$\tag{5.65}$$

Finally, we apply the Wigner–Eckart theorem once more and rewrite (5.65) as

$$\langle \gamma j_1 j_2 j m | X(k,q) | \gamma' j_1' j_2' j' m' \rangle$$

$$= \sum_{\gamma''} \sum_{m_1, m_2} \sum_{m_1', m_2'} \sum_{q_1, q_2} (-1)^{k_1 - k_2 + q + j_1 - j_2 + m + j_1' - j_2' + m'}$$

$$\times \left[(2j+1)(2j'+1)(2k+1) \right]^{\frac{1}{2}} \begin{pmatrix} j_1 & j_2 & j \\ m_1 & m_2 & -m \end{pmatrix}$$

$$\times \begin{pmatrix} j_1' & j_2' & j' \\ m_1' & m_2' & -m' \end{pmatrix} \begin{pmatrix} k_1 & k_2 & k \\ q_1 & q_2 & -q \end{pmatrix}$$

$$\times (-1)^{j_1 - m_1 + j_2 - m_2} \begin{pmatrix} j_1 & k_1 & j_1' \\ -m_1 & q_1 & m_1' \end{pmatrix} \begin{pmatrix} j_2 & k_2 & j_2' \\ -m_2 & q_2 & m_2' \end{pmatrix}$$

$$\times \langle \gamma j_1 \| T^{k_1} \| \gamma'' j_1' \rangle \langle \gamma'' j_2 \| U^{k_2} \| \gamma' j_2' \rangle \tag{5.66}$$

We have two expressions [Eqs. (5.64) and (5.66)] for the general matrix elements of $X(k,q)$. We equate them and multiply both sides by

$$(-1)^{j-m} \begin{pmatrix} j & k & j' \\ -m & q & m' \end{pmatrix}$$

and sum over m, m', and q. With the help of Eq. (2.32) we obtain

$$\langle \gamma j_1 j_2 j \| X^k \| \gamma' j_1' j_2' j' \rangle$$

$$= \sum_{\gamma''} \sum_{m,m',q} \sum_{m_1,m_1'} \sum_{m_2,m_2'} \sum_{q_1,q_2} (-1)^{k_1-k_2+j_1'-j_2'+j+q+m'-m_1-m_2}$$

$$\times \begin{pmatrix} j_1 & j_2 & j \\ m_1 & m_2 & -m \end{pmatrix} \begin{pmatrix} j_1' & j_2' & j' \\ m_1' & m_2' & -m' \end{pmatrix} \begin{pmatrix} k_1 & k_2 & k \\ q_1 & q_2 & -q \end{pmatrix}$$

$$\times \begin{pmatrix} j_1 & k_1 & j_1' \\ -m_1 & q_1 & m_1' \end{pmatrix} \begin{pmatrix} j_2 & k_2 & j_2' \\ -m_2 & q_2 & m_2' \end{pmatrix} \begin{pmatrix} j & k & j' \\ -m & q & m' \end{pmatrix}$$

$$\times \left[(2j+1)(2j'+1)(2k+1) \right]^{\frac{1}{2}}$$

$$\times \langle \gamma j_1 \| T^{k_1} \| \gamma'' j_1' \rangle \langle \gamma'' j_2 \| U^{k_2} \| \gamma' j_2' \rangle \tag{5.67}$$

We have a sum over six 3-j symbols, which has the distinct smell of a 9-j symbol. Using either graphical methods[15] or the symmetries of the 3-j symbols to remove the phase factor, we obtain the result

$$\langle \gamma j_1 j_2 j \| X^k \| \gamma' j_1' j_2' j' \rangle = \left[(2j+1)(2j'+1)(2k+1) \right]^{\frac{1}{2}} \begin{Bmatrix} j_1 & j_1' & k_1 \\ j_2 & j_2' & k_2 \\ j & j' & k \end{Bmatrix}$$

$$\times \sum_{\gamma''} \langle \gamma j_1 \| T^{k_1} \| \gamma'' j_1' \rangle \langle \gamma'' j_2 \| U^{k_2} \| \gamma' j_2' \rangle \tag{5.68}$$

Everything now becomes a special case of Eq. (5.68)!

In particular, let us examine the matrix elements of the zero-rank compound tensor $X(0,0)$. Putting $k = 0$ into the 9-j symbol of Eq. (5.68) causes it to vanish unless $j = j'$ and $k_1 = k_2$. We conclude that a scalar operator (as one would find as terms in the Hamiltonian of a system) is diagonal in the total angular momentum. Moreover, such scalar operators can be constructed only from the contraction of tensor operators of the same rank. Similar conclusions have been reached previously. However, the advantage of Eq. (5.68) is that when $k = 0$, it reduces to a 6-j symbol, giving us an explicit expression for the reduced matrix element of a compound scalar operator:

$$\langle \gamma j_1 j_2 j \| X^0 \| \gamma' j_1' j_2' j' \rangle = \langle \gamma j_1 j_2 j \| [T^k \otimes U^k]_0^{(0)} \| \gamma' j_1' j_2' j' \rangle$$

$$= (-1)^{k+j_2+j+j_1'} (2k+1)^{-\frac{1}{2}} (2j+1)^{\frac{1}{2}} \begin{Bmatrix} j_1 & j_2 & j \\ j_2' & j_1' & k \end{Bmatrix}$$

$$\times \sum_{\gamma''} \langle \gamma j_1 \| T^k \| \gamma'' j_1' \rangle \langle \gamma'' j_2 \| U^k \| \gamma' j_2' \rangle \tag{5.69}$$

Often, as we remarked before, $\mathbf{T}^k \cdot \mathbf{U}^k$ is used in place of $[T^k \otimes U^k]_0^{(0)}$ as the scalar operator. Using Eq. (5.48), we rewrite Eq. (5.69) as

$$\langle \gamma j_1 j_2 j \| \mathbf{T}^k \cdot \mathbf{U}^k \| \gamma' j_1' j_2' j' \rangle = (-1)^{j_1' + j_2 + j} (2j + 1)^{\frac{1}{2}} \left\{ \begin{array}{ccc} j_1 & j_2 & j \\ j_2' & j_1' & k \end{array} \right\}$$

$$\times \sum_{\gamma''} \langle \gamma j_1 \| T^k \| \gamma'' j_1' \rangle \langle \gamma'' j_2 \| U^k \| \gamma' j_2' \rangle \quad (5.70)$$

Because of frequent occurrence of these scalar operators, we write out Eq. (5.64) in full for this case:

$$\langle \gamma j_1 j_2 j m | \mathbf{T}^k \cdot \mathbf{U}^k | \gamma' j_1' j_2' j' m' \rangle \quad (5.71)$$

$$= \delta_{jj'} \delta_{mm'} (-1)^{j_1' + j_2 + j} \left\{ \begin{array}{ccc} j_1 & j_2 & j \\ j_2' & j_1' & k \end{array} \right\} \sum_{\gamma''} \langle \gamma j_1 \| T^k \| \gamma'' j_1' \rangle \langle \gamma'' j_2 \| U^k \| \gamma' j_2' \rangle$$

We have placed a box around Eq. (5.71) because it is so useful.

Next, we examine the special case where $X(k, q)$ acts on only the variables in space 1 or 2. For example, suppose that $X(k, q)$ behaves as only $T(k_1, q_1)$. This is the same as setting $U(k_2, q_2)$ to a constant times the identity operator, that is, to $U(0, 0)$. By putting $k_2 = 0$ in Eq. (5.68), we find that once again the $9\text{-}j$ symbol collapses to a $6\text{-}j$ symbol and we must have $j_2 = j_2'$ and $k = k_1$. We find

$$\langle \gamma j_1 j_2 j \| T^{k_1} \| \gamma' j_1' j_2' j' \rangle \quad (5.72)$$

$$= \delta_{j_2 j_2'} (-1)^{j_1 + j_2 + j' + k_1} [(2j' + 1)(2j + 1)]^{\frac{1}{2}} \left\{ \begin{array}{ccc} j_1 & j & j_2 \\ j' & j_1' & k_1 \end{array} \right\} \langle \gamma j_1 \| T^{k_1} \| \gamma' j_1' \rangle$$

Similarly, if $X(k, q)$ acts only on space 2, that is, if $X(k, q) = U(k_2, q_2)$, then

$$\langle \gamma j_1 j_2 j \| U^{k_2} \| \gamma' j_1' j_2' j' \rangle \quad (5.73)$$

$$= \delta_{j_1 j_1'} (-1)^{j_1 + j_2' + j + k_2} [(2j' + 1)(2j + 1)]^{\frac{1}{2}} \left\{ \begin{array}{ccc} j_2 & j & j_1 \\ j' & j_2' & k_2 \end{array} \right\} \langle \gamma j_2 \| U^{k_2} \| \gamma' j_2' \rangle$$

Equations (5.72) and (5.73) are also placed in boxes because of the heavy use they get as they give us the reduced matrix elements of a single particle operator in a coupled representation. Note that the phase factors are different in these two equations because of the order of coupling of $\mathbf{j}_1$ and $\mathbf{j}_2$ to form $\mathbf{j}$.

Finally, there is a more general case, which is, however, not encountered as frequently as the preceding cases. This is the situation when the system is in a sharp angular momentum state $|\gamma jm\rangle$ that is not decomposable into subsystem states and we desire to find the matrix elements of the compound tensor $X(k, q) = [T^{k_1} \otimes T^{k_2}]_q^{(k)}$, where here $X(k, q)$ can act only on the set of variables of the system. The same treatment as that used in deriving Eq. (5.68) yields the result

$$
\langle \gamma j \| X^k \| \gamma' j' \rangle = \langle \gamma j \| \left[T^{(k_1)} \otimes T^{(k_2)} \right]^{(k)} \| \gamma' j' \rangle \tag{5.74}
$$

$$
= (-1)^{k+j+j'} (2k+1)^{\frac{1}{2}} \sum_{\gamma'' j''} \begin{Bmatrix} k_1 & k_2 & k \\ j' & j' & j'' \end{Bmatrix} \langle \gamma j \| T^{k_1} \| \gamma'' j'' \rangle \langle \gamma'' j'' \| T^{k_2} \| \gamma' j' \rangle
$$

We conclude this section by working out some examples chosen to illustrate the utility of this reference library [Eqs. (5.71)–(5.74)] that we have compiled. First, let us consider the effect of spin–orbit interaction on the term energies of a two-electron atom. Here electrons 1 and 2 have orbital angular momenta ℓ_1 and ℓ_2 and spin angular momenta s_1 and s_2. According to Russell–Saunders coupling, $\mathbf{L} = \ell_1 + \ell_2$, $\mathbf{S} = s_1 + s_2$, and $\mathbf{J} = \mathbf{L} + \mathbf{S}$. The spin–orbit Hamiltonian is a sum of one-electron operators of the form $H_{SO} = \sum_{i=1}^n \xi(r_i)\ell_i \cdot s_i$, and for the particular problem under consideration

$$
H_{SO} = \xi(r_1)\ell_1 \cdot s_1 + \xi(r_2)\ell_2 \cdot s_2 \tag{5.75}
$$

The general matrix elements of H_{SO} are given by applying Eq. (5.71):

$$
\langle {}^{2S+1}L_J M_J | H_{SO} | {}^{2S'+1}L'_{J'} M'_J \rangle
$$

$$
= \zeta_{n_1\ell_1;n'_1\ell'_1} \delta_{JJ'}\delta_{M_J M'_J}(-1)^{L'+S+J} \begin{Bmatrix} L & S & J \\ S' & L' & 1 \end{Bmatrix} \langle L \| \ell_1 \| L' \rangle \langle S \| s_1 \| S' \rangle
$$

$$
+ \zeta_{n_2\ell_2;n'_2\ell'_2} \delta_{JJ'}\delta_{M_J M'_J}(-1)^{L'+S+J} \begin{Bmatrix} L & S & J \\ S' & L' & 1 \end{Bmatrix} \langle L \| \ell_2 \| L' \rangle \langle S \| s_2 \| S' \rangle
$$

$$
\tag{5.76}
$$

where the radial integral over $\xi(r)$ has been abbreviated as

$$
\zeta_{n\ell;n'\ell'} = \int_0^\infty R_{n\ell}(r)\xi(r)R_{n'\ell'}(r)r^2 \, dr \tag{5.77}
$$

Next we apply Eqs. (5.72) and (5.73) to express the reduced matrix elements appearing in Eq. (5.76) in simpler form:

$$\langle L\|\ell_1\|L'\rangle \equiv \langle\ell_1\ell_2 L\|\ell_1\|\ell_1'\ell_2'L'\rangle$$

$$= \delta_{\ell_2\ell_2'}(-1)^{\ell_1+\ell_2+L'+1}\left[(2L'+1)(2L+1)\right]^{\frac{1}{2}}$$

$$\times\left\{\begin{matrix}\ell_1 & L & \ell_2 \\ L' & \ell_1' & 1\end{matrix}\right\}\langle\ell_1\|\ell_1\|\ell_1'\rangle$$

$$\langle S\|s_1\|S'\rangle \equiv \langle s_1 s_2 S\|s_1\|s_1's_2'S'\rangle$$

$$= \delta_{s_2 s_2'}(-1)^{s_1+s_2+S'+1}\left[(2S'+1)(2S+1)\right]^{\frac{1}{2}}$$

$$\times\left\{\begin{matrix}s_1 & S & s_2 \\ S' & s_1' & 1\end{matrix}\right\}\langle s_1\|s_1\|s_1'\rangle$$

$$\langle L\|\ell_2\|L'\rangle \equiv \langle\ell_1\ell_2 L\|\ell_2\|\ell_1'\ell_2'L'\rangle$$

$$= \delta_{\ell_1\ell_1'}(-1)^{\ell_1+\ell_2+L+1}\left[(2L'+1)(2L+1)\right]^{\frac{1}{2}}$$

$$\times\left\{\begin{matrix}\ell_2 & L & \ell_1 \\ L' & \ell_2' & 1\end{matrix}\right\}\langle\ell_2\|\ell_2\|\ell_2'\rangle$$

$$\langle S\|s_2\|S'\rangle \equiv \langle s_1 s_2 S\|s_2\|s_1's_2'S'\rangle$$

$$= \delta_{s_1 s_1'}(-1)^{s_1+s_2'+S+1}\left[(2S'+1)(2S+1)\right]^{\frac{1}{2}}$$

$$\times\left\{\begin{matrix}s_2 & S & s_1 \\ S' & s_2' & 1\end{matrix}\right\}\langle s_2\|s_2\|s_2'\rangle$$

$$(5.78)$$

The reduced matrix elements in Eq. (5.78) have been evaluated before [see Eqs. (5.22) and (5.23)]. Making all these substitutions, we find

$$\left\langle {}^{2S+1}L_J M_J\right| H_{SO} \left|{}^{2S'+1}L'_{J'} M'_J\right\rangle$$

$$= \zeta_{n_1\ell_1;n_1'\ell_1'}\,\delta_{JJ'}\delta_{M_J M_J'}\delta_{\ell_1\ell_1'}\delta_{\ell_2\ell_2'}(-1)^{S+S'+J+\ell_1+\ell_2+1}$$

$$\times\left[(2L'+1)(2L+1)(2S'+1)(2S+1)(2\ell_1+1)\ell_1(\ell_1+1)(3/2)\right]^{\frac{1}{2}}$$

$$\times \left\{ \begin{matrix} L & S & J \\ S' & L' & 1 \end{matrix} \right\} \left\{ \begin{matrix} \ell_1 & L & \ell_2 \\ L' & \ell_1 & 1 \end{matrix} \right\} \left\{ \begin{matrix} \frac{1}{2} & S & \frac{1}{2} \\ S' & \frac{1}{2} & 1 \end{matrix} \right\}$$

$$+ \zeta_{n_2 \ell_2 ; n_2' \ell_2'} \, \delta_{JJ'} \delta_{M_J M_J'} \delta_{\ell_1 \ell_1'} \delta_{\ell_2 \ell_2'} (-1)^{2S+L+L'+J+\ell_1+\ell_2+1}$$

$$\times \left[(2L'+1)(2L+1)(2S'+1)(2S+1)(2\ell_2+1)\ell_2(\ell_2+1)(3/2) \right]^{\frac{1}{2}}$$

$$\times \left\{ \begin{matrix} L & S & J \\ S' & L' & 1 \end{matrix} \right\} \left\{ \begin{matrix} \ell_2 & L & \ell_1 \\ L' & \ell_2 & 1 \end{matrix} \right\} \left\{ \begin{matrix} \frac{1}{2} & S & \frac{1}{2} \\ S' & \frac{1}{2} & 1 \end{matrix} \right\} \qquad (5.79)$$

where we have used the fact that s_1, s_2, s_1', and s_2' all equal $\frac{1}{2}$.

Examination of Eq. (5.79) permits us to make a number of remarks regarding the nature of spin–orbit interaction. We see that H_{SO} can connect states of only the same total angular momentum J and projection of the total angular momentum M_J. Moreover, H_{SO} is independent of M_J and cannot split the $(2J+1)$-fold spatial degeneracy associated with each J level. The spin–orbit operator is a one-electron operator and cannot connect the terms arising from configurations that differ by two spin-orbitals. This has the simple consequence that for the $\zeta_{n_1 \ell_1 ; n_1' \ell_1'}$ term all quantum numbers associated with electron 2 must be identical while for the $\zeta_{n_2 \ell_2 ; n_2' \ell_2'}$ term, the same condition holds for electron 1. Finally, the leading $6\text{-}j$ symbol of each term vanishes unless the following two angular momentum addition equalities hold: $\mathbf{L} + \mathbf{1} = \mathbf{L'}$ and $\mathbf{S} + \mathbf{1} = \mathbf{S'}$; in other words, $\Delta(L1L')$ and $\Delta(S1S')$ must be satisfied. This yields the selection rules for spin–orbit interaction:

$$\Delta S = 0, \pm 1, \qquad\qquad \Delta L = 0, \pm 1$$

These restrictions can be augmented further by considering parity. Since H_{SO} is invariant under spatial coordinate inversion, the spin–orbit operator can connect only levels of the same parity. Let us go one step further. Consider the form of the $\zeta_{n_i \ell_i ; n_i' \ell_i'}$ terms in Eq. (5.79) when $\ell_i = \ell_i' = 0$, where $i = 1$ or 2. The $\ell_i^{1/2}$ factor shows at once that these terms vanish. Hence ns spin–orbitals cannot contribute to the spin–orbit splitting; that is, there must be a nonzero orbital angular momentum (current loop) for the electron to have a relativistic (magnetic) interaction with its own spin. Viewed from the rest frame of the electron, the current loop results from the nucleus orbiting the electron with an effective charge which depends on the distance between the electron and the nucleus. Hence, the magnitude of the current loop increases with Z for the same ℓ value and is greater for $\ell \neq 0$ electrons which penetrate the core.

In general one must diagonalize a secular matrix with the same J value to obtain the different term energies $E(^{2S+1}L_J)$. However, when the spin–orbit interaction is small compared to the separation between different terms having the same J, we may include the effects of spin–orbit coupling by using first-order perturbation theory. For the diagonal term $\langle LSJM_J| H_{SO} |LSJM_J\rangle$ the leading $6\text{-}j$ symbol of each $\zeta_{n\ell} = \zeta_{n\ell;n\ell}$ term becomes

$$\begin{Bmatrix} L & S & J \\ S & L & 1 \end{Bmatrix} = (-1)^{S+L+J}\frac{J(J+1)-S(S+1)-L(L+1)}{2[S(S+1)(2S+1)L(L+1)(2L+1)]^{\frac{1}{2}}} \quad (5.80)$$

Furthermore, for a given $^{2S+1}L_J$ level the values of S and L are fixed as well as those of s_1, s_2, ℓ_1, and ℓ_2. Hence the diagonal term depends only on J through the $6\text{-}j$ symbol given in Eq. (5.80). Thus we can write

$$H_{SO} = A\mathbf{L}\cdot\mathbf{S} = \frac{A}{2}[J(J+1)-S(S+1)-L(L+1)] \quad (5.81)$$

where the spin–orbit constant A is the same for all J levels of a ^{2S+1}L term. This treatment yields the Landé interval rule[14]

$$E(^{2S+1}L_J) - E(^{2S+1}L_{J-1}) = AJ \quad (5.82)$$

that is, the separation is proportional to J.

It is worth remarking that when $S = 0$ (singlet term) or $L = 0$ (S term), there is only one J level associated with the ^{2S+1}L term. Spin–orbit interaction cannot split this level, but its position relative to the other levels can be shifted by the existence of nonzero off-diagonal matrix elements of H_{SO} connecting this level with other levels of the same J.

Please also note that the replacement of H_{SO} by $A\mathbf{L}\cdot\mathbf{S}$ is valid only for calculating the diagonal matrix elements of H_{SO}. For example, $\mathbf{L}\cdot\mathbf{S}$ cannot connect states of different multiplicity; that is, singlet and triplet states are mixed by spin–orbit interaction but not by $A\mathbf{L}\cdot\mathbf{S}$.

For the special case of the $(n\ell)^2$ configuration having two equivalent electrons, the diagonal elements have the form

$$\langle (n\ell)^2 \,^{2S+1}L_JM_J| H_{SO} |(n\ell)^2 \,^{2S+1}L_JM_J\rangle$$

$$= 2\zeta_{n\ell}(-1)^{2S+J+1}(2L+1)(2S+1)\left[\ell(\ell+1)(2\ell+1)(3/2)\right]^{\frac{1}{2}}$$

$$\times \begin{Bmatrix} L & S & J \\ S & L & 1 \end{Bmatrix}\begin{Bmatrix} \ell & L & \ell \\ L & \ell & 1 \end{Bmatrix}\begin{Bmatrix} \frac{1}{2} & S & \frac{1}{2} \\ S & \frac{1}{2} & 1 \end{Bmatrix} \quad (5.83)$$

We see that only one spin–orbit parameter $\zeta_{n\ell}$ characterizes all the J-level splittings of the levels arising within a ^{2S+1}L term. This conclusion is also true for the configuration $(n\ell)^p$ of p identical electrons.

As a final example, we consider the following resonance fluorescence process:

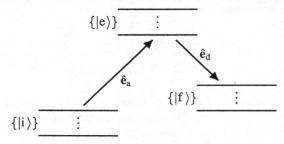

Here radiation of polarization $\hat{\mathbf{e}}_a$ is absorbed by the initial states $\{|i\rangle\}$ of an atomic or molecular system causing an electric dipole transition to the excited states $\{|e\rangle\}$, which then decay to the final states $\{|f\rangle\}$ by an electric dipole transition in which the reemitted radiation is detected with polarization $\hat{\mathbf{e}}_d$. We wish to develop a general expression for an arbitrary excitation–detection geometry and for arbitrary polarizations $\hat{\mathbf{e}}_a$ and $\hat{\mathbf{e}}_d$. Then we will specialize this to the common situation where the observation direction is at right angles to the excitation direction and the incident and detected light are both linearly polarized.

Clearly, the resonance fluorescence process is an involved one and several simplifying assumptions are required to make this problem tractable. Let all states be characterized by sharp angular momentum quantum numbers. If we assume no hyperfine structure, we may write $|i\rangle = |\alpha_i J_i M_i\rangle$, $|e\rangle = |\alpha_e J_e M_e\rangle$, and $|f\rangle = |\alpha_f J_f M_f\rangle$. We imagine at time $t = 0$ a short pulse of light is incident on our sample which is at sufficiently low pressure so that we can treat it as isolated, that is, that we can ignore the effect of collisions. We can derive an expression for the intensity at subsequent time t, denoted by $I(t)$, assuming that the following conditions are met:

1. The spectral width of the light pulse is much greater than the reciprocal of the duration of the pulse. Then the resonance fluorescence process may be treated as two independent steps, absorption followed by fluorescence.

2. The excitation process is sufficiently weak; thus the natural lifetime for the excited system to decay is much less than the average time between two successive photon absorptions by the system. This condition ensures that we can neglect optical pumping effects.

3. The duration of the pulse is short compared to characteristic "precession" times of the system, that is, compare to the reciprocal of Bohr frequencies corresponding to excited-state energy differences.

4. The initial state is isotropic, that is, has not been prepared with a preferential population in any of the M_i sublevels nor with special phase relations (coherence) among these Zeeman components.

5. The radiative lifetimes of all excited-state sublevels are the same. In what follows it will be denoted by $\tau_e = \Gamma_e^{-1}$.

Before the light pulse arrives, the wave function of the atom or molecule in some particular initial state $\alpha_i J_i M_i$ with energy $E(\alpha_i J_i M_i)$ is

$$|\psi(\alpha_i J_i M_i; t < 0)\rangle = |\alpha_i J_i M_i\rangle \exp\left[-i\omega_{M_i} t\right] \qquad (5.84)$$

where $\omega_{M_i} = E(\alpha_i J_i M_i)/\hbar$. At time $t = 0$ the light pulse with polarization $\hat{\mathbf{e}}_a$ causes excitation of the system. The wave function of the excited state is proportional to

$$|\psi(\alpha_e J_e M_e; t = 0)\rangle \propto \sum_{M_e} \langle \alpha_e J_e M_e | \hat{\mathbf{e}}_a \cdot \mathbf{r} | \alpha_i J_i M_i\rangle |\alpha_e J_e M_e\rangle \qquad (5.85)$$

where $\mathbf{r}$ is the electric dipole moment operator and $\hat{\mathbf{e}}_a$ is a unit vector having in general complex components that specify the polarization of the electric field of the incident radiation. The summation over M_e takes into account that more than one excited-state sublevel can be excited from a given ground state sublevel M_i, depending on the nature of the polarization of the incident light beam.

The excited state evolves in time according to

$$|\psi(\alpha_e J_e M_e; t)\rangle \propto \sum_{M_e} \langle \alpha_e J_e M_e | \hat{\mathbf{e}}_a \cdot \mathbf{r} | \alpha_i J_i M_i\rangle |\alpha_e J_e M_e\rangle \exp\left[-i\omega_{M_e} t - \Gamma_e t/2\right]$$

$$(5.86)$$

where $\omega_{M_e} = E(\alpha_e J_e M_e)/\hbar$ and spontaneous emission is taken into account by inclusion of the damping factor $\exp[-\Gamma_e t/2]$, where $\Gamma_e = 1/\tau_e$ is the rate of natural radiative decay. The factor of $\frac{1}{2}$ arises because the excited-state population, which is proportional to the square of the excited-state wave function, decays at the rate Γ_e. Suppose at time t that the excited state emits a photon of polarization $\hat{\mathbf{e}}_d$ as it makes a spontaneous transition to the final state $\alpha_f J_f M_f$. Then the instantaneous rate of fluorescence with that polarization is proportional to

$$I(t) \propto \sum_{M_f} |\langle \psi(\alpha_e J_e M_e; t) | \hat{\mathbf{e}}_d \cdot \mathbf{r} | \alpha_f J_f M_f\rangle|^2 \qquad (5.87)$$

where the sum is over all possible final states to which the excited state might radiate.

In the foregoing we restricted ourselves to excitation of one particular sublevel of the initial state. However, for an isotropic distribution of initial states we may include the contribution of all of them by simply rewriting Eq. (5.87) as

$$I(t) \propto \sum_{M_i} \sum_{M_f} |\langle \psi(\alpha_e J_e M_e; t) | \hat{\mathbf{e}}_d \cdot \mathbf{r} | \alpha_f J_f M_f\rangle|^2 \qquad (5.88)$$

Next we make use of the identity

$$\left| \sum_{m_a} \langle a m_a | O | b m_b \rangle \right|^2 = \sum_{m_a} \sum_{m_a'} \langle a m_a | O | b m_b \rangle^* \langle a m_a' | O | b m_b \rangle$$

$$= \sum_{m_a} \sum_{m_a'} \langle bm_b| O^\dagger |am_a\rangle \langle am_a'| O |bm_b\rangle \qquad (5.89)$$

where O is an operator and $O^\dagger$ is its adjoint. Substitution of Eq. (5.86) in (5.88) and use of Eq. (5.89) yields the result

$$I(t) \propto \sum_{M_i} \sum_{M_f} \sum_{M_e, M_e'} \langle \alpha_e J_e M_e| \hat{\mathbf{e}}_a \cdot \mathbf{r} |\alpha_i J_i M_i\rangle \langle \alpha_i J_i M_i| \hat{\mathbf{e}}_a^* \cdot \mathbf{r} |\alpha_e J_e M_e'\rangle$$

$$\times \langle \alpha_e J_e M_e'| \hat{\mathbf{e}}_d \cdot \mathbf{r} |\alpha_f J_f M_f\rangle \langle \alpha_f J_f M_f| \hat{\mathbf{e}}_d^* \cdot \mathbf{r} |\alpha_e J_e M_e\rangle$$

$$\times \exp\left[-i\left(\omega_{M_e} - \omega_{M_e'}\right)t - \Gamma_e t\right] \qquad (5.90)$$

This displays the expected (see Application 9) time behavior of an exponential decay on which is superimposed quantum beats from the interference of the indistinguishable different paths through the sublevels of the excited state that connect the same initial and final sublevels[16]. The total rate is found by integrating Eq. (5.90) over all time:

$$I \propto \int_0^\infty I(t)\,dt$$

$$\propto \sum_{M_i} \sum_{M_f} \sum_{M_e, M_e'} \frac{1}{\Gamma_e + i\left(\omega_{M_e} - \omega_{M_e'}\right)}$$

$$\times \langle \alpha_e J_e M_e| \hat{\mathbf{e}}_a \cdot \mathbf{r} |\alpha_i J_i M_i\rangle \langle \alpha_i J_i M_i| \hat{\mathbf{e}}_a^* \cdot \mathbf{r} |\alpha_e J_e M_e'\rangle$$

$$\times \langle \alpha_e J_e M_e'| \hat{\mathbf{e}}_d \cdot \mathbf{r} |\alpha_f J_f M_f\rangle \langle \alpha_f J_f M_f| \hat{\mathbf{e}}_d^* \cdot \mathbf{r} |\alpha_e J_e M_e\rangle \qquad (5.91)$$

In terms of quantum electrodynamics[17], $\hat{\mathbf{e}} \cdot \mathbf{r}$ corresponds to the photon annihilation operator and $\hat{\mathbf{e}}^* \cdot \mathbf{r}$ to the photon creation operator, that is, $\hat{\mathbf{e}}^* \cdot \mathbf{r}$ always appears in front of $|\alpha_e J_e M_e\rangle$ or $|\alpha_e J_e M_e'\rangle$, while $\hat{\mathbf{e}} \cdot \mathbf{r}$ appears in front of $|\alpha_i J_i M_i\rangle$ or $|\alpha_f J_f M_f\rangle$. Note that the denominator of Eq. (5.91) depends on the Zeeman sublevel energy splitting of the excited state and the natural radiative lifetime. Equation (5.91) was derived by Breit in 1933 and is often referred to as the *Breit level-crossing formula*[18].

Suppose that an external field (magnetic or electric) is applied along the axis of quantization (the Z axis) of the system. Then the splittings between the magnetic sublevels can be made to change. There are two limiting cases of much interest. If the excited-state magnetic sublevels are split apart much larger than the natural linewidth, that is, if $(\omega_{M_e} - \omega_{M_e'}) \gg \Gamma_e$, all terms involving $M_e \neq M_e'$ vanish and the resonance fluorescence signal is proportional to the simple expression

$$I(\text{high field limit}) \propto \sum_{M_i} \sum_{M_f} \sum_{M_e} |\langle \alpha_i J_i M_i| \hat{\mathbf{e}}_a \cdot \mathbf{r} |\alpha_e J_e M_e\rangle|^2$$

$$\times \ |\langle \alpha_e J_e M_e | \, \hat{\mathbf{e}}_d \cdot \mathbf{r} \, | \alpha_f J_f M_f \rangle|^2 \tag{5.92}$$

where each level $\alpha_e J_e M_e$ contributes *independently* (i.e., incoherently). The other extreme is represented by $(\omega_{M_e} - \omega_{M'_e}) \ll \Gamma_e$ for example, in zero external field. Then the resonance fluorescence signal is proportional to

$$I(\text{zero field}) \propto \sum_{M_i} \sum_{M_f} \sum_{M_e, M'_e} \langle \alpha_e J_e M_e | \, \hat{\mathbf{e}}_a \cdot \mathbf{r} \, | \alpha_i J_i M_i \rangle \langle \alpha_i J_i M_i | \, \hat{\mathbf{e}}_a^* \cdot \mathbf{r} \, | \alpha_e J_e M'_e \rangle$$

$$\times \ \langle \alpha_e J_e M'_e | \, \hat{\mathbf{e}}_d \cdot \mathbf{r} \, | \alpha_f J_f M_f \rangle \langle \alpha_f J_f M_f | \, \hat{\mathbf{e}}_d^* \cdot \mathbf{r} \, | \alpha_e J_e M_e \rangle$$

$$\tag{5.93}$$

which has interference terms (cross terms) among the different M_e, M'_e sublevels provided that they are coherently excited in the resonance fluorescence process (zero-field level crossing). Hence as we pass from low to high external field, this coherence is destroyed and a study of the polarization intensity as a function of field strength allows us to determine excited-state splittings with a resolution comparable to the natural linewidth.

If desired, it is actually possible to achieve resolution within the natural width! This is accomplished by biasing the detection of the fluorescence to favor those excited species that have lived longer than the radiative lifetime, for example, by having the detection system accept signal only after a delay which exceeds the radiative lifetime[19]. The narrowing of the "resonances" below the natural width does not violate the uncertainty principle because the uncertainty principle only expresses the Fourier transform pair between mean decay time and spectral linewidth. However, the gain in resolution must in practice be balanced against the loss in signal-to-noise ratio of the fluorescence, caused by the resulting decrease in the number of fluorescence events detected at longer delays.

While there are numerous uses of this coherence sub-Doppler spectroscopy, we turn our attention to the calculation of the degree of polarization in the absence of external fields. Equation (5.93) is the starting point. It may be rewritten as

$$I \propto \sum_{M_e, M'_e} A_{M_e M'_e} F_{M'_e M_e} \tag{5.94}$$

where the absorption and fluorescence matrix elements are

$$A_{M_e M'_e} = \sum_{M_i} \langle \alpha_e J_e M_e | \, \hat{\mathbf{e}}_a \cdot \mathbf{r} \, | \alpha_i J_i M_i \rangle \langle \alpha_i J_i M_i | \, \hat{\mathbf{e}}_a^* \cdot \mathbf{r} \, | \alpha_e J_e M'_e \rangle$$

$$= \langle \alpha_e J e M_e | \, (\hat{\mathbf{e}}_a \cdot \mathbf{r}) \, P_i \, (\hat{\mathbf{e}}_a^* \cdot \mathbf{r}) \, | \alpha_e J_e M'_e \rangle \tag{5.95}$$

and

$$
F_{M'_e M_e} = \sum_{M_f} \langle \alpha_e J_e M'_e | \, \hat{\mathbf{e}}_d \cdot \mathbf{r} \, | \alpha_f J_f M_f \rangle \langle \alpha_f J_f M_f | \, \hat{\mathbf{e}}_d^* \cdot \mathbf{r} \, | \alpha_e J_e M_e \rangle
$$

$$
= \langle \alpha_e J_e M'_e | \, (\hat{\mathbf{e}}_d \cdot \mathbf{r}) \, P_f \, (\hat{\mathbf{e}}_d^* \cdot \mathbf{r}) \, | \alpha_e J_e M_e \rangle \tag{5.96}
$$

respectively, and we have introduced the projection operators

$$
P_i = \sum_{M_i} | \alpha_i J_i M_i \rangle \langle \alpha_i J_i M_i | \tag{5.97}
$$

and

$$
P_f = \sum_{M_f} | \alpha_f J_f M_f \rangle \langle \alpha_f J_f M_f | \tag{5.98}
$$

The projection operators $\mathbf{P}$ are readily verified to be invariant under rotation, that is, $\mathbf{R}^{-1}\mathbf{P}\mathbf{R} = \mathbf{P}$, and we can ignore these scalars in our tensorial analysis. Hence we are concerned with operators of the form $(\hat{\mathbf{e}} \cdot \mathbf{r})(\hat{\mathbf{e}}^* \cdot \mathbf{r})$ in Eqs. (5.95) and (5.96). This has the tensor structure indicated in Eq. (5.52):

$$
(\hat{\mathbf{e}} \cdot \mathbf{r})(\hat{\mathbf{e}}^* \cdot \mathbf{r}) = \sum_{k,q} (-1)^{k-q} \left[e^{(1)} \otimes e^{*(1)} \right]_q^{(k)} \left[r^{(1)} \otimes r^{(1)} \right]_{-q}^{(k)} \tag{5.99}
$$

which conveniently separates into the scalar contraction of two tensors, the polarization tensor[20, 21]

$$
E_q^k(\hat{\mathbf{e}}, \hat{\mathbf{e}}^*) = \left[e^{(1)} \otimes e^{*(1)} \right]_q^{(k)}
$$

$$
= \sum_{\mu} (-1)^q (2k+1)^{\frac{1}{2}} e(1,\mu) e^*(1, q-\mu) \begin{pmatrix} 1 & 1 & k \\ \mu & q-\mu & -q \end{pmatrix}
$$

$$
\tag{5.100}
$$

which contains only variables concerning the electric field direction and the tensor $\left[r^{(1)} \times r^{(1)} \right]_q^{(k)}$, which contains operators that act on the atomic or molecular system. The nine components of the polarization tensor $E_q^k(\hat{\mathbf{e}}, \hat{\mathbf{e}}^*)$ for $k = 0$, 1, and 2 are readily worked out using Eqs. (5.41)–(5.44).

Let us illustrate the evaluation of the absorption matrix element $A_{M_e M'_e}$ in rather complete detail. Using Eqs. (5.99) and (5.100) and applying the Wigner–Eckart theorem, we have

$$
A_{M_e M'_e} = \langle \alpha_e J_e M_e | \sum_{k,q} (-1)^{k-q} \left[(\hat{\mathbf{e}}_a)^{(1)} \otimes (\hat{\mathbf{e}}_a^*)^{(1)} \right]_q^{(k)} P_i
$$

$$\times \left[r^{(1)} \otimes r^{(1)} \right]^{(k)}_{-q} |\alpha_e J_e M'_e\rangle$$

$$= \sum_{k,q} (-1)^{k-q} E^k_q(\hat{\mathbf{e}}_a, \hat{\mathbf{e}}^*_a) \langle \alpha_e J_e M_e | \left[r^{(1)} \otimes r^{(1)} \right]^{(k)}_{-q} P_i |\alpha_e J_e M'_e\rangle$$

$$= \sum_{k,q} (-1)^{k-q+J_e-M_e} E^k_q(\hat{\mathbf{e}}_a, \hat{\mathbf{e}}^*_a) \begin{pmatrix} J_e & k & J_e \\ -M_e & -q & M'_e \end{pmatrix}$$

$$\times \langle J_e || \left[r^{(1)} \otimes r^{(1)} \right]^{(k)} P_i || J_e \rangle \tag{5.101}$$

Next we use our trusty library of results, namely, Eq. (5.74), to express $A_{M_e M'_e}$ as

$$A_{M_e M'_e} = \sum_{k,q} (-1)^{k-q+J_e-M_e} E^k_q(\hat{\mathbf{e}}_a, \hat{\mathbf{e}}^*_a) \begin{pmatrix} J_e & k & J_e \\ -M_e & -q & M'_e \end{pmatrix}$$

$$\times (-1)^{k+J_e+J_e} (2k+1)^{\frac{1}{2}} \begin{Bmatrix} 1 & 1 & k \\ J_e & J_e & J_i \end{Bmatrix}$$

$$\times \langle J_e || r^{(1)} || J_i \rangle \langle J_i || r^{(1)} || J_e \rangle$$

$$= \sum_{k,q} (2k+1)^{\frac{1}{2}} E^k_q(\hat{\mathbf{e}}_a, \hat{\mathbf{e}}^*_a) (-1)^{J_e-M_e-q} \begin{pmatrix} J_e & k & J_e \\ -M_e & -q & M'_e \end{pmatrix}$$

$$\times (-1)^{J_e+J_i} \left| \langle J_e || r^{(1)} || J_i \rangle \right|^2 \begin{Bmatrix} 1 & 1 & k \\ J_e & J_e & J_i \end{Bmatrix} \tag{5.102}$$

where the summation over j'' in Eq. (5.74) is restricted to $j'' = J_i$ because of the presence of the projection operator P_i, and where we have also used Eq. (5.35) to express the result as involving the square of the reduced matrix element of the electric dipole operator.

By similar means,

$$F_{M'_e M_e} = \sum_{k',q'} (2k'+1)^{\frac{1}{2}} E^{k'}_{q'}(\hat{\mathbf{e}}_d, \hat{\mathbf{e}}^*_d) (-1)^{J_e-M'_e-q'} \begin{pmatrix} J_e & k' & J_e \\ -M'_e & -q' & M_e \end{pmatrix}$$

$$\times (-1)^{J_e+J_f} \left| \langle J_e || r^{(1)} || J_f \rangle \right|^2 \begin{Bmatrix} 1 & 1 & k' \\ J_e & J_e & J_f \end{Bmatrix} \tag{5.103}$$

Then Eq. (5.94) contains

$$\sum_{M_e, M'_e} (-1)^{J_e-M'_e+J_e-M_e} \begin{pmatrix} J_e & k & J_e \\ -M_e & -q & M'_e \end{pmatrix} \begin{pmatrix} J_e & k' & J_e \\ -M'_e & -q' & M_e \end{pmatrix}$$

$$= (-1)^q \delta_{kk'} \delta_{-qq'} / (2k+1) \qquad (5.104)$$

where we have used Eq. (2.32) to simplify its form. Hence

$$I \propto \left| \langle J_i \| \tau^{(1)} \| J_e \rangle \right|^2 \left| \langle J_e \| \tau^{(1)} \| J_f \rangle \right|^2$$

$$\times \sum_{k,q} (-1)^q E_q^k(\hat{\mathbf{e}}_a, \hat{\mathbf{e}}_a^*) E_{-q}^k(\hat{\mathbf{e}}_d, \hat{\mathbf{e}}_d^*) a^k(J_i, J_e, J_f) \qquad (5.105)$$

where

$$a^k(J_i, J_e, J_f) = (-1)^{J_i + J_e + J_f + J_e} \begin{Bmatrix} 1 & 1 & k \\ J_e & J_e & J_i \end{Bmatrix} \begin{Bmatrix} 1 & 1 & k \\ J_e & J_e & J_f \end{Bmatrix} \qquad (5.106)$$

The simple form of Eqs. (5.105) and (5.106) displayed above is a lovely example of the power of the tensor method. It is a key result[17, 22, 23] in calculating the resonance fluorescence intensity under general excitation–detection conditions.

Let us specialize to the case of linearly polarized excitation–detection, where $\hat{\mathbf{e}}_a$ and $\hat{\mathbf{e}}_d$ are unit vectors pointing along the polarization directions of the exciting and detected beams, taken to be the quantization axes of the tensors $E_q^k(\hat{\mathbf{e}}_a, \hat{\mathbf{e}}_a^*)$ and $E_q^k(\hat{\mathbf{e}}_d, \hat{\mathbf{e}}_d^*)$, respectively. Then the only nonvanishing components of both tensors are those for $q = 0$. It follows from Eq. (5.100) that we can write

$$E_0^k(\hat{\mathbf{e}}_a, \hat{\mathbf{e}}_a^*) = E_0^k(\hat{\mathbf{e}}_d, \hat{\mathbf{e}}_d^*) = (2k+1)^{\frac{1}{2}} \begin{pmatrix} 1 & 1 & k \\ 0 & 0 & 0 \end{pmatrix} \qquad (5.107)$$

However, the quantization axes of the two polarization tensors differ in general because $\hat{\mathbf{e}}_a$ does not necessarily coincide with $\hat{\mathbf{e}}_d$ but instead makes the angle θ. We refer the detection polarization tensor to the same axis as that of the absorption polarization tensor by carrying out the rotation $R = (0, \theta, 0)$, that is,

$$E_0^k(\hat{\mathbf{e}}_d, \hat{\mathbf{e}}_d^*) = \sum_q D_{q0}^k(0, \theta, 0) E_q^k(\hat{\mathbf{e}}_d, \hat{\mathbf{e}}_d^*)$$

$$= D_{00}^k(0, \theta, 0) E_0^k(\hat{\mathbf{e}}_d, \hat{\mathbf{e}}_d^*)$$

$$= (2k+1)^{\frac{1}{2}} \begin{pmatrix} 1 & 1 & k \\ 0 & 0 & 0 \end{pmatrix} P_k(\cos\theta) \qquad (5.108)$$

Thus for a linearly polarized excitation–detection geometry

$$I \propto \left| \langle J_i \| \tau^{(1)} \| J_e \rangle \right|^2 \left| \langle J_e \| \tau^{(1)} \| J_f \rangle \right|^2 (-1)^{J_i + J_e + J_f + J_e}$$

$$\times \sum_k (2k+1) \begin{pmatrix} 1 & 1 & k \\ 0 & 0 & 0 \end{pmatrix}^2 \begin{Bmatrix} 1 & 1 & k \\ J_e & J_e & J_i \end{Bmatrix} \begin{Bmatrix} 1 & 1 & k \\ J_e & J_e & J_f \end{Bmatrix} P_k(\cos\theta)$$

$$(5.109)$$

where the summation over k is restricted to three terms, $k = 0$, $k = 1$, and $k = 2$. Equation (5.109) shows that the angular dependence of the resonance fluorescence has a very simple form: the $k = 0$ term is isotropic, the $k = 1$ term vanishes because

$$\begin{pmatrix} 1 & 1 & 1 \\ 0 & 0 & 0 \end{pmatrix}$$

is zero, and the $k = 2$ term varies as $(3\cos^2\theta - 1)/2$. Thus by choosing θ so that the $k = 2$ also vanishes, that is, by choosing $\theta = \arccos(1/3)^{1/2} = 54.7°$, the radiation reaching the detector is *independent* of the linear polarization directions $\hat{e}_a$ and $\hat{e}_d$. This choice of θ is called the "magic angle."

Now consider the traditional case of a right-angle excitation–detection geometry in which the viewing direction is perpendicular to the direction of propagation of the linearly polarized excitation beam. It is convenient to distinguish two cases. In the first, $\hat{e}_a$ is parallel to $\hat{e}_d$, $\theta = 0°$, and the resonance fluorescence signal is denoted by $I_{\parallel}$; in the second, $\hat{e}_a$ is perpendicular to $\hat{e}_d$, $\theta = \pi/2$, and the resonance fluorescence signal is denoted by $I_{\perp}$. Then one describes the polarization by the degree of polarization

$$P = \frac{I_{\parallel} - I_{\perp}}{I_{\parallel} + I_{\perp}} \qquad (5.110)$$

or by the polarization anisotropy

$$R = \frac{I_{\parallel} - I_{\perp}}{I_{\parallel} + 2I_{\perp}} \qquad (5.111)$$

These two measures are related by

$$P = \frac{3R}{2 + R} \qquad (5.112)$$

and

$$R = \frac{2P}{3 - P} \qquad (5.113)$$

Using Eq. (5.109), we obtain for the orthogonal excitation–detection geometry ($\theta = \pi/2$) the particularly pleasing expression

$$R = \frac{\begin{Bmatrix} 1 & 1 & 2 \\ J_e & J_e & J_i \end{Bmatrix} \begin{Bmatrix} 1 & 1 & 2 \\ J_e & J_e & J_f \end{Bmatrix}}{\begin{Bmatrix} 1 & 1 & 0 \\ J_e & J_e & J_i \end{Bmatrix} \begin{Bmatrix} 1 & 1 & 0 \\ J_e & J_e & J_f \end{Bmatrix}} \qquad (5.114)$$

TABLE 5.1   Resonance Fluorescence Process; Linear Polarization

Transition	$R(J_i)$	$P(J_i)$
$(Q\uparrow, Q\downarrow)$	$\dfrac{(2J_i - 1)(2J_i + 3)}{10J_i(J_i + 1)}$	$\dfrac{(2J_i - 1)(2J_i + 3)}{8J_i^2 + 8J_i - 1}$
$(R\uparrow, R\downarrow)$	$\dfrac{(J_i + 2)(2J_i + 5)}{10(J_i + 1)(2J_i + 1)}$	$\dfrac{(J_i + 2)(2J_i + 5)}{14J_i^2 + 23J_i + 10}$
$(P\uparrow, P\downarrow)$	$\dfrac{(J_i - 1)(2J_i - 3)}{10J_i(2J_i + 1)}$	$\dfrac{(J_i - 1)(2J_i - 3)}{14J_i^2 + 5J_i + 1}$
$\left.\begin{array}{c}(R\uparrow, P\downarrow)\\ \text{or}\\ (P\uparrow, R\downarrow)\end{array}\right\}$	$\dfrac{1}{10}$	$\dfrac{1}{7}$
$(Q\uparrow, R\downarrow)$	$-\dfrac{2J_i + 3}{10J_i}$	$-\dfrac{2J_i + 3}{6J_i - 1}$
$(Q\uparrow, P\downarrow)$	$-\dfrac{2J_i - 1}{10(J_i + 1)}$	$-\dfrac{2J_i - 1}{6J_i + 7}$
$(R\uparrow, Q\downarrow)$	$-\dfrac{2J_i + 5}{10(J_i + 1)}$	$-\dfrac{2J_i + 5}{6J_i + 5}$
$(P\uparrow, Q\downarrow)$	$-\dfrac{2J_i - 3}{10J_i}$	$-\dfrac{2J_i - 3}{6J_i + 1}$

This is readily evaluated for the nine possible branches $P\uparrow$, $Q\uparrow$, or $R\uparrow$ on excitation and $P\downarrow$, $Q\downarrow$, or $R\downarrow$ on fluorescence, where $\Delta J = -1, 0$, or $+1$ is identified with P, Q, or R, respectively. (Here $\Delta J = J_e - J_i$ on excitation and $\Delta J = J_e - J_f$ on fluorescence). Algebraic expressions for the polarization are collected in Table 5.1 for the different branches by evaluating Eq. (5.114) for R and using Eq. (5.112) to reexpress R in terms of P. These expressions have been found previously[24–26] but by much more cumbersome means. As J_i increases, the results in Table 5.1 approach the classical limiting forms found previously in Application 8, specifically: $R = 2/5$ ($P = 1/2$) for $(Q\uparrow, Q\downarrow)$; $R = 1/10$ ($P = 1/7$) for $(P\uparrow, P\downarrow)$, $(P\uparrow, R\downarrow)$, $(R\uparrow, P\downarrow)$, or $(R\uparrow, R\downarrow)$; and $R = -1/5$ ($P = -1/3$) for $(Q\uparrow, P\downarrow)$, $(Q\uparrow, R\downarrow)$, $(P\uparrow, Q\downarrow)$, or $(R\uparrow, Q\downarrow)$. The effects of hyperfine structure are easily included[17, 22, 23], easy that is, when the tensor methods of this section are used!

Another variant of the resonance fluorescence process occurs when the incident radiation is circularly polarized, with either left- or right-hand helicity, and the detected light is analyzed for circular polarization[24, 27]. Here the propagation direction of the incident and detected light beams may be chosen to define the Z axis. This choice has the virtue that once again only one component of the polarization vector expressed in the spherical basis set is nonvanishing, either $e(1, 1)$ for left circularly polarized light or $e(1, -1)$ for right circularly polarized light.

Recall that the electric field vector $\mathbf{E}$ of a monochromatic wave of frequency ω and wave vector $\mathbf{k} = 2\pi/\lambda\hat{\mathbf{n}}$, where $\hat{\mathbf{n}}$ is a unit vector in the direction of motion, can be written in the form

$$\mathbf{E}(t) = A\hat{\mathbf{e}}\exp[i(\mathbf{k}\cdot\mathbf{r} - \omega t)] \qquad (5.115)$$

where A is the amplitude and $\hat{\mathbf{e}}$ is a unit vector called the *polarization vector*. Because of the transverse nature of electromagnetic waves, $\hat{\mathbf{e}}$ is perpendicular to $\hat{\mathbf{n}}$. Hence the choice of $\hat{\mathbf{n}}$ to be along the Z axis confines $\hat{\mathbf{e}}$ to lie in the XY plane. For light in a pure state of arbitrary polarization propagating along the Z axis, its polarization vector may be represented by

$$\hat{\mathbf{e}}(\beta, \delta) = \cos\beta\hat{\mathbf{X}} + e^{i\delta}\sin\beta\hat{\mathbf{Y}} \qquad (5.116)$$

where β ranges from 0 to π and δ is the phase necessary to describe elliptical polarization. For example, for linear polarization $\delta = 0$ and β is the angle between the polarization vector and the X axis. Circularly polarized light corresponds to equal components of $\hat{\mathbf{e}}$ along $\hat{\mathbf{X}}$ and $\hat{\mathbf{Y}}$ but vibrating out of phase by $90°$. We define left circularly polarized light by $\beta = \pi/4$ and $\delta = \pi/2$

$$\hat{\mathbf{e}}(\beta = \pi/4, \delta = \pi/2) = (2)^{-\frac{1}{2}}(\hat{\mathbf{X}} + i\hat{\mathbf{Y}}) \qquad (5.117)$$

and right circularly polarized light by $\beta = 3\pi/4$ and $\delta = \pi/2$

$$\hat{\mathbf{e}}(\beta = 3\pi/4, \delta = \pi/2) = (2)^{-\frac{1}{2}}(\hat{\mathbf{X}} - i\hat{\mathbf{Y}}) \qquad (5.118)$$

Thus the only nonvanishing spherical component for left circularly polarized light is $-e(1, 1)$ and for right circularly polarized light is $e(1, -1)$.

Special care must be taken in interpreting the spherical components of $\hat{\mathbf{e}}^*$. We take $e^*(1, \mu)$ to mean the $(1, \mu)$ spherical component of the $\hat{\mathbf{e}}^*$ vector. Reference to Eq. (5.116) for $\delta = \pi/2$ shows that $e(1, 1) = -(\cos\beta + \sin\beta)/\sqrt{2}$, $e(1, -1) = (\cos\beta - \sin\beta)/\sqrt{2}$, $e^*(1, 1) = -(\cos\beta - \sin\beta)/\sqrt{2}$, and $e^*(1, -1) = (\cos\beta + \sin\beta)/\sqrt{2}$. Hence

$$e^*(1, \mu) = (-1)^{\mu}e(1, -\mu) \qquad (5.119)$$

The polarization tensor $E_q^k(\hat{\mathbf{e}}, \hat{\mathbf{e}}^*)$ can then be constructed according to Eq. (5.100) with the help of Eq. (5.119). We find that for circularly polarized light the only nonvanishing components of the polarization tensor are with $q = 0$, specifically

$$E_0^k(\hat{\mathbf{e}}, \hat{\mathbf{e}}^*) = -(2k + 1)^{\frac{1}{2}}\begin{pmatrix} 1 & 1 & k \\ 1 & -1 & 0 \end{pmatrix} \qquad (5.120)$$

for left circularly polarized light and

$$E_0^k(\hat{e}, \hat{e}^*) = -(2k+1)^{\frac{1}{2}} \begin{pmatrix} 1 & 1 & k \\ -1 & 1 & 0 \end{pmatrix} \tag{5.121}$$

for right circularly polarized light.

Let $I(\text{same})$ denote the resonance fluorescence signal having the same helicity as the circular polarization of the incident beam and $I(\text{opposite})$, the opposite sense of circular polarization. It is useful to introduce the degree of circular polarization, C, which is defined in analogy to the degree of linear polarization, P, by

$$C = \frac{I(\text{same}) - I(\text{opposite})}{I(\text{same}) + I(\text{opposite})} \tag{5.122}$$

Then in the same manner as Eq. (5.109) was derived, it is found that

$$C = \frac{\begin{Bmatrix} 1 & 1 & 1 \\ J_e & J_e & J_i \end{Bmatrix} \begin{Bmatrix} 1 & 1 & 1 \\ J_e & J_e & J_f \end{Bmatrix} P_1(\cos\theta)}{\frac{2}{3}\begin{Bmatrix} 1 & 1 & 0 \\ J_e & J_e & J_i \end{Bmatrix} \begin{Bmatrix} 1 & 1 & 0 \\ J_e & J_e & J_f \end{Bmatrix} + \frac{1}{3}\begin{Bmatrix} 1 & 1 & 2 \\ J_e & J_e & J_i \end{Bmatrix} \begin{Bmatrix} 1 & 1 & 2 \\ J_e & J_e & J_f \end{Bmatrix} P_2(\cos\theta)} \tag{5.123}$$

where θ is the angle between the incident beam and the detection direction. Note that C vanishes for a right-angle excitation–detection geometry ($\theta = \pi/2$) but attains a maximum value for a collinear geometry ($\theta = 0$ or π). In particular, when the excitation beam coincides in direction with that of the fluorescence beam detected ($\theta = 0$), the degree of circular polarization is given by

$$C = \frac{3\begin{Bmatrix} 1 & 1 & 1 \\ J_e & J_e & J_i \end{Bmatrix} \begin{Bmatrix} 1 & 1 & 1 \\ J_e & J_e & J_f \end{Bmatrix}}{2\begin{Bmatrix} 1 & 1 & 0 \\ J_e & J_e & J_i \end{Bmatrix} \begin{Bmatrix} 1 & 1 & 0 \\ J_e & J_e & J_f \end{Bmatrix} + \begin{Bmatrix} 1 & 1 & 2 \\ J_e & J_e & J_i \end{Bmatrix} \begin{Bmatrix} 1 & 1 & 2 \\ J_e & J_e & J_f \end{Bmatrix}} \tag{5.124}$$

Closed algebraic expressions and limiting forms for large J_i are listed in Table 5.2 for the nine possible resonance fluorescence branches.

For an isotropic sample, measurement of the degree of circular polarization in molecular resonance fluorescence generally requires the resolution of individual rotational branches in both the excitation and fluorescence steps. This is a consequence of the fact that P and R branch transitions usually have the same strength from a given excited level, and the sums $C(R\uparrow, R\downarrow) + C(R\uparrow, P\downarrow)$ and $C(P\uparrow, R\downarrow) + C(P\uparrow, P\downarrow)$ are approximately zero for large J_i values. Moreover, resonance fluorescence branches involving $Q\uparrow$ or $Q\downarrow$ also make a negligible contribution to C in the high J_i limit. For an oriented sample, nonzero values of C generally require only the resolution of individual rotational branches either in the excitation or the fluorescence step.

TABLE 5.2 Resonance Fluorescence Process; Circular Polarization

Transition	$C(J_i)$	$C(J_i \to \infty)$
$(Q\uparrow, Q\downarrow)$	$\dfrac{5}{8J_i^2 + 8J_i - 1}$	0
$(R\uparrow, R\downarrow)$	$\dfrac{5(J_i + 2)(2J_i + 1)}{14J_i^2 + 23J_i + 10}$	$\dfrac{5}{7}$
$(P\uparrow, P\downarrow)$	$\dfrac{5(J_i - 1)(2J_i + 1)}{14J_i^2 + 5J_i + 1}$	$\dfrac{5}{7}$
$(R\uparrow, P\downarrow)$ or $(P\uparrow, R\downarrow)$	$-\dfrac{5}{7}$	$-\dfrac{5}{7}$
$(Q\uparrow, R\downarrow)$	$\dfrac{5}{6J_i - 1}$	0
$(Q\uparrow, P\downarrow)$	$-\dfrac{5}{6J_i + 7}$	0
$(R\uparrow, Q\downarrow)$	$-\dfrac{5}{6J_i + 5}$	0
$(P\uparrow, Q\downarrow)$	$-\dfrac{5}{6J_i + 1}$	0

We have assumed that the light beam is in a pure polarization state. This need not be the case, and a beam of photons is said to be in a mixed state if it is not possible to describe the beam in terms of a single state vector, as in Eq. (5.116). A mixed state might result from two (or more) light sources emitting independently with no definite phase relation between the sources. An example is a beam of totally unpolarized light that may be regarded as equal incoherent combinations of two beams of opposite polarization, that is, one beam of horizontal linear polarization with another beam of vertical linear polarization having the same intensity or one beam of left circular polarization with another beam of right circular polarization having the same intensity. Mixed photon states are most conveniently treated by use of density matrix methods[28].

In the foregoing we have also assumed that the ground-state distribution of J_i directions is random. This need not be the case, especially if the molecules are formed in some directional process such as in photodissociation or in molecular beam scattering either with other molecules or with surfaces. The determination of the angular momentum distribution by use of the resonance fluorescence process has been considered elsewhere in some detail[29–32].

NOTES AND REFERENCES

1. Wigner and Racah were the first to introduce irreducible tensor operators [see E. P. Wigner, "On the Matrices which Reduce the Kronecker Products of Representations of S. R. Groups," in *Quantum Theory of Angular Momentum*, L. C. Biedenharn and H. van Dam, eds. (Academic Press, New York, 1965); E. P. Wigner, *Am. J. Math.*, **63**, 57 (1941); G. Racah, *Phys. Rev.*, **61**, 186 (1942); **62**, 438 (1942); **63**, 367 (1943)]. The latter three papers by Racah are classics in the field of atomic spectroscopy because they first illustrated the power of these tensor methods compared to previous treatments. For an excellent introduction to this topic, see B. L. Silver, *Irreducible Tensor Methods* (Academic Press, New York, 1976), which contains many worked-out examples, particularly in regard to applications to atomic systems in spherical and point group environments. Other extremely useful references are U. Fano and G. Racah, *Irreducible Tensorial Sets* (Academic Press, New York, 1959); B. R. Judd, *Operator Techniques in Atomic Spectroscopy* (McGraw-Hill, New York, 1963); B. R. Judd, *Angular Momentum Theory for Diatomic Molecules* (Academic Press, New York, 1975); and I. I. Sobel'man, *An Introduction to the Theory of Atomic Spectra* (Pergamon Press, New York, 1972). Finally, for all you might want to know (and probably more!), there is the two-volume set: L. C. Biedenharn and J. D. Louck, *Angular Momentum in Quantum Physics* (Addison-Wesley, Reading, MA, 1981); *The Racah–Wigner Algebra in Quantum Theory* (Addison-Wesley, Reading, MA, 1981).

2. R. F. Curl, Jr. and J. L. Kinsey, *J. Chem. Phys.*, **35**, 1758 (1961).

3. J. M. Brown and B. J. Howard, *Molec. Phys.*, **31**, 1517 (1976); **32**, 1197 (1976).

4. E. P. Wigner, *Group Theory* (Academic Press, New York, 1959); C. Eckart, *Rev. Mod. Phys.*, **2**, 305 (1930).

5. D. M. Brink and G. R. Satchler, *Angular Momentum* (Clarendon Press, Oxford, 1979).

6. A. R. Edmonds, *Angular Momentum in Quantum Mechanics* (Princeton University Press, Princeton, NJ, 1974).

7. S. Devons and J. B. Goldfarb, "Angular Correlations," in *Encyclopedia of Physics* Vol. XLII, (Springer-Verlag, Berlin, 1957), pp. 362–554.

8. L. C. Biedenharn, in *Nuclear Spectroscopy*, F. Ajzenberg-Selove, ed. (Academic Press, New York, 1960), Part B, pp. 732–810.

9. H. Feshbach, in *Polarization Phenomena in Nuclear Reactions*, H. H. Barschall and W. Haeberli, eds. (University of Wisconsin Press, Madison, WI, 1971), pp. 29–38.

10. R. M. Steffen and K. Alder, in *The Electromagnetic Interaction in Nuclear Spectroscopy* (North-Holland, Amsterdam, 1975), Chapters 12 and 13.

11. M. Peshkin, *Adv. Chem. Phys.*, **18**, 1 (1970).

12. R. N. Dixon, *J. Chem. Phys.*, **85**, 1866 (1986).

13. D. A. Case and D. R. Herschbach, *Molec. Phys.*, **30**, 1537 (1975); D. A. Case and D. R. Herschbach, *J. Chem. Phys.*, **64**, 4212 (1976); G. M. McClelland and D. R. Herschbach, *J. Phys. Chem.*, **83**, 1445 (1979); J. D. Barnwell, J. G. Loeser, and D. R. Herschbach, *J. Phys. Chem.*, **87**, 2781 (1983).

14. E. U. Condon and G. H. Shortley, *The Theory of Atomic Spectra* (Cambridge University Press, 1935); J. S. Griffith, *The Theory of Transition-Metal Ions* (Cambridge University Press, 1964).

15. Graphical methods are readily adapted to the evaluation of spherical tensor matrix elements; see for example I. Lindgren and J. Morrison, *Atomic Many-Body Theory* (Springer-Verlag, Berlin, 1982), Chapters 3 and 4. For example, the Wigner–Eckart theorem becomes

$$\langle \alpha' j' m' | T(k, q) | \alpha j m \rangle = \langle \alpha' j' \| T^k \| \alpha j \rangle$$

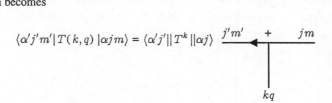

16. For small molecules, like diatomics and triatomics, the excited state rotational level J_e is almost always unique since only one such rotational transition $J_i \rightarrow J_e$ can occur for a short pulse of light with a small frequency spread. For large molecular systems, it is possible to prepare a coherent distribution of excited rotational levels with a short pulse of light. This results in more than one indistinguishable path from J_i to J_f and causes the appearance of rotational quantum beats. See P. M. Felker, J. S. Baskin, and A. H. Zewail, *J. Phys. Chem.*, **90**, 724 (1986); J. S. Baskin, P. M. Felker, and A. H. Zewail, *J. Chem. Phys.*, **84**, 4708 (1986); P. M. Felker and A. H. Zewail, *J. Chem. Phys.*, **86**, 2460, 2483 (1987); and N. F. Scherer, L. R. Khundkar, T. S. Rose, and A. H. Zewail, *J. Phys. Chem.*, **91**, 6478 (1987).

17. A. Omont, *Progr. Quantum Electronics*, **5**, 69 (1977).

18. G. Breit, *Rev. Mod. Phys.*, **5**, 91 (1933). See also P. Franken, *Phys. Rev.*, **121**, 508 (1961) and M. E. Rose and L. Carovillano, *Phys. Rev.*, **122**, 1185 (1961).

19. P. Schenck, R. C. Hilbern, and H. Metcalf, *Phys. Rev. Lett.*, **31**, 189 (1973).

20. M. I. Dyakonov, *Sov. Phys. JETP*, **20**, 1484 (1965).

21. H. Kretzen and H. Walther, *Phys. Lett.*, **27A**, 718 (1968).

22. G. Gouedard and J. C. Lehmann, *J. Phys.*, **34**, 693 (1973); M. Broyer, G. Gouedard, J. C. Lehmann, and J. Vigué, "Optical Pumping of Molecules," *Adv. Atomic Molec. Phys.*, **12**, 165 (1976).

23. R. Luypaert and J. Van Craen, *J. Phys. B*, **10**, 3627 (1977).

24. P. P. Feofilov, *The Physical Basis of Polarized Emission* (Consultants Bureau, New York, 1961).

25. R. N. Zare, *J. Chem. Phys.*, **45**, 4510 (1966).

26. G. W. Loge and C. S. Parmenter, *J. Chem. Phys.*, **74**, 29 (1981).

27. H. Katô, S. R. Jeyes, A. J. McCaffery, and M. D. Rowe, *Chem. Phys. Lett.*, **39**, 573 (1976); S. R. Jeyes, A. J. McCaffery, M. D. Rowe, P. A. Madden, and H. Katô, *Chem. Phys. Lett.*, **47**, 550 (1977); R. Clark and A. J. McCaffery, *Molec. Phys.*, **35**, 609, 617 (1978); S. R. Jeyes, A. J. McCaffery, and M. D. Rowe, *Molec. Phys.*, **36**, 845, 1865 (1978); M. D. Rowe and A. J. McCaffery, *Chem. Phys.*, **34**, 81 (1978); ibid., **43**, 35 (1979); J. McCormack, A. J. McCaffery, and M. D. Rowe, *Chem. Phys.*, **48**, 121 (1980); A. J. McCaffery, *Specialist Periodical Reports*, "Gas Kinetics and Energy Transfer," **4**, 47 (1981); B. J. Whitaker and A. J. McCaffery, *Chem. Phys. Lett.*, **86**, 185 (1982); ibid., *J. Chem. Phys.*, **78**, 3256 (1983); A. J. Bain and A. J. McCaffery, *J. Chem. Phys.*, **83**, 2627 (1985).

28. K. Blum, *Density Matrix Theory and Applications* (Plenum Press, New York, 1981).

29. D. A. Case, G. M. McClelland, and D. R. Herschbach, *Molec. Phys.*, **35**, 541 (1978).

30. C. H. Greene and R. N. Zare, *J. Chem. Phys.*, **78**, 6741 (1983).

31. R. I. Altkorn and R. N. Zare, *Ann. Rev. Phys. Chem.*, **35**, 265 (1984).

32. A. C. Kummel, G. O. Sitz, and R. N. Zare, *J. Chem. Phys.*, **85**, 6874 (1986); *J. Chem. Phys.* (in press).

PROBLEM SET 3

1. Use angular momentum coupling algebra to prove the vector identity

$$\mathbf{a} \cdot (\mathbf{b} \times \mathbf{c}) = (\mathbf{a} \times \mathbf{b}) \cdot \mathbf{c} \tag{5.125}$$

2. Derive Eq. (5.123) from Eq. (5.122). Offer an interpretation why $C(J_i) \to 0$ as $J_i \to \infty$ for the resonance fluorescence process involving $Q\uparrow$ or $Q\downarrow$ branches.

3. Explore an alternative derivation of Eq. (5.105), which expresses the resonance fluorescence intensity in the absence of an external field. Start with Eq. (5.93), and write

$$\hat{\mathbf{e}}_a \cdot \mathbf{r} = \sum_{\mu} (-1)^{\mu} (e_a)_{-\mu} \, r_{\mu}^1$$

$$\hat{\mathbf{e}}_a^* \cdot \mathbf{r} = \sum_{\mu'} (-1)^{\mu'} (e_a^*)_{-\mu'} \, r_{\mu'}^1$$

$$\hat{\mathbf{e}}_d \cdot \mathbf{r} = \sum_{\nu} (-1)^{\nu} (e_d)_{-\nu} \, r_{\nu}^1$$

$$\hat{\mathbf{e}}_d^* \cdot \mathbf{r} = \sum_{\nu'} (-1)^{\nu'} (e_d^*)_{-\nu'} \, r_{\nu'}^1 \tag{5.126}$$

Apply the Wigner–Eckart theorem to show that

$$I \propto \sum_{\substack{\mu,\mu' \\ \nu,\nu'}} (-1)^{\mu+\mu'+\nu+\nu'} (e_a)_{-\mu} (e_a^*)_{-\mu'} (e_d)_{-\nu} (e_d^*)_{-\nu'}$$

$$\times \left| \langle \alpha_e J_e || \, r^{(1)} \, || \alpha_i J_i \rangle \right|^2 \left| \langle \alpha_e J_e || \, r^{(1)} \, || \alpha_f J_f \rangle \right|^2$$

$$\times (-1)^{J_i - J_e + J_f - J_e} S(J_i, J_e, J_f; \mu, \mu', \nu, \nu') \tag{5.127}$$

where

$$S(J_i, J_e, J_f; \mu, \mu', \nu, \nu')$$

$$= \sum_{\substack{M_i, M_e \\ M_e', M_f}} (-1)^{J_e - M_e + J_i - M_i + J_e - M_e' + J_f - M_f} \begin{pmatrix} J_e & 1 & J_i \\ -M_e & \mu & M_i \end{pmatrix}$$

$$\times \begin{pmatrix} J_i & 1 & J_e \\ -M_i & \mu' & M_e' \end{pmatrix} \begin{pmatrix} J_e & 1 & J_f \\ -M_e' & \nu & M_f \end{pmatrix} \begin{pmatrix} J_f & 1 & J_e \\ -M_f & \nu' & M_e \end{pmatrix}$$

$$(5.128)$$

Use the results of Problem Set 2, part 3 to reexpress the summation S as

$$S = \sum_{k,q} (-1)^{q+J_f-J_e+J_f-J_e} (2k+1) \begin{pmatrix} 1 & 1 & k \\ \mu & \mu' & q \end{pmatrix} \begin{pmatrix} 1 & 1 & k \\ \nu & \nu' & -q \end{pmatrix}$$

$$\times \begin{Bmatrix} 1 & 1 & k \\ J_e & J_e & J_i \end{Bmatrix} \begin{Bmatrix} 1 & 1 & k \\ J_e & J_e & J_f \end{Bmatrix} \tag{5.129}$$

Show that on substitution this yields the expression for the resonance fluorescence intensity given by Eqs. (5.105) and (5.106).

4. A second-rank Cartesian tensor T_{AB} ($A = X, Y, Z$ and $B = X, Y, Z$) has $3^2 = 9$ independent components. Decompose T_{AB} into its irreducible spherical tensor components. Show that T_{AB} is composed of a zero-rank irreducible tensor, proportional to $T_{XX} + T_{YY} + T_{ZZ} = \text{Tr}[T]$ and having one independent component, a first-rank irreducible tensor, proportional to the antisymmetric tensor $\frac{1}{2}(T_{AB} - T_{BA})$ and having three independent components, and a second-rank irreducible tensor, proportional to the symmetric traceless tensor $\frac{1}{2}(T_{AB} + T_{BA}) - \frac{1}{3}\delta_{AB}\text{Tr}[T]$ and having five independent components. Present a general procedure for decomposing an arbitrary Nth-rank Cartesian tensor into its spherical irreducible components. *Hint*: Any Cartesian tensor may be written in a spherical basis using the correspondence between first rank Cartesian tensors T_X, T_Y, T_Z and first rank spherical tensors $T_1^{(1)}, T_0^{(1)}, T_{-1}^{(1)}$:

$$\begin{pmatrix} T_1^{(1)} \\ T_0^{(1)} \\ T_{-1}^{(1)} \end{pmatrix} = \begin{pmatrix} -\frac{1}{\sqrt{2}} & -\frac{i}{\sqrt{2}} & 0 \\ 0 & 0 & 1 \\ \frac{1}{\sqrt{2}} & -\frac{i}{\sqrt{2}} & 0 \end{pmatrix} \begin{pmatrix} T_X \\ T_Y \\ T_Z \end{pmatrix} \tag{5.130}$$

or

$$T_\mu^{(1)} = \sum_A U_{\mu A} T_A \tag{5.131}$$

with the inverse relation

$$\begin{pmatrix} T_X \\ T_Y \\ T_Z \end{pmatrix} = \begin{pmatrix} -\frac{1}{\sqrt{2}} & 0 & \frac{1}{\sqrt{2}} \\ \frac{i}{\sqrt{2}} & 0 & \frac{i}{\sqrt{2}} \\ 0 & 1 & 0 \end{pmatrix} \begin{pmatrix} T_1^{(1)} \\ T_0^{(1)} \\ T_{-1}^{(1)} \end{pmatrix} \tag{5.132}$$

or

$$T_A = \sum_\mu (U^{-1})_{A\mu} T_\mu^{(1)} = \sum_\mu U_{\mu A}^\dagger T_\mu^{(1)} \tag{5.133}$$

Thus the AB element of a second-rank Cartesian tensor T_{AB} is written in a spherical basis as

$$T_{AB} = \sum_{\mu,\nu} U_{\mu A}^{\dagger} U_{\nu B}^{\dagger} T_{\mu}^{(1)} T_{\nu}^{(1)} \tag{5.134}$$

The product $T_{\mu}^{(1)} T_{\nu}^{(1)}$ is not irreducible but transforms under rotation according to

$$\mathbf{R} T_{\mu}^{(1)} T_{\nu}^{(1)} \mathbf{R}^{-1} = \mathbf{R} T_{\mu}^{(1)} \mathbf{R}^{-1} \, \mathbf{R} T_{\nu}^{(1)} \mathbf{R}^{-1}$$

$$= \sum_{\mu',\nu'} D_{\mu'\mu}^{1}(R) D_{\nu'\nu}^{1}(R) \, T_{\mu'}^{(1)} T_{\nu'}^{(1)}$$

$$= \sum_{\mu',\nu'} \sum_{j} \langle 1\mu', 1\nu' | jm' \rangle \langle 1\mu, 1\nu | jm \rangle$$

$$\times D_{m'm}^{j}(R) T_{\mu'}^{(1)} T_{\nu'}^{(1)} \tag{5.135}$$

and it follows that

$$\mathbf{R} T_{AB} \mathbf{R}^{-1}$$

$$= \sum_{j} \sum_{\mu,\nu} U_{\mu A}^{\dagger} U_{\nu B}^{\dagger} \sum_{\mu',\nu'} \langle 1\mu', 1\nu' | jm' \rangle \langle 1\mu, 1\nu | jm \rangle \, D_{m'm}^{j}(R) T_{\mu'}^{(1)} T_{\nu'}^{(1)}$$

$$= \sum_{j} \sum_{m,m'} C_{jm} D_{m'm}^{j}(R) T_{m'}^{(j)} \tag{5.136}$$

Then solve for the coefficients C_{jm} in the expansion. The generalization to higher-order Cartesian tensors relies on repeated application of the procedure shown above. The explicit decomposition of a third-rank Cartesian tensor into spherical irreducible components is given in B. Dick, *Chem. Phys.*, **96**, 199 (1985).

APPLICATION 11
THE ENERGY LEVELS OF A TWO-VALENCE-ELECTRON ATOM
REVISITED

A. Work out the term energies for the p^2 configuration, using spherical tensor
methods:

$$E(^{2S+1}L) = \langle \ell_1 = 1 \; \ell_2 = 1 \; SM_SLM_L | \frac{e^2}{r_{12}} | \ell_1 = 1 \; \ell_2 = 1 \; SM_SLM_L \rangle \quad (1)$$

$$= \sum_k \langle p^2 SM_SLM_L | C^k(1) \cdot C^k(2) | p^2 SM_SLM_L \rangle$$

$$\times e^2 \int_0^\infty r_1^2 \, dr_1 \int_0^\infty r_2^2 \, dr_2 \, \frac{r_<^k}{r_>^{k+1}} \left[R_p(1) R_p(2) \right]^2 \quad (2)$$

$$= \sum_k (-1)^L \begin{Bmatrix} 1 & 1 & k \\ 1 & 1 & L \end{Bmatrix} \langle 1 \| C^k \| 1 \rangle^2 F^k \quad (3)$$

where we have made use of the expansion

$$\frac{e^2}{r_{12}} = e^2 \sum_k \frac{r_<^k}{r_>^{k+1}} P_k(\cos \Theta) = e^2 \sum_k \frac{r_<^k}{r_>^{k+1}} C^k(1) \cdot C^k(2) \quad (4)$$

Show in particular that

$$E(^1S) = F^0 + \tfrac{2}{5} F^2$$

$$E(^3P) = F^0 - \tfrac{1}{5} F^2$$

$$E(^1D) = F^0 + \tfrac{1}{25} F^2 \quad (5)$$

It is traditional to follow the practice of Condon and Shortley and to define
$F_0 = F^0$ and $F_2 = (1/25) F^2$ so that

$$E(^1S) = F_0 + 10 F_2$$

$$E(^3P) = F_0 - 5 F_2$$

$$E(^1D) = F_0 + F_2 \quad (6)$$

B. By considering only the electrostatic interaction, the resulting energy expression does not contain J, and hence the different J levels $^{2S+1}L_J$ of the term ^{2S+1}L have the same energy. However, the inclusion of the spin–orbit interaction term

$$H_{SO} = \sum_i \xi(r_i) \boldsymbol{\ell}_i \cdot \mathbf{s}_i \tag{7}$$

in the Hamiltonian removes this degeneracy. Find the $E(^{2S+1}L_J)$ treating H_{SO} as a perturbation; that is, use only first-order perturbation theory.
Set up, but do not evaluate, the exact expressions for $E(^{2S+1}L_J)$ when the spin–orbit interaction is not small compared to the separation between term energies. Express your energy level expressions in terms of the spin–orbit parameter

$$\zeta_{n\ell} = \int R_{n\ell}(r)\,\xi(r)\,R_{n\ell}(r)\,r^2\,dr \tag{8}$$

It may be shown that

$$E(^1S_0) = F_0 + \tfrac{5}{2}F_2 - \tfrac{1}{2}\zeta_{np} + \left[\tfrac{225}{4}F_2^2 + \tfrac{15}{2}F_2\zeta_{np} + \tfrac{9}{4}\zeta_{np}^2\right]^{\frac{1}{2}}$$

$$E(^3P_0) = F_0 + \tfrac{5}{2}F_2 - \tfrac{1}{2}\zeta_{np} - \left[\tfrac{225}{4}F_2^2 + \tfrac{15}{2}F_2\zeta_{np} + \tfrac{9}{4}\zeta_{np}^2\right]^{\frac{1}{2}}$$

$$E(^3P_1) = F_0 - 5F_2 - \tfrac{1}{2}\zeta_{np}$$

$$E(^1D_2) = F_0 - 2F_2 + \tfrac{1}{4}\zeta_{np} + \left[9F_2^2 - \tfrac{3}{2}F_2\zeta_{np} + \tfrac{9}{16}\zeta_{np}^2\right]^{\frac{1}{2}}$$

$$E(^3P_2) = F_0 - 2F_2 + \tfrac{1}{4}\zeta_{np} - \left[9F_2^2 - \tfrac{3}{2}F_2\zeta_{np} + \tfrac{9}{16}\zeta_{np}^2\right]^{\frac{1}{2}}$$

$$\tag{9}$$

As ζ_{np} approaches 0, the energy levels follow LS coupling; if we keep only terms first order in ζ_{np}, the energy level spacings follow the Landé interval rule; and if ζ_{np}/F_2 tends to infinity, the energy levels follow jj coupling.

C. Fluorescent lighting utilizes the Hg $6s\,6p\ ^3P_1^\circ \rightarrow 6s^2\ ^1S_0$ UV transition at 253.7 nm to cause a phosphor on the inside walls of a fluorescent lamp to glow "white." The (strongly allowed) resonant transition of Hg corresponds to the transition $6s\,6p\ ^1P_1^\circ \rightarrow 6s^2\ ^1S_0$ at 184.9 nm. Present arguments to explain why the Hg 253.7 nm line has significant intensity although singlet–triplet electric dipole transitions are forbidden in the LS coupling limit.

APPLICATION 12
DIRECTIONAL CORRELATIONS;
BREAKUP OF A LONG-LIVED COMPLEX

Consider a reactive scattering process

$$A + BC \rightarrow AB + C$$

As a first approximation, we regard the reagents and products to be spinless particles possessing no internal angular momenta. Let $\hat{\mathbf{k}}$ and $\hat{\mathbf{k}}'$ specify the respective initial and final relative velocity directions in the center-of-mass frame, and let $\boldsymbol{\ell}$ and $\boldsymbol{\ell}'$ be the corresponding external orbital angular momenta whose directions are perpendicular to $\hat{\mathbf{k}}$ and $\hat{\mathbf{k}}'$, respectively, and whose magnitudes are given by the product of the reduced mass, the impact parameter, and the relative velocity magnitude of the collision partners. We denote the total angular momentum of the scattering system by $\mathbf{j}$. Conservation of angular momentum implies that

$$\mathbf{j} = \boldsymbol{\ell} = \boldsymbol{\ell}' \tag{1}$$

The direction correlation[1, 2] between $\hat{\mathbf{k}}$ and $\hat{\mathbf{k}}'$, which is proportional to the differential cross section, depends only on the included angle θ, where

$$\cos \theta = \hat{\mathbf{k}} \cdot \hat{\mathbf{k}}' \tag{2}$$

that is, we may write the differential cross section as the function $I(\theta)$. Next we expand $I(\theta)$ in a complete set of Legendre polynomials:

$$I(\theta) = \frac{1}{4\pi} \sum_{\nu=0}^{\infty} (2\nu + 1) A_\nu P_\nu(\cos \theta) \tag{3}$$

We may solve for the expansion coefficients A_ν by multiplying both sides of Eq. (3) by $P_{\nu'}(\cos \theta)$ and integrating over all solid angle elements $d\Omega = \sin \theta \, d\theta \, d\phi$. With the help of Eq. (1.59) we find that

$$A_\nu = \int_0^{2\pi} d\phi \int_0^\pi I(\theta) P_\nu(\cos \theta) \sin \theta \, d\theta$$

$$= \left\langle P_\nu(\cos \theta) \right\rangle_{\text{av}} = \left\langle P_\nu(\hat{\mathbf{k}} \cdot \hat{\mathbf{k}}') \right\rangle_{\text{av}} \tag{4}$$

Here the expansion coefficient A_ν is interpreted as the value of $P_\nu(\hat{\mathbf{k}} \cdot \hat{\mathbf{k}}')$ averaged over all possible configurations of the scattering system.

We choose to define the z axis to be along $\boldsymbol{\ell}$. We denote the polar angles of $\hat{\mathbf{k}}$ by (θ_1, ϕ_1) and those of $\hat{\mathbf{k}}'$ by (θ_2, ϕ_2). Use of the spherical harmonic addition theorem [Eq. (3.85)], allows us to rewrite Eq. (4) as

$$\langle P_\nu(\hat{\mathbf{k}} \cdot \hat{\mathbf{k}}') \rangle_{\text{av}} = \frac{4\pi}{2\nu + 1} \left\langle \sum_q Y_{\nu q}^*(\theta_1, \phi_1) Y_{\nu q}(\theta_2, \phi_2) \right\rangle_{\text{av}} \tag{5}$$

So far these results are general. Now we introduce the condition that the reaction, which is characterized by the angular momentum $\boldsymbol{\ell}$, is long-lived. If the time between formation and breakup of the complex is long compared to the rotational period of the complex, $\hat{\mathbf{k}}'$ will be uniformly distributed about $\boldsymbol{\ell}' = \boldsymbol{\ell}$. This has the consequence that the angle in the xy plane conjugate to $\boldsymbol{\ell}$ is uniformly distributed; that is, the collision complex breaks up with a random value of $\phi_1 - \phi_2$.

A. Show that

$$\langle P_\nu(\hat{\mathbf{k}} \cdot \hat{\mathbf{k}}') \rangle_{\text{av}} = \left\langle P_\nu(\hat{\mathbf{k}} \cdot \hat{\boldsymbol{\ell}}) P_\nu(\hat{\mathbf{k}}' \cdot \hat{\boldsymbol{\ell}}) \right\rangle_{\text{av}} \tag{6}$$

and hence that

$$A_\nu = [P_\nu(0)]^2 \tag{7}$$

For such a long-lived collision complex it follows that

$$I(\theta) = \frac{1}{4\pi} \sum_\nu (2\nu + 1)[P_\nu(0)]^2 P_\nu(\cos\theta) \tag{8}$$

B. Explain why this result implies that the product angular distribution has forward–backward symmetry in the center-of-mass frame, that is, is symmetric about $\theta = \pi/2$.

C. Show that $I(\theta)$ is proportional to $1/\sin\theta$, that is, peaks at the poles and fans out at the equator. *Hint*: Expand $1/\sin\theta$ in a Legendre series and compare to Eq. (8).

D. We remove the restriction that the reactants and products are spinless particles. The total angular momentum of the collision complex is then

$$\mathbf{j} = \boldsymbol{\ell} + \mathbf{s} = \boldsymbol{\ell}' + \mathbf{s}' \tag{9}$$

where $\mathbf{s} = \mathbf{J}_A + \mathbf{J}_{BC}$ and $\mathbf{s}' = \mathbf{J}_{AB} + \mathbf{J}_C$ denote the entrance and exit channel spins (angular momenta). In the classical limit each angular momentum value may be associated with a definite direction in space, specified by its corresponding unit vector. Successive applications of the spherical harmonic addition theorem give

$$\left\langle P_\nu(\hat{\mathbf{k}} \cdot \hat{\mathbf{k}}') \right\rangle_{av} = \left\langle P_\nu(\hat{\mathbf{k}} \cdot \hat{\mathbf{j}}) P_\nu(\hat{\mathbf{j}} \cdot \hat{\mathbf{k}}') \right\rangle_{av}$$

$$= \left\langle P_\nu(\hat{\mathbf{k}} \cdot \hat{\boldsymbol{\ell}}) P_\nu(\hat{\boldsymbol{\ell}} \cdot \hat{\mathbf{j}}) P_\nu(\hat{\mathbf{j}} \cdot \hat{\boldsymbol{\ell}}') P_\nu(\hat{\boldsymbol{\ell}}' \cdot \hat{\mathbf{k}}') \right\rangle_{av}$$

$$= \left\langle P_\nu(\hat{\mathbf{k}} \cdot \hat{\boldsymbol{\ell}}) P_\nu(\hat{\mathbf{k}}' \cdot \hat{\boldsymbol{\ell}}') P_\nu(\hat{\boldsymbol{\ell}} \cdot \hat{\mathbf{j}}) P_\nu(\hat{\mathbf{j}} \cdot \hat{\boldsymbol{\ell}}') \right\rangle_{av} \qquad (10)$$

where we assume that directional properties of the reagent and product trajectories are uncorrelated except by the conservation of total angular momentum (definition of a separable long-lived collision complex[2]). Here the first line of Eq. (10) is a consequence of the random distribution of the dihedral angle $\phi_{kjk'}$, and the second line is a consequence of the random distribution of $\phi_{k\ell j}$ and $\phi_{k'\ell'j}$, which results because **s** is assumed randomly oriented with respect to **k** and **s'** is unobserved. Note that an unobserved direction is equivalent to a random one since each leads to an unweighted average over the relevant dihedral angle. Show once again that only even values of ν contribute to $I(\theta)$, that is, that $I(\theta)$ has forward–backward scattering symmetry.

E. The preceding arguments are semiclassical in nature in that they ignore the difficulty of specifying the magnitude of $\boldsymbol{\ell}$ and its exact direction. A fully quantum mechanical treatment shows that we must replace the Legendre function of the angle between a quantized angular momentum $\hat{\boldsymbol{\ell}}$ and an unquantized direction $\hat{\mathbf{k}}$ by an appropriate Clebsch–Gordan coefficient

$$P_\nu(\hat{\boldsymbol{\ell}} \cdot \hat{\mathbf{k}}) \sim \langle \ell 0, \nu 0 | \ell 0 \rangle \qquad (11)$$

for large ℓ, and the Legendre function for the angle between the two quantized angular momenta $\hat{\boldsymbol{\ell}}$ and $\hat{\mathbf{j}}$ by an appropriate 6-j symbol

$$P_\nu(\hat{\boldsymbol{\ell}} \cdot \hat{\mathbf{j}}) \sim (-1)^{\ell+s+j}[(2\ell+1)(2j+1)]^{\frac{1}{2}} \begin{Bmatrix} \ell & \nu & \ell \\ j & s & j \end{Bmatrix} \qquad (12)$$

for large ℓ and j. The derivations of Eqs. (11) and (12) may be found in P. J. Brussaard and H. A. Tolhoek, *Physica*, **23**, 955 (1957). The complete expression for the differential cross section becomes

$$I(\theta) = \frac{1}{4\pi} \sum_\nu (2\nu+1)(-1)^{\ell+s+j+\ell'+s'+j}$$

$$\times [(2\ell+1)(2j+1)(2\ell'+1)(2j'+1)]^{\frac{1}{2}}$$

$$\times \langle \ell 0, \nu 0 | \ell 0 \rangle \langle \ell' 0, \nu 0 | \ell' 0 \rangle \begin{Bmatrix} \ell & \nu & \ell \\ j & s & j \end{Bmatrix} \begin{Bmatrix} \ell' & \nu & \ell' \\ j & s' & j \end{Bmatrix} P_\nu(\cos\theta)$$

$$(13)$$

An important consequence of the quantization is that the index ν is restricted to a finite number of even values. What conditions can be placed on the maximum value of ν?

These same considerations apply to the directional correlation of photons emitted from an excited state, such as the angular correlation of successive gamma rays from an excited atomic nucleus[3] or the angular correlation of two fluorescence photons successively emitted from an electronically excited atomic or molecular state. We consider the process $j_i \rightarrow j$ followed by $j \rightarrow j_f$, where j_i, j, and j_f are the angular momentum quantum numbers of the initial, intermediate, and final states of the system, respectively. We denote the two directions as $\hat{\mathbf{k}}_i$ and $\hat{\mathbf{k}}_f$ and suppose that in the first transition the emitted photon has the intrinsic angular momentum $\boldsymbol{\ell}_i$ and in the second transition $\boldsymbol{\ell}_f$. We also suppose that the intermediate state is not disturbed (perturbed) between the time it is formed and the time it decays.

Any correlation in which only two directions of motion are measured (called a *direction–direction correlation*) must be expressible as a Legendre series in the angle θ_{if} between the two directions $\hat{\mathbf{k}}_i$ and $\hat{\mathbf{k}}_f$. In analogy to Eqs. (3) and (4) we may write

$$W(\theta_{if}) = \frac{1}{4\pi} \sum_{\nu} (2\nu + 1) B_\nu P_\nu(\cos\theta_{if}) \tag{14}$$

where

$$B_\nu = \left\langle P_\nu(\hat{\mathbf{k}}_i \cdot \hat{\mathbf{k}}_f) \right\rangle_{av} \tag{15}$$

Because the polarizations of the two photons are not measured, we may regard $\boldsymbol{\ell}_i$ and $\boldsymbol{\ell}_f$ to be randomly and uniformly oriented. Repeated application of the spherical harmonic addition theorem yields

$$B_\nu = \left\langle P_\nu(\hat{\mathbf{k}}_i \cdot \hat{\mathbf{j}}) P_\nu(\hat{\mathbf{j}} \cdot \hat{\mathbf{k}}_f) \right\rangle_{av}$$

$$= \left\langle P_\nu(\hat{\mathbf{k}}_i \cdot \hat{\boldsymbol{\ell}}_i) P_\nu(\hat{\boldsymbol{\ell}}_i \cdot \hat{\mathbf{j}}) P_\nu(\hat{\mathbf{j}} \cdot \hat{\boldsymbol{\ell}}_f) P_\nu(\hat{\boldsymbol{\ell}}_f \cdot \hat{\mathbf{k}}_f) \right\rangle_{av}$$

$$= P_\nu(\hat{\mathbf{k}}_i \cdot \hat{\boldsymbol{\ell}}_i) P_\nu(\hat{\mathbf{k}}_f \cdot \hat{\boldsymbol{\ell}}_f) \left\langle P_\nu(\hat{\boldsymbol{\ell}}_i \cdot \hat{\mathbf{j}}) P_\nu(\hat{\boldsymbol{\ell}}_f \cdot \hat{\mathbf{j}}) \right\rangle_{av} \tag{16}$$

The above gives the general form of the direction–direction angular distribution as more rigorous arguments show[3]. Let us specialize this to the emission of two electric dipole photons. In this case $\ell_i = \ell_f = 1$. Because the electric vector is transverse to the propagation direction of the radiation, the projection of ℓ on $\mathbf{k}$ can have only the values plus or minus one (net helicity of unity). We use semiclassical arguments to deduce the quantum transcription of Eq. (16). In particular [in analogy to Eq. (11)]

$$P_\nu(\hat{\boldsymbol{\ell}}_i \cdot \hat{\mathbf{k}}_i) \sim \langle \ell_i\, 1, \nu 0 | \ell_i - 1 \rangle \tag{17}$$

and [in analogy to Eq. (12)]

$$P_\nu(\hat{\boldsymbol{\ell}}_i \cdot \hat{\mathbf{j}}) \sim (-1)^{\ell_i + j_i + j}[(2\ell_i + 1)(2j + 1)]^{\frac{1}{2}} \begin{Bmatrix} \ell_i & \nu & \ell_i \\ j & j_i & j \end{Bmatrix} \tag{18}$$

with corresponding expressions for $P_\nu(\hat{\boldsymbol{\ell}}_f \cdot \hat{\mathbf{k}}_f)$ and $P_\nu(\hat{\boldsymbol{\ell}}_f \cdot \hat{\mathbf{j}})$. Thus for the successive emission of two electric dipole photons the direction–direction angular correlation has the form

$$W(\theta_{if}) = [1 + a_2 P_2(\cos\theta_{if})]/4\pi \tag{19}$$

where a_2 depends on j_i, j, j_f as

$$a_2 = (-1)^{j_i + 2j + j_f} \begin{Bmatrix} 1 & 1 & 2 \\ j & j & j_i \end{Bmatrix} \begin{Bmatrix} 1 & 1 & 2 \\ j & j & j_f \end{Bmatrix} \tag{20}$$

up to a normalization constant. Hence a measurement of a_2 can be used to deduce the value of j in the unobserved intermediate state!

A simple example[4] serves to illustrate Eq. (19). We consider the radiative cascade process ${}^1S \to {}^1P \to {}^1S$ in an atom corresponding to $j_i = 0$, $j = 1$, and $j_f = 0$. In the first transition ($j_i \to j$) the three possibilities $\Delta m = 0$, $\Delta m = +1$, and $\Delta m = -1$ are equally probable. However, a $\Delta m = 0$ transition in $j_i \to j$ must be followed by a $\Delta m = 0$ transition in $j \to j_f$. Similarly, a $\Delta m = \pm 1$ transition in $j_i \to j$ must be followed by a $\Delta m = \mp 1$ transition in $j \to j_f$.

The directional photon distribution in each of these transitions has a dipole antenna pattern. Specifically, for a $\Delta m = 0$ transition the photon angular distribution is $\sin^2\theta = 1 - \cos^2\theta$, whereas for a $\Delta m = \pm 1$ transition the photon angular distribution is $\frac{1}{2}(1 + \cos^2\theta)$, where θ is measured from the quantization axis. Hence the radiation in the $j_i = 0 \to j = 1$ transition is isotropic when summed over all possible Δm components weighted by their transition probabilities (which should be no surprise since the state $j_i = 0$, $m_i = 0$ is also isotropic).

The direction of the quantization axis may be arbitrarily chosen without any loss of generality. In particular, we choose the z axis to lie along the direction of the first photon; that is, $\mathbf{z}$ is parallel to $\hat{\mathbf{k}}_i$. For this choice no $\Delta m = 0$ transition can occur since the photon angular distribution associated with the $\Delta m = 0$ transition has a node along the z axis; that is, for $\Delta m = 0$ transitions no radiation is emitted in the z direction. Hence the first transition is of the $\Delta m = \pm 1$ type, and this must be followed by a second transition of the $\Delta m = \mp 1$ type. Therefore, it follows that

$$W(\theta_{if}) = \frac{3}{16\pi}\left(1 + \cos^2\theta_{if}\right) \tag{21}$$

where the normalization constant is chosen so that the total emission probability of the second photon is unity. According to Eq. (21), the direction–direction correlation for the emission of two successive electric dipole photons is twice as large when the two photons are detected in the same or opposite directions ($\theta_{if} = 0$ or π) than when detected at right angles ($\theta_{if} = \pi/2$).

F. Yang[5] has used group theoretic arguments to prove a general theorem about angular correlations between successive photons; namely, $W(\theta_{if})$ is an even power of $\cos \theta_{if}$, and the highest power in this polynomial is smaller than or equal to $2\ell_i$, $2\ell_f$, or $2j$, whichever is the smallest. Show that the semiclassical arguments presented in the preceding paragraphs lead to the same conclusion as Yang's theorem.

NOTES AND REFERENCES

1. L. C. Biedenharn, in *Nuclear Spectroscopy*, F. Azjenberg-Selove, ed., Academic Press, New York, 1960, Part B, p. 732.

2. G. M. McClelland and D. R. Herschbach, *J. Phys. Chem.*, **83**, 1445 (1979). These authors show that even if a complex survives for many rotational periods, the product angular distribution need not display forward–backward symmetry unless the complex is separable. This replaces the "beguiling but misbegotten conclusion" that all long-lived collision complexes yield forward–backward scattering symmetry. Unfortunately, examples of nonseparable long-lived complexes appear not to have been observed.

3. There is an extensive literature on correlations in nuclear cascade processes, starting with the pioneering work of D. R. Hamilton, *Phys. Rev.*, **58**, 122 (1940), who treated γ-ray correlations using time-dependent perturbation theory. See also G. Goertzel, *Phys. Rev.*, **70**, 897 (1946), who extended Hamilton's treatment and D. L Falkoff and G. E. Uhlenbeck, *Phys. Rev.*, **79**, 323 (1950), who generalized this treatment to successive radiations of arbitrary multipole moments. The first use of tensor operators is to be found in G. Racah, *Phys. Rev.*, **84**, 910 (1951); for a review, see L. C. Biedenharn and M. E. Rose, *Rev. Mod. Phys.*, **25**, 729 (1953). However, angular correlations are most generally treated by using the powerful density matrix formalism; see S. Devon and L. J. B. Goldfarb, "Angular Correlations," in *Encyclopedia of Physics*, Vol. XLII, Springer-Verlag, Berlin, 1957, pp. 362–554, as well as R. M. Steffen and K. Alder, in *The Electromagnetic Interaction in Nuclear Spectroscopy*, North-Holland, Amsterdam, 1975, pp. 505–643.

4. This example is taken from J. M. Blatt and V. F. Weisskopf, *Theoretical Nuclear Physics*, Wiley, New York, 1952, pp. 635–639.

5. C. N. Yang, *Phys. Rev.*, **74**, 764 (1948).

APPLICATION 13
ORIENTATION AND ALIGNMENT

Inelastic and reactive scattering, as well as photodissociation and photoionization processes, typically produce anisotropic distributions of the angular momentum of the collision products and photofragments. We consider how to describe the *angular momentum polarization*. One natural choice is to make an expansion of the angular momentum distribution into multipole moments. Here the monopole term is proportional to the population. All odd multipoles (dipole, octupole, etc.) describe the *orientation* of the angular momentum **J**, that is, which way **J** preferentially points regarded as a collection of single-headed arrows, and all even multipoles (quadrupole, hexadecapole, etc.) describe the *alignment* of **J**, that is, the spatial distribution of **J** regarded as a collection of double-headed arrows. If all multipoles (except the monopole) vanish, the spatial distribution of **J** is said to be isotropic; that is, **J** is uniformly distributed in space. In the $|JM\rangle$ representation, orientation corresponds to a preferential population of some $|JM\rangle$ versus some $|J-M\rangle$ level causing the system to have a net helicity or spin. In the same representation, alignment corresponds to a preferential population in the combined $|JM\rangle$ and $|J-M\rangle$ levels compared to the combined $|JM'\rangle$ and $|J-M'\rangle$ levels. Note that a system can exhibit both orientation and alignment.

To carry out the multipole moment expansion of **J**, we introduce the spherical tensor operators $J_q^{(k)}$. In accord with Eq. (5.15)

$$J_1^{(1)} = -(2)^{-\frac{1}{2}}(J_x + iJ_y) = -(2)^{-\frac{1}{2}}J_+$$

$$J_0^{(1)} = J_z$$

$$J_{-1}^{(1)} = (2)^{-\frac{1}{2}}(J_x - iJ_y) = (2)^{-\frac{1}{2}}J_- \tag{1}$$

It follows from Eqs. (5.6) and (5.7) that

$$\left[J_z, J_q^{(k)}\right] = qJ_q^{(k)} \tag{2}$$

$$\left[J_\pm, J_q^{(k)}\right] = \left[k(k+1) - q(q\pm 1)\right]^{\frac{1}{2}} J_{q\pm1}^{(k)} \tag{3}$$

A. Prove that

$$J_{\pm k}^{(k)} = \left[J_{\pm1}^{(1)}\right]^k \tag{4}$$

Hint: Contract $J_{\pm1}^{(1)}$ and $J_{\pm1}^{(1)}$ to form the spherical tensor $J_{\pm2}^{(2)}$ and make an inductive argument.

Application of Eq. (3) to Eq. (4) can be used to generate the $J_q^{(k)}$ for all q. We list for reference purposes the $J_q^{(k)}$ for $k = 0$ through 4 in terms of $J_\pm$, J_z, and $\mathbf{J}^2$:

$$J_0^{(0)} = 1 \tag{5}$$

$$J_{\pm 1}^{(1)} = \mp (2)^{-\frac{1}{2}} J_\pm \tag{6}$$

$$J_0^{(1)} = J_z \tag{7}$$

$$J_{\pm 2}^{(2)} = \frac{1}{2} J_\pm^2 \tag{8}$$

$$J_{\pm 1}^{(2)} = \mp \frac{1}{2} J_\pm (2 J_z \pm 1) \tag{9}$$

$$J_0^{(2)} = (6)^{-\frac{1}{2}} (3 J_z^2 - \mathbf{J}^2) \tag{10}$$

$$J_{\pm 3}^{(3)} = \mp (2)^{-\frac{3}{2}} J_\pm^3 \tag{11}$$

$$J_{\pm 2}^{(3)} = \frac{1}{2} (3)^{\frac{1}{2}} J_\pm^2 (J_z \pm 1) \tag{12}$$

$$J_{\pm 1}^{(3)} = \mp \frac{1}{20} (30)^{\frac{1}{2}} J_\pm (5 J_z^2 - \mathbf{J}^2 \pm 5 J_z + 2) \tag{13}$$

$$J_0^{(3)} = (10)^{-\frac{1}{2}} (5 J_z^2 - 3 \mathbf{J}^2 + 1) J_z \tag{14}$$

$$J_{\pm 4}^{(4)} = \frac{1}{4} J_\pm^4 \tag{15}$$

$$J_{\pm 3}^{(4)} = \mp (2)^{-\frac{3}{2}} J_\pm^3 (2 J_z \pm 3) \tag{16}$$

$$J_{\pm 2}^{(4)} = \frac{1}{2} (7)^{-\frac{1}{2}} J_\pm^2 (7 J_z^2 - \mathbf{J}^2 \pm 14 J_z + 9) \tag{17}$$

$$J_{\pm 1}^{(4)} = \mp \frac{1}{2} (14)^{-\frac{1}{2}} J_\pm (14 J_z^3 - 6 \mathbf{J}^2 J_z \pm 21 J_z^2 \mp 3 \mathbf{J}^2 + 19 J_z \pm 6) \tag{18}$$

$$J_0^{(4)} = \frac{1}{2} (70)^{-\frac{1}{2}} (35 J_z^4 - 30 \mathbf{J}^2 J_z^2 + 3 \mathbf{J}^4 + 25 J_z^2 - 6 \mathbf{J}^2) \tag{19}$$

B. Use Eqs. (2) and (3) to derive $J_{\pm 1}^{(2)}$ and $J_0^{(2)}$ from $J_{\pm 2}^{(2)}$.

C. The angular momentum tensors satisfy the relation

$$\left(J_q^{(k)}\right)^\dagger = (-1)^q J_{-q}^{(k)} \tag{20}$$

By explicit reference to Eqs. (8), (9), and (10), show that $(J_{+2}^{(2)})^\dagger = J_{-2}^{(2)}$, $(J_{+1}^{(2)})^\dagger = -J_{-1}^{(2)}$, and $(J_0^{(2)})^\dagger = J_0^{(2)}$. *Hint*: Recall that $(AB)^\dagger = B^\dagger A^\dagger$, where A and B are two operators.

We are interested in calculating the expectation values of the operators $J_q^{(k)}$. This motivates a more general consideration of what is the minimum set of information required to calculate the mean value of any operator Q for our system "prepared" by collision with another system or a photon. This question is most readily answered by using *density matrix methods*[1–3].

We suppose that our system is in a definite angular momentum state. If it is in a *pure state*, then its state vector $|\psi\rangle$ can be expanded in terms of the complete set of eigenvectors $|JM\rangle$

$$|\psi\rangle = \sum_M a_M |JM\rangle \tag{21}$$

where

$$\sum_M |a_M|^2 = 1 \tag{22}$$

The mean value of the operator Q is given by

$$\langle Q \rangle = \frac{\langle \psi | Q | \psi \rangle}{\langle \psi | \psi \rangle} = \sum_{M',M} a_{M'}^* a_M \langle JM' | Q | JM \rangle = \sum_{M',M} a_{M'}^* a_M Q_{M'M} \tag{23}$$

In the general case our system is in a *mixed state*, which is represented by an incoherent superposition of a number of pure states $|\psi^{(i)}\rangle$ with statistical weights $W^{(i)}$

$$|\psi\rangle = \sum_i W^{(i)} |\psi^{(i)}\rangle \tag{24}$$

where

$$\sum_i W^{(i)} = 1 \tag{25}$$

in other words, in this ensemble $W^{(i)}$ of the states are in $|\psi^{(i)}\rangle$. Then

$$\langle Q \rangle = \sum_i W^{(i)} \langle Q \rangle_i = \sum_i \sum_{M',M} W^{(i)} a_{M'}^{(i)*} a_M^{(i)} Q_{M'M} \tag{26}$$

The density matrix is defined as

$$\rho_{MM'} = \sum_i W^{(i)} a_{M'}^{(i)*} a_M^{(i)} \tag{27}$$

so that the expectation value of Q becomes

$$\langle Q \rangle = \sum_{M',M} \rho_{MM'} Q_{M'M} = \text{Tr}[\rho Q] \tag{28}$$

Here $\text{Tr}[A]$ indicates the trace of the matrix A, that is, the sum of the diagonal elements of A.

D. Prove that

$$\text{Tr}[\rho] = 1 \tag{29}$$

E. Prove that ρ is Hermitian, that is,

$$\langle JM'|\rho|JM \rangle^* = \langle JM|\rho|JM' \rangle \tag{30}$$

Clearly, a complete knowledge of ρ suffices for calculation of the expectation value of any arbitrary operator. We now consider the number of independent parameters needed to specify a given density matrix. Of course, this depends on the number of orthogonal states $|JM\rangle$ over which the sum in Eq. (21) ranges, but for angular momentum $\mathbf{J}$ this sum is restricted to $2J + 1$. For this case ρ is a $(2J + 1)$ by $(2J + 1)$ square matrix with $(2J + 1)^2$ complex elements corresponding to $2(2J + 1)^2$ real parameters. The hermiticity condition given in Eq. (30) shows that the number of real parameters is $(2J + 1)^2$, and the normalization condition given in Eq. (29) reduces this number to $(2J + 1)^2 - 1 = 4J(J + 1)$.

F. Show that for a system in a pure state of definite angular momentum $\mathbf{J}$ the number of real parameters necessary to specify the density matrix is $4J$.

The diagonal and off-diagonal elements of the density matrix have special significance. The diagonal element $\langle JM|\rho|JM \rangle = \sum_i W^{(i)} |a_M|^2$ is the probability of finding the system in the state $|JM\rangle$; that is, the diagonal elements of ρ represent the populations in the set of orthogonal eigenstates. However, the off-diagonal elements of ρ vanish unless some two eigenstates $|JM\rangle$ and $|JM'\rangle$ are in phase so that their ensemble average is nonzero; that is, the off-diagonal elements of ρ represent coherent preparation of the system when expressed in the $|JM\rangle$ representation.

Note that if a pure state in one representation is transformed to another representation, then in general the density matrix has both diagonal and off-diagonal elements. For a mixed state, the density matrix also has both diagonal and off-diagonal elements in general, but there is a difference in that the mixed state has lost phase information through the combination. Only for pure states is $\text{Tr}[\rho^2] = \text{Tr}[\rho] = 1$; for mixed states $\text{Tr}[\rho^2]$ is less than 1 and in the chaotic limit becomes $1/(2J+1)$, where the dimensionality is $2J+1$. To prove this, recall that the trace of a matrix equals the sum of its eigenvalues λ_i and that the eigenvalues of the density matrix correspond to populations and hence are positive definite. Then we may write $\text{Tr}[\rho] = \sum_i \lambda_i = 1$ and $\text{Tr}[\rho^2] = \sum_i \lambda_i^2$. Because $\sum_i \lambda_i^2 \leq (\sum_i \lambda_i)^2$, we conclude that $\text{Tr}[\rho^2] \leq 1$. The minimum value for $\text{Tr}[\rho^2]$ occurs in the uniform mixed state for which each eigenvalue λ_i equals $1/(2J+1)$. The pure states and the uniform mixed state of an ensemble are at opposite extremes with respect to the extent of order. A pure state represents a completely ordered ensemble, whereas the mixed state, which is uniform, is in a state of maximum disorder, that is, complete chaos. A more familiar measure of the lack of information or disorder is the entropy S, defined as $S = -\text{Tr}[\rho \ln \rho]$. Then for a pure state $S = 0$, whereas for a completely chaotic mixed state $S = \ln(2J+1)$.

We return to the problem of describing the orientation and alignment of a system in a definite state of angular momentum. This is accomplished by calculating the mean value of the angular momentum tensor components $J_q^{(k)}$. Hence the multipole moments of the system are proportional to

$$\langle J_q^{(k)} \rangle = \text{Tr}\left[\rho J_q^{(k)}\right] = \sum_{M',M} \rho_{MM'} \langle J M' | J_q^{(k)} | J M \rangle$$

$$\equiv \langle (J | J_q^{(k)} | J) \rangle \tag{31}$$

where the brackets $\langle \cdots \rangle$ in the last line of Eq. (31) remind us that this is an ensemble average. The multipole moments $\langle J_q^{(k)} \rangle$ contain all the dynamical information that can be learned about the system as to how the angular momentum vector points in space. As Eq. (31) shows, its full calculation requires a complete knowledge of the system's density matrix, a daunting task in general!

In many collision experiments we have axial symmetry, and when this is the case calculation of the $\langle J_q^{(k)} \rangle$ is simplified. For example, there is cylindrical symmetry along the beam axis of beam–gas scattering experiments (involving isotropic projectiles incident on isotropic targets), along the relative velocity vector of two crossed (isotropic) beams, along the propagation direction of a beam of unpolarized light incident on an isotropic gas sample, or along the electric vector of a beam of linearly polarized light incident on an isotropic gas sample. In such cases it is natural to choose the Z axis to lie along the axis of cylindrical symmetry. It then follows that the physical properties of the ensemble are independent of the choice of the X

and Y axes. Thus the density matrix is diagonal in M in this coordinate frame, and Eq. (31) takes on the particularly simple form

$$\langle (J | J_q^{(k)} | J) \rangle = \sum_M \rho_{MM} \langle JM | J_q^{(k)} | JM \rangle$$

$$= \sum_M \rho_{MM} (-1)^{J-M} \begin{pmatrix} J & k & J \\ -M & q & M \end{pmatrix} \langle J \| J^{(k)} \| J \rangle \qquad (32)$$

where the second line of Eq. (32) was obtained from the first by use of the Wigner–Eckart theorem [Eq. (5.14)]. Reference to the 3-j symbol in Eq. (32) shows that it vanishes for all nonzero values of q. Hence we conclude that in a system having axial symmetry the only nonvanishing multipole moments are $\langle J_0^{(k)} \rangle$ (provided the quantization axis coincides with the symmetry axis). Conversely, the existence of a $q \neq 0$ multipole moment implies that the system was prepared through a collision process not having axial symmetry.

For reference purposes we list the reduced matrix elements $\langle J \| J^{(k)} \| J \rangle$ for $k = 0$ through $k = 4$, on the basis of the convention of Eq. (5.14):

$$\langle J \| J^{(0)} \| J \rangle = (2J + 1)^{\frac{1}{2}} \qquad (33)$$

$$\langle J \| J^{(1)} \| J \rangle = [J(J + 1)(2J + 1)]^{\frac{1}{2}} \qquad (34)$$

$$\langle J \| J^{(2)} \| J \rangle = \left[\frac{J(J + 1)(2J - 1)(2J + 1)(2J + 3)}{6} \right]^{\frac{1}{2}} \qquad (35)$$

$$\langle J \| J^{(3)} \| J \rangle = \left[\frac{(J - 1)J(J + 1)(J + 2)(2J - 1)(2J + 1)(2J + 3)}{10} \right]^{\frac{1}{2}} \qquad (36)$$

$$\langle J \| J^{(4)} \| J \rangle =$$

$$\left[\frac{(J - 1)J(J + 1)(J + 2)(2J - 3)(2J - 1)(2J + 1)(2J + 3)(2J + 5)}{70} \right]^{\frac{1}{2}}$$

$$(37)$$

G. Verify Eq. (35).

H. Show that an axially symmetric system that also possesses the additional symmetry of a plane of reflection perpendicular to the axial symmetry axis has only nonvanishing multipole moments with even rank k, that is, only shows alignment. It follows that photodissociation or photoionization of an isotropic gas sample by a beam of plane polarized or unpolarized light can only produce photofragments that are aligned but not oriented. Similarly, beam–gas or beam–beam scattering in which a beam of isotropic particles impinges on an isotropic sample or beam of atoms or molecules can only produce products that are aligned (in the coordinate frame in which the quantization axis coincides with the axis of cylindrical symmetry).

I. To make the foregoing less abstract, consider a $J = 2$ system having axial symmetry. Let the magnetic sublevel populations $N(J, M)$ be

$$N(2, 2) \quad = \tfrac{1}{3}$$
$$N(2, 1) \quad = \tfrac{1}{6}$$
$$N(2, 0) \quad = 0$$
$$N(2, -1) \quad = \tfrac{1}{6}$$
$$N(2, -2) \quad = \tfrac{1}{3}$$

Find all the nonvanishing multipole moments $\langle J_q^{(k)} \rangle$ of this system.

For many problems it is convenient to introduce the dipolar orientation vector $O_q^{(1)}$ and the quadrupolar alignment tensor $A_q^{(2)}$. Of course, $O_q^{(1)}$ is proportional to $\langle J_q^{(1)} \rangle$ and $A_q^{(2)}$ to $\langle J_q^{(2)} \rangle$. Specifically, they are defined[4] by

$$O_q^{(1)}(J) = [J(J + 1)]^{-\frac{1}{2}} \operatorname{Re} \langle (J| J_q^{(1)} |J) \rangle \tag{38}$$

and

$$A_q^{(2)}(J) = (6)^{\frac{1}{2}} [J(J + 1)]^{-1} \operatorname{Re} \langle (J| J_q^{(2)} |J) \rangle \tag{39}$$

In the high J limit, where $\cos \theta = M/[J(J + 1)]^{1/2}$, we find that $O_0^{(1)}$ equals the ensemble average of $P_1(\cos \theta)$ and its value ranges from 1 to -1 and that $A_0^{(2)}$ becomes the ensemble average of $2 P_2(\cos \theta)$ and its value ranges from 2 to -1.

J. Use Eq. (39) to find the value of the alignment tensor component $A_0^{(2)}$ for the system described in part I. What is the average [root-mean-square (RMS)] value of the angle that $\mathbf{J}$ makes with the quantization axis?

As a further illustration of the power of these methods, we consider at some length the orientation and alignment produced by photoionization[5]. First we introduce the angular momentum persona:

$\mathbf{j}_0$ — angular momentum of target state before photoelectron ejection

1 — angular momentum of photon causing electric-dipole-allowed transition

s_e — spin angular momentum of photoelectron ($s_e = \frac{1}{2}$)

ℓ_e — orbital angular momentum of photoelectron ($\ell_e = 0, 1, 2, \ldots$)

$\mathbf{j}_e$ — total angular momentum of photoelectron ($\mathbf{j}_e = \boldsymbol{\ell}_e + \mathbf{s}_e$)

$\mathbf{j}_1$ — angular momentum of target state after photoelectron ejection

We wish to determine the spatial distribution of $\mathbf{j}_1$.

The total angular momentum of the system $\mathbf{j}$ is given by $\mathbf{j}_0 + \mathbf{1}$ prior to photoionization and by $\mathbf{j}_e + \mathbf{j}_1$ after photoionization. Conservation of angular momentum implies

$$\mathbf{j} = \mathbf{j}_0 + \mathbf{1} = \mathbf{j}_e + \mathbf{j}_1 \tag{40}$$

However, it is much more convenient to characterize the photoionization process not by the total angular momentum but by $\mathbf{j}_t$, the unobserved angular momentum transferred from the target to the photofragments:

$$\mathbf{j}_t = \mathbf{j}_e - \mathbf{j}_0 = \mathbf{1} - \mathbf{j}_1 \tag{41}$$

When so defined, $\mathbf{j}_t^2$ commutes with the angular momenta associated with the observed anisotropy, and hence the contributions to the anisotropy from each $\mathbf{j}_t$ add incoherently[6].

The photoionization cross section $\sigma(j_1 m_1)$ for leaving the target in the state $|j_1 m_1\rangle$ is proportional to the square of the dipole moment matrix element, averaged over all initial orientations of the target, and summed over all possible (undetected) photoelectron final states:

$$\sigma(j_1 m_1) = K_0 \sum_{m_0} \sum_{j_e, m_e} \left| \langle j_1 m_1, j_e m_e | r_q^{(1)} | j_0 m_0 \rangle \right|^2$$

$$= K_0 \sum_{m_0} \sum_{j_e, m_e} \left| \sum_{j,m} \langle j_1 m_1, j_e m_e | j m \rangle \langle j m | r_q^{(1)} | j_0 m_0 \rangle \right|^2$$

$$= K_0 \sum_{m_0} \sum_{j_e, m_e} \sum_{j,m} \sum_{j',m'} \langle j_1 m_1, j_e m_e | j' m' \rangle \langle j_1 m_1, j_e m_e | j m \rangle$$

$$\times \langle j' m' | r_q^{(1)} | j_0 m_0 \rangle^* \langle j m | r_q^{(1)} | j_0 m_0 \rangle \tag{42}$$

In Eq. (42) K_0 is a proportionality constant and the index q describes the polarization state of the dipole photon ($q = 0$ corresponds to linearly polarized light, and $q = \pm 1$ corresponds to left- and right-handed circularly polarized light, respectively).

K. Show that

$$\sigma(j_1 m_1) = K_0 \sum_{j_e} \sum_{j,j'} [(2j+1)(2j'+1)]^{\frac{1}{2}} \langle j' \| r^{(1)} \| j_0 \rangle^* \langle j \| r^{(1)} \| j_0 \rangle$$

$$\times \sum_{j_t} (2j_t + 1) \left\{ \begin{matrix} j_1 & j_e & j' \\ j_0 & 1 & j_t \end{matrix} \right\} \left\{ \begin{matrix} j_1 & j_e & j \\ j_0 & 1 & j_t \end{matrix} \right\}$$

$$\times \begin{pmatrix} j_1 & 1 & j_t \\ m_1 & -q & q - m_1 \end{pmatrix}^2 \tag{43}$$

from which it follows that

$$\sigma(j_1 m_1) = \sum_{j_t} \sigma(j_t) \langle j_1 m_1, j_t \, q - m_1 | 1 q \rangle^2 \tag{44}$$

that is, the cross section $\sigma(j_1 m_1)$ has the form of an incoherent summation over the angular momentum transfer quantum number. *Hint*: Use the Wigner–Eckart theorem and the identity (4.16).

L. We wish to calculate the orientation and alignment

$$O_0^{(1)}(j_1) = \frac{\sum_{m_1} \sigma(j_1 m_1) m_1 [j_1(j_1+1)]^{-\frac{1}{2}}}{\sum_{m_1} \sigma(j_1 m_1)} \tag{45}$$

and

$$A_0^{(2)}(j_1) = \frac{\sum_{m_1} \sigma(j_1 m_1) \left[3 m_1^2 - j_1(j_1+1) \right] / [j_1(j_1+1)]}{\sum_{m_1} \sigma(j_1 m_1)} \tag{46}$$

and find the contribution for each angular momentum transfer continuum channel. With the help of Eq. (44), we may rewrite these expressions as

$$O_0^{(1)}(j_1) = \frac{\sum_{j_t} \sigma(j_t) O_0^{(1)}(j_1, j_t; q)}{\sum_{j_t} \sigma(j_t)} \tag{47}$$

and

$$A_0^{(2)}(j_1) = \frac{\sum_{j_t} \sigma(j_t) A_0^{(2)}(j_1, j_t; q)}{\sum_{j_t} \sigma(j_t)} \tag{48}$$

Find expressions for $O_0^{(1)}(j_1, j_t; q)$ and $A_0^{(2)}(j_1, j_t; q)$, and show that

$$O_0^{(1)}(j_1, j_t; q) = -\frac{1}{2}q \left[j_1/(j_1 + 1) \right]^{\frac{1}{2}} \qquad j_t = j_1 + 1$$

$$O_0^{(1)}(j_1, j_t; q) = \frac{1}{2}q \left[j_1(j_1 + 1) \right]^{-\frac{1}{2}} \qquad j_t = j_1$$

$$O_0^{(1)}(j_1, j_t; q) = \frac{1}{2}q \left[(j_1 + 1)/j_1 \right]^{\frac{1}{2}} \qquad j_t = j_1 - 1 \qquad (49)$$

and that

$$A_0^{(2)}(j_1, j_t; q = 0) = -\frac{2}{5} + \frac{3}{5(j_1 + 1)} \qquad j_t = j_1 + 1$$

$$A_0^{(2)}(j_1, j_t; q = 0) = \frac{4}{5} - \frac{3}{5j_1(j_1 + 1)} \qquad j_t = j_1$$

$$A_0^{(2)}(j_1, j_t; q = 0) = -\frac{2}{5} - \frac{3}{5j_1} \qquad j_t = j_1 - 1 \qquad (50)$$

Hint: Use the results of Eqs. (1.97)–(1.99) of Problem Set 1.

A number of conclusions follow from Eqs. (49) and (50). For example, $O_0^{(1)}(j_1, j_t; q)$ vanishes for $q = 0$, which implies that interaction of the (isotropic) target atom or molecule with linearly polarized light cannot produce orientation of the target following photoionization. Moreover, $O_0^{(1)}(j_1, j_t; q)$ changes sign if the sign of the incident photon circular polarization is changed. Unpolarized light may be regarded as an equal weighting of $q = +1$ and $q = -1$. Therefore, interaction of the (isotropic) target with unpolarized light also cannot produce orientation in the target.

The orientation and alignment of the target state following photoionization can be expressed in terms of the contribution from three angular momentum transfer continuum channels $j_t = j_1 + 1$, $j_t = j_1$, and $j_t = j_1 - 1$. Let the fractional contribution of each channel be denoted by $f(j_t)$, where

$$\sum_{j_t} f(j_t) = 1 \qquad (51)$$

Then the orientation and alignment of the final state of the target may be written as

$$O_0^{(1)}(j_1) = \sum_{j_t} f(j_t) O_0^{(1)}(j_1, j_t; q) \qquad (52)$$

and

$$A_0^{(2)}(j_1) = \sum_{j_t} f(j_t) A_0^{(2)}(j_1, j_t; q) \tag{53}$$

Hence a measurement of $O_0^{(1)}(j_1)$ and $A_0^{(2)}(j_1)$ provides sufficient information to extract all the continuum channel contributions $f(j_t)$. Equations (51)–(53) completely disentangle the dynamics of the photoionization process from geometric terms involving the angular momentum of the resulting target state and the polarization state of the absorbed photon.

We return once more to the description of orientation and alignment in a state of definite angular momentum in order to develop a different point of view. Following Fano[2], we introduce the operators

$$T(J)_{KQ} = \sum_{M',M} (-1)^{J-M} \langle JM', J-M|KQ\rangle |JM'\rangle \langle JM| \tag{54}$$

which have the matrix elements

$$\langle JM'|T(J)_{KQ}|JM\rangle = (-1)^{J-M} \langle JM', J-M|KQ\rangle \tag{55}$$

The nonvanishing of the Clebsch–Gordan coefficient $\langle JM', J-M|KQ\rangle$ in the definition of $T(J)_{KQ}$ shows that K is restricted to integral values extending from 0 to $2J$ and that Q ranges from $-K$ to K in integral steps.

M. Consider a rotation $\mathbf{R}$ acting on $T(J)_{KQ}$. Show that

$$\mathbf{R}\, T(J)_{KQ} \mathbf{R}^{-1} = \sum_{Q'} D_{Q'Q}^K(R) T(J)_{KQ'} \tag{56}$$

from which it follows that $T(J)_{KQ}$ is a spherical tensor operator of rank K and component Q. The $T(J)_{KQ}$ are called *state multipoles*, a designation whose appropriateness will become evident as we proceed.

Clearly, the $T(J)_{KQ}$ span the $(2J+1)$ by $(2J+1)$ space of the density matrix:

$$\rho = \sum_{M',M} \langle JM'|\rho|JM\rangle |JM'\rangle \langle JM| \tag{57}$$

Equation (54) is readily inverted by multiplying both sides by $\langle JN', J-N|KQ\rangle$ and summing over all values of K and Q:

$$|JN'\rangle \langle JN| = \sum_{K,Q} (-1)^{J-N} \langle JN', J-N|KQ\rangle T(J)_{KQ} \tag{58}$$

Specifically we may decompose ρ as

$$\rho = \sum_{K,Q} \rho_{KQ} T_{KQ} \tag{59}$$

N. Invert Eq. (59) to show that

$$\rho_{KQ} = \mathrm{Tr}\left[T_{KQ}^{\dagger} \rho\right] = \left\langle T_{KQ}^{\dagger} \right\rangle \tag{60}$$

Hint: Use the orthonormality properties of the Clebsch–Gordan coefficients to prove that

$$\mathrm{Tr}\left[T_{K'Q'}^{\dagger} T_{KQ}\right] = \delta_{K'K} \delta_{Q'Q} \tag{61}$$

where $\langle JM| T(J)_{KQ}^{\dagger} |JM'\rangle = \langle JM'| T(J)_{KQ} |JM\rangle^{*}$.

The state multipoles T_{KQ} transform as the spherical harmonics Y_{KQ} and the expansion coefficients $\rho_{KQ} = \langle T_{KQ}^{\dagger}\rangle$ must be interpreted as the multipole moments of the state (up to some undetermined constant that depends on the normalization chosen). Substitution of Eq. (58) in Eq. (57) and comparison with Eq. (59) shows that

$$\left\langle T_{KQ}^{\dagger} \right\rangle = \sum_{M',M} (-1)^{J-M'} \langle JM', J-M| KQ\rangle \langle JM'| \rho |JM\rangle \tag{62}$$

Alternatively, Eq. (62) can be inverted to yield the expression

$$\langle JM'| \rho |JM\rangle = \sum_{K,Q} (-1)^{J-M'} \langle JM', J-M| KQ\rangle \left\langle T(J)_{KQ}^{\dagger} \right\rangle \tag{63}$$

Hence the two descriptions of a system in terms of density matrix elements and its multipole moments are equivalent. To determine one is to determine the other.

However, we have previously shown that the $\langle J_q^{(k)}\rangle$ also describe the multipole moments of the system. Clearly, the $\langle J_q^{(k)}\rangle$ and the $\langle T(J)_{KQ}^{\dagger}\rangle$ must be related. Indeed, this must be the case because both transform as Y_{KQ}. Hence to determine $\langle J_q^{(k)}\rangle$ is equivalent to determining the state multipoles of a system.

We establish the proportionality constant. Application of the Wigner–Eckart theorem to Eq. (55) shows that $\langle J|| T^K ||J\rangle = (2K+1)^{1/2}$. Therefore

$$\frac{\langle JM| T_{KQ}^{\dagger} |JM\rangle}{\langle J|| T^K ||J\rangle} = \frac{\langle JM| J_q^{(k)} |JM\rangle}{\langle J|| J^{(k)} ||J\rangle} \tag{64}$$

and it follows that

$$\left\langle T(J)_{KQ}^{\dagger} \right\rangle = \left\langle J_q^{(k)} \right\rangle \frac{\langle J \| T^K \| J \rangle}{\langle J \| J^{(k)} \| J \rangle} \tag{65}$$

that is,

$$\left\langle T(J)_{00}^{\dagger} \right\rangle = \frac{1}{(2J+1)^{\frac{1}{2}}}$$

$$\left\langle T(J)_{1Q}^{\dagger} \right\rangle = \left[\frac{3}{J(J+1)(2J+1)} \right]^{\frac{1}{2}} \left\langle J_q^{(1)} \right\rangle$$

$$\left\langle T(J)_{2Q}^{\dagger} \right\rangle = \left[\frac{30}{J(J+1)(2J-1)(2J+1)(2J+3)} \right]^{\frac{1}{2}} \left\langle J_q^{(2)} \right\rangle \tag{66}$$

and so on.

O. Consider an electronically excited atom or molecule with initial angular momen-
    tum quantum number J_i that makes a spontaneous electric dipole transition to
    a final state characterized by the angular momentum quantum number J_f. Let
    the M_i population of the initial state be distributed according to the probability
    law

$$P(J_i, M_i) = \frac{1}{2J_i + 1} \left\{ 1 + O_0^{(1)}(J_i) \frac{M_i}{[J_i(J_i+1)]^{\frac{1}{2}}} \right.$$

$$\left. + A_0^{(2)}(J_i) \left[\frac{3M_i^2 - J_i(J_i+1)}{J_i(J_i+1)} \right] \right\} \tag{67}$$

that is, the initial state has cylindrical symmetry with an orientation characterized
by the value $O_0^{(1)}(J_i)$ and an alignment characterized by the value $A_0^{(2)}(J_i)$.
Present arguments to indicate that the probability $P(J_f, M_f)$ of populating the
final state J_f, M_f by the radiative transition is

$$P(J_f, M_f) = \sum_{M_i} P(J_i, M_i) \langle J_i M_i, 1 M_f - M_i | J_f M_f \rangle^2 \tag{68}$$

Explicit evaluation of Eq. (68) shows that[7]

$$P(J_f, M_f) = \frac{1}{2J_f + 1} \left\{ 1 + O_0^{(1)}(J_f) \frac{M_f}{[J_f(J_f+1)]^{\frac{1}{2}}} \right.$$

$$\left. + A_0^{(2)}(J_f) \left[\frac{3M_f^2 - J_f(J_f+1)}{J_f(J_f+1)} \right] \right\} \tag{69}$$

where

$$O_0^{(1)}(J_f) = C^{(1)}(J_i, J_f) O_0^{(1)}(J_i) \tag{70}$$

and

$$A_0^{(2)}(J_f) = C^{(2)}(J_i, J_f) A_0^{(2)}(J_f) \tag{71}$$

The values of the proportionality constants are for $J_f = J_i - 1$

$$C^{(1)} = 1 - \frac{1}{(J_f + 1)^2} \tag{72}$$

$$C^{(2)} = 1 - \frac{3}{(J_f + 1)(2J_f + 3)} \tag{73}$$

for $J_f = J_i$

$$C^{(1)} = 1 - \frac{1}{J_f(J_f + 1)} \tag{74}$$

$$C^{(2)} = 1 - \frac{3}{J_f(J_f + 1)} \tag{75}$$

and for $J_f = J_i + 1$:

$$C^{(1)} = 1 \tag{76}$$

$$C^{(2)} = 1 - \frac{3}{J_f(2J_f - 1)} \tag{77}$$

Hence the orientation and alignment of the final state is proportional to the orientation and alignment of the initial state following a spontaneous electric dipole transition from J_i to J_f; moreover, for J values large compared to the angular momentum carried off by the photon, $C^{(1)}(J_i, J_f)$ and $C^{(2)}(J_i, J_f)$ both approach unity; that is, the degree of orientation and alignment is unchanged by the spontaneous electric dipole transition.

As a concluding example, consider the time evolution of a composite system having angular momenta $\mathbf{J}$ and $\mathbf{I}$. Here we have in mind a molecule with rotational angular momentum $\mathbf{J}$ and nuclear spin angular momentum $\mathbf{I}$. We suppose that our system initially has no preference for how $\mathbf{J}$ or $\mathbf{I}$ points in space. Suppose at time $t = 0$ that a photon or particle collision causes $\mathbf{J}$ to become oriented or aligned but does not act on $\mathbf{I}$. Consequently, the system acquires through the collision process a multipole moment of its $\mathbf{J}$ distribution, $\langle J_q^{(k)}(t = 0)\rangle = \langle J_q^{(k)}\rangle$ but an isotropic distribution of its $\mathbf{I}$ distribution $\langle I_0^{(0)}\rangle$. After the collision, $\mathbf{J}$ and $\mathbf{I}$ couple to form the resultant $\mathbf{F}$, which commutes with the system Hamiltonian $\mathcal{H}$. According to the vector model, the coupling of $\mathbf{J}$ and $\mathbf{I}$ causes $\mathbf{J}$ to precess about $\mathbf{F}$. Hence we expect that the presence of hyperfine structure, even if unresolved, causes $\langle J_q^{(k)}(t) \rangle$ to be less than $\langle J_q^{(k)}(0) \rangle$, that is, depolarizes the multipole moments of $\mathbf{J}$. We wish to determine the time dependence of $\langle J_q^{(k)}(t) \rangle$ and make these comments quantitative.

Using the Wigner–Eckart theorem we replace the matrix elements of one spherical tensor for those of another with the same rank and component:

$$\langle (J| J_q^{(k)}(t) |J) \rangle = \langle (J| J_q^{(k)}(0) |J) \rangle \frac{\langle J|| J^{(k)}(t) ||J \rangle}{\langle J|| J^{(k)} ||J \rangle} \tag{78}$$

Thus the time dependence of $\langle (J| J_q^{(k)}(t) |J) \rangle$ is contained entirely in the time dependence of the reduced matrix element $\langle J|| J^{(k)}(t) ||J \rangle$. We use Eq. (5.68) to evaluate this. Let

$$X_q^{(k)}(t) = \left[J_q^{(k)}(t) \otimes I_0^{(0)} \right]_q^{(k)} \tag{79}$$

be a spherical tensor of rank k and component q constructed from the angular momentum tensor operators in J space and I space. Then in the coupled representation

$$\langle JIF'|| X^{(k)}(t) ||JIF \rangle$$

$$= \langle JIF'|| \left[J_q^{(k)}(t) \otimes I_0^{(0)} \right]^k ||JIF \rangle$$

$$= [(2F' + 1)(2F + 1)(2k + 1)]^{\frac{1}{2}} \begin{Bmatrix} J & J & k \\ I & I & 0 \\ F' & F & k \end{Bmatrix}$$

$$\times \langle J|| J^{(k)}(t) ||J \rangle \langle I|| I^{(0)} ||I \rangle \tag{80}$$

We solve Eq. (80) for $\langle J|| J^{(k)}(t) ||J \rangle$ by using the orthonormality properties of the 9-j symbols:

$$\langle J|| J^{(k)}(t) ||J \rangle$$

$$= \sum_{F',F} \frac{\langle JIF'|| X^{(k)}(t) ||JIF \rangle}{\langle I|| I^{(0)} ||I \rangle} \begin{Bmatrix} J & J & k \\ I & I & 0 \\ F' & F & k \end{Bmatrix} (2k + 1)^{\frac{1}{2}} \tag{81}$$

But $X^{(k)}(t)$ is related to $X^{(k)}(t = 0)$ by a unitary transformation involving the time evolution operator $\mathbf{U}(t)$ [see Eq. (3.16)]

$$X^{(k)}(t) = e^{-i\mathcal{H}t/\hbar} X^{(k)}(0) e^{i\mathcal{H}t/\hbar} \tag{82}$$

where $\mathcal{H}$ is the system Hamiltonian, which is diagonal in the coupled representation with eigenvalues E_F. Thus it is readily shown that

$$\langle J \| J^{(k)}(t) \| J \rangle = \langle J \| J^{(k)} \| J \rangle \, G^{(k)}(t) \tag{83}$$

where

$$G^{(k)}(t) = \sum_{F,F'} e^{i(E_{F'}-E_F)t/\hbar}$$

$$\times (2F'+1)(2F+1)(2k+1) \left\{ \begin{array}{ccc} J & J & k \\ F' & F & k \\ I & I & 0 \end{array} \right\}^2$$

$$= \sum_{F,F'} e^{i(E_{F'}-E_F)t/\hbar} \frac{(2F'+1)(2F+1)}{2I+1} \left\{ \begin{array}{ccc} F' & F & k \\ J & J & I \end{array} \right\}^2 \tag{84}$$

Under interchange of F' and F, the imaginary terms are odd and thus cancel in the sum. Hence

$$G^{(k)}(t) = \sum_{F,F'} \cos[(E_{F'} - E_F)t/\hbar] \frac{(2F'+1)(2F+1)}{2I+1} \left\{ \begin{array}{ccc} F' & F & k \\ J & J & I \end{array} \right\}^2 \tag{85}$$

Substitution of Eq. (83) into Eq. (78) shows that

$$\langle (J | J_q^{(k)}(t) | J) \rangle = \langle (J | J_q^{(k)} | J) \rangle \, G^{(k)}(t) \tag{86}$$

We see that the interaction between **J** and **I** does not mix multipole moments of **J** but causes their value in general to be reduced. If the hyperfine splitting is small, the $F' \neq F$ terms oscillate rapidly about zero and the time-averaged value of $G^{(k)}(t)$ is given by the $F' = F$ terms in Eq. (85), namely,

$$\langle G^{(k)}(t) \rangle_{av} = \frac{1}{2I+1} \sum_F (2F+1)^2 \left\{ \begin{array}{ccc} F & F & k \\ J & J & I \end{array} \right\}^2 \tag{87}$$

NOTES AND REFERENCES

1. K. Blum, *Density Matrix Theory and Applications*, Plenum Press, New York, 1981.

2. U. Fano, *Rev. Mod. Phys.*, **29**, 74 (1957).

3. D. ter Haar, *Rep. Progr. Phys.*, **24**, 304 (1961).

4. The definition of $A_q^{(2)}(J)$ is from U. Fano and J. H. Macek, *Rev. Mod. Phys.*, **45**, 553 (1973). The definition of $O_q^{(1)}(J)$ is $[J(J+1)]^{1/2}$ times larger than that of Fano and Macek.

5. C. H. Greene and R. N. Zare, *Ann. Rev. Phys. Chem.*, **33**, 119 (1982).

6. The theory of angular momentum transfer is developed in U. Fano and D. Dill, *Phys. Rev.*, **A6**, 185 (1972); D. Dill and U. Fano, *Phys. Rev. Lett.*, **29**, 1203 (1972); D. Dill, *Phys. Rev.*, **A7**, 1976 (1973); D. Dill and J. L. Dehmer, *J. Chem. Phys.*, **61**, 692 (1974); J. L. Dehmer and D. Dill, *J. Chem. Phys.*, **65**, 5327 (1976); D. Dill, S. Wallace, J. Siegel, and J. L. Dehmer, *Phys. Rev. Lett.*, **41**, 1230 (1978); ibid., **42**, 411 (1979). See also U. Fano and A. R. P. Rau, *Atomic Collisions and Spectra*, Academic Press, Orlando, FL, 1986.

7. R. Bersohn and S. H. Lin, *Adv. Chem. Phys.*, **16**, 67 (1969).

APPLICATION 14
NUCLEAR QUADRUPOLE INTERACTIONS

Armed with a reference library for evaluating spherical tensor matrix elements, let us put these to use in one important application, the interaction of the electric quadrupole moment of the nucleus with the electric field gradient at the nucleus caused by all electronic (and other nuclear) charges in the atom (or molecule)[1–6]. Specifically, we will examine how nuclear quadrupole interactions affect the appearance of pure rotational spectra of molecules and how such spectra can provide information about the electronic charge distribution in the molecule. The rotational spectrum of the linear triatomic molecule, cyanogen chloride (ClCN), is chosen as an example.

The rotational energy levels of a rigid rotor have the form

$$E(J) = BJ(J + 1) \tag{1}$$

where B is the rotational constant and J the rotational quantum number. For electric-dipole-allowed rotational transitions, $\Delta J = \pm 1$ and the so-called pure rotation spectrum consists of a series of lines located at

$$h\nu = E(J + 1) - E(J) = B(J + 1)(J + 2) - BJ(J + 1) = 2B(J + 1) \tag{2}$$

These considerations predict a simple spectrum of equally spaced lines having a frequency separation equal to twice the rotational constant.

Indeed this is what is observed for $^{35}Cl\,^{12}C\,^{14}N$ and $^{37}Cl\,^{12}C\,^{14}N$ under low resolution, each molecule having a slightly different rotational constant because of the differing masses of the ^{35}Cl and ^{37}Cl isotopes. However, under higher resolution the pure rotation spectrum is actually far from simple! Figure 1 shows the $^{35}ClCN$ microwave spectrum (the $^{37}ClCN$ microwave spectrum looks qualitatively very similar) taken at two different pressures, 50 and 1 mTorr, using Stark modulated phase sensitive detection to yield first-derivative lineshapes. These have been kindly recorded for us by Professor John S. Muenter, Department of Chemistry, University of Rochester, who suggested this example. The static electric field is parallel to the electric field of the microwave radiation, and in the absence of nuclear quadrupole interaction, Figure 1 corresponds to a $J = 0, M = 0$ to $J = 1, M = 0$ transition. Clearly, there is not one line, as predicted by Eq. (2), but several closely spaced lines. Why? Has the rigid rotor model failed?

The answer is that our treatment is incomplete; it does not consider the interaction of the rotational angular momentum of the molecule with the spins of the nuclei. In ClCN, Cl has $I = \frac{3}{2}$ (which holds for ^{35}Cl and ^{37}Cl) N has $I = 1$ (which holds for ^{14}N), and C has $I = 0$ (which holds for ^{12}C). Figure 1 shows that the $J = 0$ to $J = 1$ transition of ClCN has three lines for just the $^{35}ClCN$ species spread out over 35 MHz. Under higher resolution two of these lines split up into triplets and one into a doublet with these new splittings ranging from 1.2 MHz to 50 kHz. So there are eight lines instead of just one.

The $J = 0$ level does not split into different components because, as will be shown below, a rotational level with zero rotational angular momentum cannot couple with nuclear spins. For the $J = 1$ level, the major coupling is with the Cl nucleus and the more minor coupling is with the N nucleus. The vector sum of $\mathbf{J}$ and $\mathbf{I}$ is traditionally labeled $\mathbf{F}$. In the case of a single nuclear spin, $\mathbf{F}$ is the total angular momentum. If there are two nuclear spins, as in ClCN, we need to introduce a coupling scheme. We label the two nuclear spins $\mathbf{I}_1$ and $\mathbf{I}_2$. Then $\mathbf{J} + \mathbf{I}_1 = \mathbf{F}_1$ and $\mathbf{F}_1 + \mathbf{I}_2 = \mathbf{F}$. We call $\mathbf{F}_1$ the intermediate angular momentum and $\mathbf{F}$ still the total. We have two possible coupling schemes, depending on whether we associate $\mathbf{I}_1$ with the nuclear spin of Cl or N. For ClCN, the interaction between $\mathbf{J}$ and the spin on the Cl nucleus is dominant and we take $I_1 = \frac{3}{2}$ and $I_2 = 1$. Thus in the $J = 1$ level, F_1 can be $\frac{1}{2}, \frac{3}{2}$, or $\frac{5}{2}$ corresponding to the three lines seen in Figure 1 at lower resolution. Coupling $F_1 = \frac{1}{2}$ with $I_2 = 1$ gives $F = \frac{1}{2}$ or $\frac{3}{2}$, coupling $F_1 = \frac{3}{2}$ with $I_2 = 1$ gives $F = \frac{1}{2}, \frac{3}{2}$, or $\frac{5}{2}$, and coupling $F_1 = \frac{5}{2}$ with $I_2 = 1$ gives $F = \frac{3}{2}, \frac{5}{2}$ or $\frac{7}{2}$. Hence the $J = 1$ level is split into three levels by the Cl nuclear spin angular momentum (upper trace of Figure 1) which is what is observed at normal sample pressure since pressure broadening is sufficiently large to obscure the smaller splittings from N. However, at reduced pressure (lower trace of Figure 1) the N nuclear spin angular momentum causes each of the three levels to split further, one into a doublet (unresolved) and the other two into triplets. The $J = 0$ to $J = 1$ transition of ClCN has been studied by Lafferty, Lide, and Toth[7] and we list below the line frequencies associated with (F_1, F) for the $J = 1$ level of ^{35}ClCN:

ν (MHz)	(F_1, F)	ν (MHz)	(F_1, F)
11,924.33	$\frac{3}{2}, \frac{1}{2}$	11,946.06	$\frac{5}{2}, \frac{7}{2}$
11,924.89	$\frac{3}{2}, \frac{5}{2}$	11,946.32	$\frac{5}{2}, \frac{3}{2}$
11,925.57	$\frac{3}{2}, \frac{3}{2}$	11,962.56	$\frac{1}{2}, \frac{1}{2}$
11,945.27	$\frac{5}{2}, \frac{5}{2}$	11,962.56	$\frac{1}{2}, \frac{3}{2}$

The last two transitions are calculated to be only 50 kHz apart and were not resolved.

A. Even these elementary considerations suffice to show that the coupling of $\mathbf{J}$ with $\mathbf{I}($Cl$)$ is dominant in ClCN. Discuss how the appearance of the $J = 0$ to $J = 1$ transition of ClCN would differ from that in Figure 1 if the coupling of $\mathbf{J}$ with the nitrogen nuclear spin dominated. Would the total number of lines change? Of course, both coupling schemes provide a complete set of basis functions for describing the nuclear quadrupole couplings and would give the same answer when all interactions are fully considered. However, as is often the case, one coupling scheme gives a better zero-order approximation to the observed splittings. We turn next to the form of the coupling between $\mathbf{J}$ and $\mathbf{I}$ that causes the hyperfine structure seen in Figure 1.

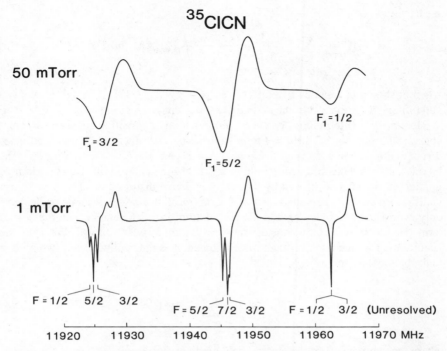

FIGURE 1   Microwave spectrum of the ^{35}ClCN $J = 0, M = 0 \rightarrow J = 1, M = 0$ rotational transition, taken by J. S. Muenter at two different pressures.

The classical expression for the electrostatic interaction of a nuclear charge density $\rho_n(\mathbf{r}_n)$ with position vector $\mathbf{r}_n$ and an electron charge density $\rho_e(\mathbf{r}_e)$ with position vector $\mathbf{r}_e$ is

$$W = -\int_n \rho_n(\mathbf{r}_n) \left[\int_e \frac{\rho_e(\mathbf{r}_e)}{|\mathbf{r}_n - \mathbf{r}_e|} d^3 r_e \right] d^3 r_n \tag{3}$$

where the center of the nucleus is taken as the origin of the coordinate system. For simplicity, we ignore other nuclei in the molecule (or surroundings) and assume that the charge distribution of the electrons is *entirely outside* the nuclear charge distribution, that is, $r_e > r_n$. Making use of the expansion (see Application 11)

$$\frac{1}{|\mathbf{r}_e - \mathbf{r}_n|} = \sum_{k=0}^{\infty} \frac{r_n^k}{r_e^{k+1}} P_k(\cos \Theta)$$

$$= \sum_{k=0}^{\infty} \frac{r_n^k}{r_e^{k+1}} \mathbf{C}^{(k)}(e) \cdot \mathbf{C}^{(k)}(n) \tag{4}$$

we have

$$W = -\sum_{k=0}^{\infty} \int \frac{\rho_e \mathbf{C}^{(k)}(e) d^3 r_e}{r_e^{k+1}} \int \rho_n r_n^k \mathbf{C}^{(k)}(n) d^3 r_n \tag{5}$$

where it should be recalled that $C_q^{(k)}(\theta,\phi) = [4\pi/(2k+1)]^{1/2} Y_{kq}(\theta,\phi)$ and (e) and (n) stand for the electronic and nuclear coordinates, respectively.

Here the $k = 0$ term represents the Coulomb term of attraction between two point charges of separation r_e, the $k = 1$ term vanishes because nuclei in their ground states do not possess permanent electric dipole moments, and the $k = 2$ term represents the interaction of the electric quadrupole moment of the nucleus with the electric field gradient at the nucleus caused by the electron charge distribution. The $k = 3$ term vanishes for the same reason as the $k = 1$ term (recall that nondegenerate nuclei have only even electric moments), and the $k = 4$ term is much smaller than the $k = 2$ term. Hence we need only consider the $k = 2$ term in what follows. We write the Hamiltonian for the nuclear quadrupole interaction as the contraction of two second rank spherical tensors

$$H_Q = \mathbf{V}^{(2)} \cdot \mathbf{Q}^{(2)} = \sum_{q=-2}^{2} (-1)^q V_q^{(2)} Q_{-q}^{(2)} \tag{6}$$

where the components of the electric field gradient tensor are

$$V_q^{(2)} = \int \left[\rho_e(\mathbf{r}_e) r_e^{-3} \right] C_q^{(2)}(\theta_e, \phi_e) d^3 r_e \tag{7}$$

and those of the nuclear quadrupole tensor are

$$Q_q^{(2)} = \int \left[r_n^2 \rho_n(\mathbf{r}_n) \right] C_q^{(2)}(\theta_n, \phi_n) d^3 r_n \tag{8}$$

To make Eqs. (7) and (8) into valid quantum mechanical expressions, we need to replace the charge densities by expectation values of the appropriate operators. We begin with the electric field gradient at the nucleus. For a nucleus at the center of the coordinate system and an electron located on the Z axis a distance r_e away, the electric field gradient can be written in terms of the potential $V = e/r_e$ as

$$q_{ZZ} = \frac{\partial^2 V}{\partial Z^2} = \frac{2e}{r_e^3} \tag{9}$$

If the electron is located off the Z axis by an angle θ_e, then

$$q_{ZZ} = \frac{\partial^2 V}{\partial Z^2} = e \left(\frac{3\cos^2\theta_e - 1}{r_e^3} \right) \tag{10}$$

and the generalization of Eq. (10) to an electron charge density distribution $\rho_e(\mathbf{r}_e)$ is

$$q_{ZZ} = \int \rho_e(\mathbf{r}_e) \left(\frac{3 \cos^2 \theta_e - 1}{r_e^3} \right) d^3 r_e \tag{11}$$

The quantum mechanical transcription of Eq. (11) is readily obtained. We define the electric field gradient coupling constant as

$$q_{ZZ} = 2 \langle \alpha J J | V_0^{(2)} | \alpha J J \rangle \tag{12}$$

where $|\alpha J J\rangle$ denotes the rigid rotor wave function with $\mathbf{J}$ making a maximum projection on the space-fixed Z axis of $M_J = J$.

B. Show that q_{ZZ} vanishes for $J = 0$ or $J = \frac{1}{2}$ and show that for $J \geq 1$

$$\langle \alpha J \| V^{(2)} \| \alpha J \rangle = \frac{1}{2} q_{ZZ} \left[\frac{(2J+1)(2J+2)(2J+3)}{(2J)(2J-1)} \right]^{\frac{1}{2}} \tag{13}$$

In a similar manner it is traditional to define the electric quadrupole moment of the nucleus as

$$eQ = \int r_n^2 \rho_n(\mathbf{r}_n)(3 \cos^2 \theta_n - 1) d^3 r_n \tag{14}$$

and just as in Eq. (12) to write

$$eQ = 2 \langle II | Q_0^{(2)} | II \rangle \tag{15}$$

where $|II\rangle$ is the nuclear spin function with $\mathbf{I}$ making a maximum projection on the space-fixed Z axis of $M_I = I$. In exact analogy with Eq. (13) it follows that the nuclear electric quadrupole moment vanishes for $I = 0$ or $I = \frac{1}{2}$, and for $I \geq 1$

$$\langle I \| Q^{(2)} \| I \rangle = \frac{1}{2} eQ \left[\frac{(2I+1)(2I+2)(2I+3)}{(2I)(2I-1)} \right]^{\frac{1}{2}} \tag{16}$$

Reference to Eq. (14) shows the nuclear quadrupole moment takes on a positive value if the nuclear charge distribution is prolate (elongated), takes on a negative value if the nuclear charge distribution is oblate (flattened), and vanishes for a spherical nuclear charge distribution as in the states $I = 0$ and $I = \frac{1}{2}$.

C. Consider the coupled states $|JIFM_F\rangle$. Find an expression for the nuclear quadrupole interaction energy using first-order perturbation theory. *Hint*: Show that

$$\Delta E = \langle \alpha JIFM_F | \mathbf{V}^{(2)} \cdot \mathbf{Q}^{(2)} | \alpha JIFM_F \rangle$$

$$= \frac{1}{2} eq_{ZZ} Q \left[\frac{\frac{3}{4} C(C+1) - I(I+1)J(J+1)}{I(2I-1)J(2J-1)} \right] \tag{17}$$

where

$$C = F(F+1) - J(J+1) - I(I+1) \tag{18}$$

The factor $eq_{ZZ}Q$ is called the *quadrupole coupling constant*. If q_{ZZ} can be calculated from the electronic structure, then a measurement of the nuclear quadrupole coupling constant allows a determination of the nuclear constant Q.

Equation (17) is a celebrated result that can be applied to both atoms and molecules. It assumes that the quadrupole splitting is much smaller than the spacing between different J levels. When this is not the case, we must solve a secular determinant involving interacting levels with the same F but different J. This introduces matrix elements of the form $\langle J'IFM_F | \mathbf{V}^{(2)} \cdot \mathbf{Q}^{(2)} | JIFM_F \rangle$ and can cause levels with $\Delta J = 0, \pm 1$, and ± 2 to interact. Fortunately for ClCN, the $J = 0$ to $J = 1$ rotational spacing ($\sim 12,000$ MHz) is so much larger than the hyperfine splittings of the $J = 1$ level (~ 25 MHz) that we need not go beyond first-order perturbation theory.

However, when applying Eq. (17) to molecules we encounter the awkward problem that the electric field gradient coupling constant q_{ZZ} is referenced to the space-fixed Z axis, causing the value of q_{ZZ} to depend not only on the electron charge distribution in the molecule but also on an average of the orientation of the molecule with respect to $\mathbf{J}$. Consequently, for molecular applications it is more convenient to express the electric field gradient coupling constant in terms of the principal inertial axis system fixed in the molecule since the field gradients with respect to the molecule-fixed system are constant independent of the rotational state. For a linear molecule, the principal-axis system has its z axis along the internuclear axis and we can relate $q_{zz} \equiv q$ to q_{ZZ} by

$$q_{ZZ} = q \int Y_{JJ}^* C_0^{(2)} Y_{JJ} \, d\Omega = q \left(\frac{4\pi}{5} \right)^{\frac{1}{2}} \int Y_{JJ}^* Y_{20} Y_{JJ} \, d\Omega = \frac{-qJ}{2J+3} \tag{19}$$

In the high J limit, $q_{ZZ} = -\frac{1}{2}q$ since classically the internuclear axis is perpendicular to $\mathbf{J}$ and the expectation value of $(3\cos^2\theta_e - 1)/2$ becomes $-\frac{1}{2}$. Hence the first-order nuclear quadrupole energy for a linear molecule is given by

$$\Delta E = -\frac{1}{2}eqQ\left[\frac{\frac{3}{4}C(C+1) - I(I+1)J(J+1)}{I(2I-1)(2J+3)(2J-1)}\right] \tag{20}$$

D. Estimate the value of eqQ for ^{35}Cl in ^{35}ClCN from the data provided. Show that this is approximately -83 MHz. *Hint*: Use Eq. (20) to evaluate the matrix elements $\langle FM_F|\,H_Q\,|FM_F\rangle$, and show that $\langle\frac{1}{2}M_F|\,H_Q\,|\frac{1}{2}M_F\rangle = -eqQ/4$, $\langle\frac{3}{2}M_F|\,H_Q\,|\frac{3}{2}M_F\rangle = eqQ/5$, and $\langle\frac{5}{2}M_F|\,H_Q\,|\frac{5}{2}M_F\rangle = -eqQ/20$.

This value of $eqQ(^{35}\text{Cl}) = -83$ MHz in ClCN immediately allows us to draw some conclusions concerning the nature of the C–Cl bond in the molecule. Contributions to q in a molecule may arise from valence electrons of the nucleus under study, distortion of the closed shells of electrons around this nucleus, and charge distributions associated with nearby atoms. Because q refers to the electric field gradient at the nucleus, vanishes for spherically symmetric charge distributions, and depends inversely on the third power of the charge separation, the overwhelmingly dominant contribution to q is from the nature of the valence electrons. For pure s valence electrons $q = 0$ (again because the charge distribution has spherical symmetry), while for pure p valence electrons the value of q is large because the electron density at the nucleus is rapidly changing. Therefore, eqQ provides an excellent measure of the p character associated with the bonding involving the nucleus under study. Note that for an isolated free Cl atom (pure p valence electrons) $eqQ(^{35}\text{Cl}) = -110$ MHz, while for a Cl$^-$ anion (closed valence shell) $eqQ(^{35}\text{Cl}) = 0$. Thus, for example, in KCl, $eqQ(^{35}\text{Cl})$ is less than 0.04 MHz while in CH$_3$Cl, FCl, and ICl, $eqQ(^{35}\text{Cl})$ is -75 MHz, -146 MHz, and -83 MHz, respectively. Hence we conclude that the C–Cl bonding in ClCN is covalent rather than ionic and that CN behaves as a pseudo-halogen. This is one example of the rich chemical information available from high-resolution studies of molecular hyperfine structure.

As Figure 1 shows, the pure rotation spectrum of ClCN contains more information, namely, the "super hyperfine splittings" caused by ^{14}N with $I = 1$. We treat this problem by working in the $|(JI_1)F_1I_2FM_F\rangle$ coupling scheme where $\mathbf{J} + \mathbf{I}_1 = \mathbf{F}_1$ and $\mathbf{F}_1 + \mathbf{I}_2 = \mathbf{F}$ and set up the secular determinant involving all levels with the same F but different F_1. We now need to evaluate matrix elements of the form

$$\langle(JI_1)F_1'I_2FM_F|\,H_Q\,|(JI_1)F_1I_2FM_F\rangle$$

$$= \langle(JI_1)F_1'I_2FM_F|\,\mathbf{V}^{(2)}(^{35}\text{Cl})\cdot\mathbf{Q}^{(2)}(^{35}\text{Cl})$$

$$+ \mathbf{V}^{(2)}(^{14}\text{N})\cdot\mathbf{Q}^{(2)}(^{14}\text{N})\,|(JI_1)F_1I_2FM_F\rangle \tag{21}$$

For ClCN $J = 1$ there are eight hyperfine levels and the secular determinant is 8×8. However, by taking advantage of the fact that F is a good quantum number, we can factor this problem into a 2×2 block for $F = \frac{1}{2}$, a 3×3 for $F = \frac{3}{2}$, another 2×2 for $F = \frac{5}{2}$, and a 1×1 for $F = \frac{7}{2}$. Each of these can be diagonalized individually since there are no off-diagonal elements connecting these subblocks. Thus by assigning the spectrum and measuring at least some of the small splittings, it is possible to determine the $J = 0$ to $J = 1$ rotational spacing ($2B$) and the two nuclear quadrupole coupling constants $eqQ(^{35}\text{Cl})$ and $eqQ(^{14}\text{N})$. In a serious spectroscopic study one would make a least-squares fit to the line positions in which B, $eqQ(^{35}\text{Cl})$, and $eqQ(^{14}\text{N})$ were adjustable parameters. In this manner Lafferty, Lide, and Toth found for $^{35}\text{Cl}^{12}\text{C}^{14}\text{N}$ $eqQ(^{35}\text{Cl}) = -83.39 \pm 0.20$ MHz and $eqQ(^{14}\text{N}) = -3.37 \pm 0.26$ MHz.

E. Set up, but do not solve, the 2×2 secular determinant for the $F = \frac{1}{2}$ block. *Hint*: Evaluate Eq. (21) to show that the nonvanishing matrix elements in this block are

$$\left\langle F_1 = \tfrac{1}{2},\ I_2 = 1,\ F = \tfrac{1}{2} \middle| H_Q \middle| F_1 = \tfrac{1}{2},\ I_2 = 1,\ F = \tfrac{1}{2} \right\rangle$$

$$= \frac{-eqQ(^{35}\text{Cl})}{4}$$

$$\left\langle F_1 = \tfrac{1}{2},\ I_2 = 1,\ F = \tfrac{1}{2} \middle| H_Q \middle| F_1 = \tfrac{3}{2},\ I_2 = 1,\ F = \tfrac{1}{2} \right\rangle$$

$$= \frac{(5)^{\frac{1}{2}} eqQ(^{14}\text{N})}{20}$$

$$= \left\langle F_1 = \tfrac{3}{2},\ I_2 = 1,\ F = \tfrac{1}{2} \middle| H_Q \middle| F_1 = \tfrac{1}{2},\ I_2 = 1,\ F = \tfrac{1}{2} \right\rangle$$

$$\left\langle F_1 = \tfrac{3}{2},\ I_2 = 1,\ F = \tfrac{1}{2} \middle| H_Q \middle| F_1 = \tfrac{3}{2},\ I_2 = 1,\ F = \tfrac{1}{2} \right\rangle$$

$$= \frac{eqQ(^{35}\text{Cl}) + eqQ(^{14}\text{N})}{5} \tag{22}$$

NOTES AND REFERENCES

1. C. H. Townes and A. L. Schawlow, *Microwave Spectroscopy*, McGraw-Hill, New York, 1955, Chapter 9.

2. J. E. Wollrab, *Rotational Spectra and Molecular Structure*, Academic Press, New York, 1967, Chapter 5.

3. E. A. C. Lucken, *Nuclear Quadrupole Coupling Constants*, Academic Press, New York, 1969.

4. W. Gordy and R. L. Cook, *Microwave Molecular Spectra*, Wiley-Interscience, New York, 1970.

5. R. L. Cook and F. C. De Lucia, *Am. J. Phys.*, **39**, 1433 (1971).

6. H. W. Kroto, *Molecular Rotation Spectra*, Wiley, New York, 1975.

7. W. J. Lafferty, D. R. Lide, and R. A. Toth, *J. Chem. Phys.*, **43**, 2063 (1965). There appears to be a typographical error for the assignment of the 11924.89 MHz line for which F is listed as $\frac{3}{2}$ rather than $\frac{5}{2}$.

Chapter 6

ENERGY-LEVEL STRUCTURE
AND WAVE FUNCTIONS
OF A RIGID ROTOR

Although no real molecule is a rigid rotor, the deviations from rigid rotor behavior are often so small that nonrigid effects (vibration–rotation interaction) can be described quite accurately by perturbation theory. Consequently, solutions to the rigid rotor form the basis for more sophisticated treatments, and it is incumbent upon us to examine the quantum mechanical description of a rotating body all of whose subparts are at fixed angles and distances from one another. This serves as the starting point in discussing the rotational energy levels of polyatomic molecules and the radiative transitions they undergo. The rigid rotor problem was first solved in the late 1920s and early 1930s by several investigators[1–10], but the original literature is difficult to read because these authors could not avail themselves of what has now become standard angular momentum machinery. After a brief review of the classical motion of rigid bodies, we follow closely the expositions of King, Hainer, and Cross[11–14] and of Van Winter[15].

6.1 MOMENTS OF INERTIA: CLASSIFICATION OF RIGID ROTORS BY TOP TYPES

Figure 6.1 shows two reference frames for describing the n particles of a rigid body, a coordinate frame located in the laboratory (space-fixed frame), and a coordinate frame located at the center of mass of the rigid body. The position vectors $\mathbf{r}_i$ and $\bar{\mathbf{r}}_i$ to the ith particle (having mass m_i) originate from the center of mass and the arbitrary lab-frame origin, respectively. They are related by

$$\bar{\mathbf{r}}_i = \mathbf{R} + \mathbf{r}_i \tag{6.1}$$

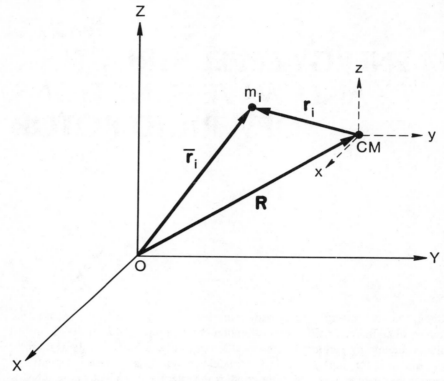

FIGURE 6.1   Laboratory and center-of-mass coordinate frames for a rigid body consisting of $i = 1, 2, \ldots, n$ particles of mass m_i having the position vector $\mathbf{r}_i$ in the center-of-mass frame and $\bar{\mathbf{r}}_i$ in the lab frame. The vector $\mathbf{R}$ is the position vector to the center of mass in the lab frame. The XYZ and xyz axes are chosen to be parallel to each other.

where $\mathbf{R}$ is the position vector from the arbitrary origin to the center of mass. The total mass M is obtained by summing over all n particles

$$M = \sum_i m_i \tag{6.2}$$

By definition, the center-of-mass position vectors satisfy the condition

$$\sum_i m_i \mathbf{r}_i = 0 \tag{6.3}$$

and it follows that

$$\sum_i m_i \bar{\mathbf{r}}_i = \sum_i m_i (\mathbf{R} + \mathbf{r}_i) = \mathbf{R} \sum_i m_i = M\mathbf{R} \tag{6.4}$$

that is,

$$\mathbf{R} = \frac{1}{M} \sum_i m_i \bar{\mathbf{r}}_i \tag{6.5}$$

The angular momentum of the ith particle about the origin of the lab frame is given by $\bar{\mathbf{r}}_i \times \bar{\mathbf{p}}_i$, where $\bar{\mathbf{p}}_i = m_i \bar{\mathbf{v}}_i = d/dt(m_i \bar{\mathbf{r}}_i)$ is the linear momentum measured in the same frame and $\bar{\mathbf{v}}_i = \mathbf{V} + \mathbf{v}_i$ is the velocity. Hence the angular momentum with respect to the lab frame of the rigid body is

$$\bar{\mathbf{J}} = \sum_i \bar{\mathbf{r}}_i \times \bar{\mathbf{p}}_i = \sum_i m_i(\mathbf{R} + \mathbf{r}_i) \times (\mathbf{V} + \mathbf{v}_i)$$

$$= \mathbf{R} \times \mathbf{V} \sum_i m_i + \mathbf{R} \times \frac{d}{dt} \sum_i m_i \mathbf{r}_i - \mathbf{V} \times \sum_i m_i \mathbf{r}_i + \sum_i m_i(\mathbf{r}_i \times \mathbf{v}_i)$$

$$= \mathbf{R} \times M\mathbf{V} + \sum_i (\mathbf{r}_i \times m_i \mathbf{v}_i) = \mathbf{R} \times M\mathbf{V} + \mathbf{J} \tag{6.6}$$

Thus the total angular momentum in the lab frame $\bar{\mathbf{J}}$ is the sum of the total angular momentum in the center-of-mass frame $\mathbf{J}$ and the center-of-mass angular momentum contribution $\mathbf{R} \times \mathbf{P}$ as if all the mass were concentrated at the center of mass. In particular, let us choose to use a coordinate frame in which the center of mass is at rest. Then

$$\bar{\mathbf{J}} = \mathbf{J} = \sum_i m_i(\mathbf{r}_i \times \mathbf{v}_i) \tag{6.7}$$

For a rigid rotor, all particles have the same angular velocity $\boldsymbol{\omega}$ about an instantaneous axis of rotation passing through the center of mass. The velocity of the ith particle is related to $\boldsymbol{\omega}$ by

$$\mathbf{v}_i = \boldsymbol{\omega} \times \mathbf{r}_i \tag{6.8}$$

Substitution of Eq. (6.8) into (6.7) and use of the vector identity $\mathbf{A} \times (\mathbf{B} \times \mathbf{C}) = \mathbf{B}(\mathbf{A} \cdot \mathbf{C}) - \mathbf{C}(\mathbf{A} \cdot \mathbf{B})$ gives

$$\mathbf{J} = \sum_i m_i[\mathbf{r}_i \times (\boldsymbol{\omega} \times \mathbf{r}_i)]$$

$$= \sum_i m_i[\boldsymbol{\omega}\, r_i^2 - \mathbf{r}_i(\mathbf{r}_i \cdot \boldsymbol{\omega})] \tag{6.9}$$

whose Cartesian components are

$$J_x = \left[\sum_i m_i \left(r_i^2 - x_i^2 \right) \right] \omega_x - \left[\sum_i m_i x_i y_i \right] \omega_y - \left[\sum_i m_i x_i z_i \right] \omega_z$$

$$J_y = -\left[\sum_i m_i x_i y_i\right]\omega_x + \left[\sum_i m_i\left(r_i^2 - y_i^2\right)\right]\omega_y - \left[\sum_i m_i y_i z_i\right]\omega_z$$

$$J_z = -\left[\sum_i m_i x_i z_i\right]\omega_x - \left[\sum_i m_i y_i z_i\right]\omega_y + \left[\sum_i m_i\left(r_i^2 - z_i^2\right)\right]\omega_z$$

(6.10)

In matrix notation, Eq. (6.10) may be rewritten as

$$\mathbf{J} = \underset{\sim}{\mathbf{I}}\cdot\boldsymbol{\omega}$$

(6.11)

or

$$\begin{pmatrix} J_x \\ J_y \\ J_z \end{pmatrix} = \begin{pmatrix} I_{xx} & I_{xy} & I_{xz} \\ I_{yx} & I_{yy} & I_{yz} \\ I_{zx} & I_{zy} & I_{zz} \end{pmatrix}\begin{pmatrix} \omega_x \\ \omega_y \\ \omega_z \end{pmatrix}$$

(6.12)

where $\underset{\sim}{\mathbf{I}}$ is called the *inertia tensor* (a second-rank Cartesian tensor) whose elements are

$$I_{jj} = \sum_i m_i\left(r_i^2 - r_{ij}^2\right), \qquad I_{jk} = I_{kj} = -\sum_i m_i r_{ij} r_{ik}$$

(6.13)

where j, k represent x, y, z. Note that if the moment of inertia tensor is computed with respect to the lab frame, we replace $\mathbf{r}_i$ by $\bar{\mathbf{r}}_i - \mathbf{R}$ in Eq. (6.13) and obtain the expression

$$I_{jj} = \sum_i m_i\left(\bar{r}_i^2 - \bar{r}_{ij}^2\right) - M\left(R^2 - R_j^2\right)$$

$$I_{jk} = I_{kj} = -\sum_i m_i \bar{r}_{ij}\bar{r}_{ik} + MR_j R_k$$

(6.14)

In Eqs. (6.13) and (6.14) the diagonal elements I_{jj} are known as *moments of inertia* and the off-diagonal elements, as *products of inertia*. The moment of inertia tensor $\underset{\sim}{\mathbf{I}}$ in Eq. (6.11) must be regarded as acting on the vector $\boldsymbol{\omega}$ and not on the coordinate system. The vectors $\mathbf{J}$ and $\boldsymbol{\omega}$ are two physically different vectors having different dimensions. Unlike the rotation operator $\mathbf{R}$, $\underset{\sim}{\mathbf{I}}$ has dimensions (g cm^2) and does not satisfy the conditions of a unitary transformation.

From Eq. (6.13) or (6.14) we see that the components of the inertia tensor $\underset{\sim}{\mathbf{I}}$ are real and symmetric so that the tensor is self-adjoint or Hermitian. Consequently, there exists a particular orientation of the rigid body (transformation of the coordinates) for which the inertia tensor is a 3×3 diagonal matrix. The diagonal elements are called the *principal moments of inertia*, and the axis system is termed the *principal axis system.* The three principal moments of inertia are designated I_{aa}, I_{bb}, I_{cc}, which are the roots of the determinantal equation

$$\begin{vmatrix} I_{xx} - \lambda & I_{xy} & I_{xz} \\ I_{yx} & I_{yy} - \lambda & I_{yz} \\ I_{zx} & I_{zy} & I_{zz} - \lambda \end{vmatrix} = 0 \qquad (6.15)$$

By convention, I_{aa} is the smallest moment while I_{cc} is the largest:

$$I_{aa} \leq I_{bb} \leq I_{cc} \qquad (6.16)$$

With respect to the principal axes, the components of $\mathbf{J}$ involve only the corresponding components of $\boldsymbol{\omega}$:

$$J_a = I_{aa}\omega_a, \qquad J_b = I_{bb}\omega_b, \qquad J_c = I_{cc}\omega_c \qquad (6.17)$$

Molecules as rigid rotors are classified in terms of their three principal moments according to the following types:

$I_{aa} = 0, \; I_{bb} = I_{cc}$	linear
$I_{aa} < I_{bb} = I_{cc}$	prolate symmetric top
$I_{aa} = I_{bb} < I_{cc}$	oblate symmetric top
$I_{aa} = I_{bb} = I_{cc}$	spherical top
$I_{aa} < I_{bb} < I_{cc}$	asymmetric top

For example, CO_2 is a linear top, CH_3Cl is a prolate symmetric top (football-like, American style), $CHCl_3$ is an oblate symmetric top (frisbee-like), CCl_4 is a spherical top, and CH_2Cl_2 is an asymmetric top (which happens to be a "near" prolate top, i.e., $I_{bb} \approx I_{cc}$).

In the center-of-mass frame, the rotational kinetic energy is

$$T = \tfrac{1}{2} \sum_i m_i v_i^2 = \tfrac{1}{2} \sum_i m_i \mathbf{v}_i \cdot (\boldsymbol{\omega} \times \mathbf{r}_i) = \tfrac{1}{2}\boldsymbol{\omega} \cdot \sum_i m_i(\mathbf{r}_i \times \mathbf{v}_i) = \tfrac{1}{2}\boldsymbol{\omega} \cdot \mathbf{J} \quad (6.18)$$

where use is made of the vector identity $\mathbf{A} \cdot (\mathbf{B} \times \mathbf{C}) = \mathbf{B} \cdot (\mathbf{C} \times \mathbf{A})$. In the principal-axis system, Eq. (6.18) may be written with the help of Eq. (6.17)

$$T = \tfrac{1}{2}(\omega_a J_a + \omega_b J_b + \omega_c J_c)$$

$$= \tfrac{1}{2}\left(\omega_a^2 I_{aa} + \omega_b^2 I_{bb} + \omega_c^2 I_{cc}\right)$$

$$= \frac{J_a^2}{2 I_{aa}} + \frac{J_b^2}{2 I_{bb}} + \frac{J_c^2}{2 I_{cc}}$$

$$= \left(A J_a^2 + B J_b^2 + C J_c^2\right) \qquad (6.19)$$

where A, B, and C are rotational constants, which in units of Hertz (Hz) are

$$A = \frac{\hbar}{4\pi I_{aa}}, \qquad B = \frac{\hbar}{4\pi I_{bb}}, \qquad C = \frac{\hbar}{4\pi I_{cc}} \qquad (6.20)$$

Note that the inequality $I_{aa} \leq I_{bb} \leq I_{cc}$ implies that

$$A \geq B \geq C \qquad (6.21)$$

Thus there are three mutually perpendicular directions for which the moment of inertia is a maximum or a minimum. If $I_{jj}^{-1/2}$ is plotted against the respective principal axes, an ellipsoid is obtained, the so-called ellipsoid of inertia. If the molecule has symmetry, the direction of one or more of the principal axes can easily be found since axes of symmetry are always principal axes and a plane of symmetry is always perpendicular to a principal axis. Molecules having a threefold or higher axis of rotational symmetry are symmetric tops, while the presence of more than one such axis causes the molecule to be a spherical top. Molecules without a threefold or higher axis of rotation cannot be symmetric tops, although many molecules with less symmetry are near-symmetric tops. It is interesting to note that prolate or near-prolate tops are more common than the oblate or near-oblate variety because in most molecules the light hydrogen atoms are located farthest from the center of mass.

6.2 CLASSICAL MOTION OF A FREE TOP

The Euler angles θ, ϕ, χ describe the position of the triad of orthogonal unit vectors $\hat{\mathbf{x}}, \hat{\mathbf{y}}, \hat{\mathbf{z}}$ fixed in a rigid body and turning about a common origin O relative to the triad of orthogonal unit vectors $\hat{\mathbf{X}}, \hat{\mathbf{Y}}, \hat{\mathbf{Z}}$ fixed in space. The motion of the body is determined when θ, ϕ, χ are known as functions of the time t. This motion can also be described by the angular velocity $\boldsymbol{\omega}(t)$, where, referred to the body-fixed frame,

$$\boldsymbol{\omega} = \omega_x \hat{\mathbf{x}} + \omega_y \hat{\mathbf{y}} + \omega_z \hat{\mathbf{z}} \qquad (6.22)$$

We seek expressions for $\omega_x, \omega_y, \omega_z$ in terms of θ, ϕ, χ and their rates of change[16]. In what follows we make extensive use of Eq. (3.43). The angular velocity $\dot{\theta}$ is along the line of nodes; its components are

$$\dot{\theta}_x = \dot{\theta} \sin \chi$$

$$\dot{\theta}_y = \dot{\theta} \cos \chi$$

$$\dot{\theta}_z = 0 \qquad (6.23)$$

The angular velocity $\dot{\phi}$ is along the Z axis; its components are

$$\dot{\phi}_x = -\dot{\phi} \sin \theta \cos \chi$$

$$\dot{\phi}_y = \dot{\phi} \sin \theta \sin \chi$$

$$\dot{\phi}_z = \dot{\phi} \cos \theta \qquad (6.24)$$

Finally, the angular velocity $\dot{\chi}$ is along the z axis; its components are

$$\dot{\chi}_x = 0$$

$$\dot{\chi}_y = 0$$

$$\dot{\chi}_z = \dot{\chi} \qquad (6.25)$$

Hence, collecting the components along each axis, we find

$$\omega_x = -\dot{\phi} \sin \theta \cos \chi + \dot{\theta} \sin \chi$$

$$\omega_y = \dot{\phi} \sin \theta \sin \chi + \dot{\theta} \cos \chi$$

$$\omega_z = \dot{\phi} \cos \theta + \dot{\chi} \qquad (6.26)$$

Let us choose the xyz axes to be the principal axes of the rigid body. Then the rotational kinetic energy is given by

$$T = \tfrac{1}{2} \left(I_x \omega_x^2 + I_y \omega_y^2 + I_z \omega_z^2 \right) \qquad (6.27)$$

Let us further specialize this to a symmetric top for which $I_x = I_y \neq I_z$. Then substitution of Eq. (6.26) into Eq. (6.27) yields

$$T = \tfrac{1}{2} I_x \left(\dot{\phi}^2 \sin^2 \theta + \dot{\theta}^2 \right) + \tfrac{1}{2} I_z \left(\dot{\phi} \cos \theta + \dot{\chi} \right)^2 \qquad (6.28)$$

Let us choose the Z axis of the space-fixed frame to be along the direction of the total angular momentum $\mathbf{L}$ of the top. Then the components of $\mathbf{L}$ in the body-fixed frame are

$$L_x = I_x \omega_x = I_x \left(-\dot{\phi} \sin \theta \cos \chi + \dot{\theta} \sin \chi \right)$$

$$L_y = I_y \omega_y = I_y \left(\dot{\phi} \sin \theta \sin \chi + \dot{\theta} \cos \chi \right)$$

$$L_z = I_z \omega_z = I_z \left(\dot{\phi} \cos \theta + \dot{\chi} \right) \qquad (6.29)$$

However, we also have

$$L_x = \Phi_{xZ} L = -L \sin \theta \cos \chi$$

$$L_y = \Phi_{yZ} L = L \sin \theta \sin \chi$$

$$L_z = \Phi_{zZ} L = L \cos \theta \qquad (6.30)$$

Hence

$$I_x \left(-\dot{\phi} \sin \theta \cos \chi + \dot{\theta} \sin \chi \right) = -L \sin \theta \cos \chi \qquad (6.31)$$

$$I_y \left(\dot{\phi} \sin \theta \sin \chi + \dot{\theta} \cos \chi \right) = L \sin \theta \sin \chi \qquad (6.32)$$

$$I_z \left(\dot{\phi} \cos \theta + \dot{\chi} \right) = L \cos \theta \qquad (6.33)$$

Because the line of nodes is perpendicular to the Z axis, the component of $\mathbf{L}$ along the line of nodes vanishes:

$$L_N = 0 \qquad (6.34)$$

This component can be resolved into components along x and y as

$$L_N = L_x \cos \left(\frac{\pi}{2} - \chi \right) + L_y \cos \chi$$

$$= L_x \sin \chi + L_y \cos \chi$$

$$= I_x \dot{\theta} \qquad (6.35)$$

where we have replaced L_x and L_y by the expressions given in Eq. (6.29). Combining Eqs. (6.34) and (6.35), we have the result

$$\dot{\theta} = 0 \qquad (6.36)$$

provided I_x is not equal to zero.

On substitution of Eq. (6.36) into Eq. (6.31) or Eq. (6.32), we find the condition for the rate of change of ϕ with time

$$\dot{\phi} = \frac{L}{I_x} \qquad (6.37)$$

On substitution of Eq. (6.37) into Eq. (6.33), the condition for the rate of change of χ with time becomes

$$\dot{\chi} = \left(\frac{L}{I_z} - \frac{L}{I_x} \right) \cos \theta \qquad (6.38)$$

Equations (6.36)–(6.38) admit of a very simple interpretation. Equation (6.36) shows that the angle between the figure axis of the top and the direction of $\mathbf{L}$ is constant; Eq. (6.37) shows that the figure axis precesses about $\mathbf{L}$ at a constant angular velocity L/I_x, while Eq. (6.38) shows that at the same time the top rotates about its figure axis at a constant angular velocity, $(L/I_z - L/I_x) \cos \theta$.

The preceding analysis leads to a picture of two cones rolling without slipping about one another. At any instant $\mathbf{L}$, $\boldsymbol{\omega}$, and $\hat{z}$ are coplanar. The angle between $\mathbf{L}$

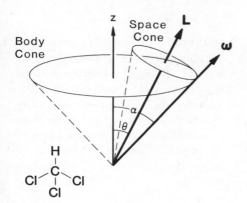

FIGURE 6.2   The classical motion of an oblate symmetric top $(I_z > I_x = I_y)$ is represented by a body cone in contact with a fixed space cone with the space cone lying inside the body cone. The body cone rolls at a uniform rate without slipping about the space cone. The line of contact is the instantaneous axis of rotation, and the common vertex of the two cones is located at the center of mass of the top.

(the Z axis) and the figure axis of the top (the z axis) is θ. Let α be the angle that $\boldsymbol{\omega}$ makes with respect to $\hat{z}$ so that

$$\tan \alpha = \frac{\left(\omega_x^2 + \omega_y^2\right)^{\frac{1}{2}}}{\omega_z} \tag{6.39}$$

Similarly, we have

$$\tan \theta = \frac{\left(L_x^2 + L_y^2\right)^{\frac{1}{2}}}{L_z}$$

$$= \frac{I_x \left(\omega_x^2 + \omega_y^2\right)^{\frac{1}{2}}}{I_z \omega_z} \tag{6.40}$$

Thus

$$\tan \theta = \left(\frac{I_x}{I_z}\right) \tan \alpha \tag{6.41}$$

We distinguish two cases: if $I_z > I_x$, then $\alpha > \theta$. The axis of rotation makes a constant angle $(\alpha - \theta)$ with respect to $\mathbf{L}$ so that the axis of rotation is a generator of a right circular cone, fixed in space, of angle $(\alpha - \theta)$. But the axis of rotation $\boldsymbol{\omega}$ makes a constant angle α with the figure axis and traces out in the body a right circular cone of angle α. Thus the motion is given by the rolling of a cone of angle α, which is fixed in the body, on a cone of angle $(\alpha - \theta)$, which is fixed in space. The smaller cone touches the larger cone internally (see Figure 6.2). If $I_z < I_x$, it follows that $\alpha < \theta$. A cone of angle α, fixed in the body, rolls on a cone of angle $(\theta - \alpha)$ fixed in space. Then the smaller cone touches the larger cone externally (see Figure 6.3).

In each case the figure axis describes a cone of angle θ about $\mathbf{L}$. It is seen from both figures that viewed from the molecular frame, $\mathbf{L}$ and $\boldsymbol{\omega}$ continuously change position.

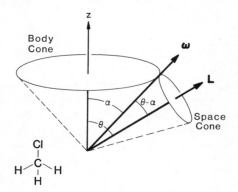

FIGURE 6.3　The classical motion of a prolate symmetric top ($I_z < I_x = I_y$) is represented by a body cone in contact with a fixed space cone with the space cone lying outside the body cone. The body cone rolls at a uniform rate without slipping about the space cone. The line of contact is the instantaneous axis of rotation, and the common vertex of the two cones is located at the center of mass of the top.

Let us return to the general case of an asymmetric top with $I_{aa} \leq I_{bb} \leq I_{cc}$. From the conservation of energy and angular momentum, we have

$$I_{aa}\omega_a^2 + I_{bb}\omega_b^2 + I_{cc}\omega_c^2 = 2E \tag{6.42}$$

and

$$I_{aa}^2\omega_a^2 + I_{bb}^2\omega_b^2 + I_{cc}^2\omega_c^2 = \mathbf{L} \cdot \mathbf{L} = L^2 \tag{6.43}$$

where E is the rotational kinetic energy and L^2 is the square of the magnitude of the angular momentum of the top. In what follows we outline a construction due to Poinsot for describing the top motion.

Let the vector $\boldsymbol{\omega}$ be represented by a line segment OQ, drawn from the center of mass O to the terminus Q. Let OP be a line drawn from O along the direction of $\mathbf{L}$. Draw the perpendicular QN of OQ on OP (see Figure 6.4). Then $ON = \boldsymbol{\omega} \cdot \hat{\mathbf{L}} = \boldsymbol{\omega} \cdot \mathbf{L}/L = 2E/L$. Thus line segment ON is a constant (*invariable line*) and point N is fixed during the motion of the top. Therefore, the plane through N perpendicular to $\mathbf{L}$ is a fixed plane, called the *invariable plane*. The terminus of $\boldsymbol{\omega}$ moves on the invariable plane.

To an observer riding in the top frame, the unit vectors $\hat{\mathbf{x}}, \hat{\mathbf{y}}, \hat{\mathbf{z}}$, taken along the principal axes a, b, c, are fixed, but both $\boldsymbol{\omega}$ and $\mathbf{L}$ are changing direction in time. Let

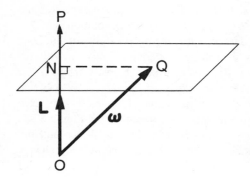

FIGURE 6.4　The invariable line and the invariable plane.

Moment of Inertia Ellipsoid

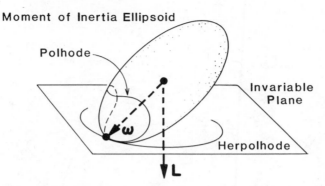

FIGURE 6.5 Motion of the moment of inertia ellipsoid relative to the invariable plane. Reprinted with permission from H. Goldstein, *Classical Mechanics*, 2nd ed., Addison-Wesley, Reading, MA, 1953, p. 206, Fig. 5-4.

the terminus Q have the body-fixed coordinate (x, y, z). Then Eqs. (6.42) and (6.43) become

$$\left(\frac{I_{aa}}{2E}\right) x^2 + \left(\frac{I_{bb}}{2E}\right) y^2 + \left(\frac{I_{cc}}{2E}\right) z^2 = 1 \qquad (6.44)$$

and

$$\left(\frac{I_{aa}}{L}\right)^2 x^2 + \left(\frac{I_{bb}}{L}\right)^2 y^2 + \left(\frac{I_{cc}}{L}\right)^2 z^2 = 1 \qquad (6.45)$$

These two equations are ellipsoids; the first of these two ellipsoids, called the *Poinsot ellipsoid*, has the same axes as the moment of inertia ellipsoid but is scaled by the factor $(2E)^{-1}$. To an observer moving with the top, the terminus of the angular velocity $\boldsymbol{\omega}$ describes a curve that is the intersection of these two ellipsoids, fixed in the body. As we have seen, the invariable plane touches the Poinsot ellipsoid at the terminus of the angular velocity vector. Hence we can picture the motion of the asymmetric top as being such that the moment of inertia ellipsoid rolls without slipping on the invariable plane, with the center of the ellipsoid maintained at a constant height above the plane (see Figure 6.5). The curve traced out by the point of contact on the moment of inertia ellipse is called the *polhode*; the curve traced out on the invariable plane is called the *herpolhode*. In the special case of a symmetric top, the moment of inertia ellipsoid is an ellipsoid of revolution so that the polhodes degenerate to circles about the symmetry axis; that is, $\boldsymbol{\omega}$ moves on the surface of a cone.

Let us return to Eqs. (6.42) and (6.43) but rewrite them in terms of the components of **L**:

$$\frac{L_a^2}{2EI_{aa}} + \frac{L_b^2}{2EI_{bb}} + \frac{L_c^2}{2EI_{cc}} = 1 \qquad (6.46)$$

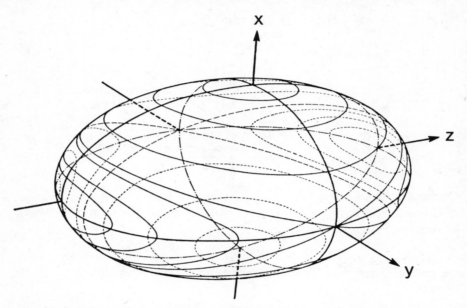

FIGURE 6.6   Some possible paths of **L** on the kinetic energy ellipsoid fixed in the body axes. Adapted from L. D. Landau and E. M. Lifschitz, *Mechanics*, 2nd ed., Pergamon Press, New York, 1969, p. 117, Fig. 51.

$$\frac{L_a^2}{L^2} + \frac{L_b^2}{L^2} + \frac{L_c^2}{L^2} = 1 \qquad (6.47)$$

Equation (6.46) is the equation of an ellipsoid with semiaxes $(2\,EI_{aa})^{1/2}$, $(2\,EI_{bb})^{1/2}$, and $(2\,EI_{cc})^{1/2}$; Eq. (6.47) is the equation of a sphere of radius L. When **L** moves relative to the principal axes of inertia, its terminus traces out the intersection of these two surfaces. These paths change as the magnitude of **L** varies from $(2\,EI_{aa})^{1/2}$ to $(2\,EI_{cc})^{1/2}$ for a given value of E. When L^2 is only slightly greater than $2\,EI_{aa}$, the sphere intersects the ellipsoid in two small closed curves around the a axis. When L^2 increases, these curves become larger, and for $L^2 = 2\,EI_{bb}$ they become two plane curves (ellipses) that intersect at the poles of the ellipsoid on the b axis. As L^2 increases further, two separate closed paths appear again, but now around the poles on the c axis.

Because these paths are closed, the motion of **L** relative to the top axes is periodic. However, there is a sharp difference in the nature of the paths near the various poles. Around a and c the paths lie close together, but around b the paths make great excursions before returning. This difference corresponds to a difference in the stability of top rotations about its three principal axes. Rotation about c and a, corresponding to the largest and the smallest moments of inertia, are stable in that slight deviations produce a motion close to the original one. However, rotation about b, corresponding to the intermediate moment of inertia, is unstable[17]. This

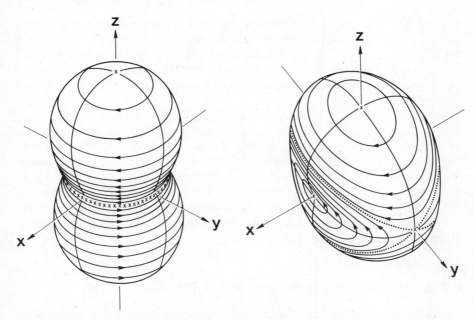

FIGURE 6.7   Rotational energy surfaces for two types of molecules: (a) rigid prolate symmetric top RE surface; (b) rigid asymmetric top RE surface. Contour lines represent angular
momentum trajectories; dotted lines, separatrix curves that separate different types of trajectories; dashed lines, meridians of spherical energy surfaces that intersect the highest-energy
trajectories. The yz meridian is a tunneling path. Adapted from W. G. Harter and C. W. Patterson, *J. Chem. Phys.*, **80**, 4241 (1984), Fig. 1.

can be demonstrated very easily by spinning a book (taped shut) about its three axes.
A molecular analog may be the behavior of some polyatomic molecules in which,
with increasing internal excitation, vibration–rotation interaction alters sufficiently
the moments of inertia to cause large changes in the molecule's rotational motion[18].
The time dependence of θ, ϕ, χ can be solved analytically for an asymmetric top, but
the solutions are quite complex (see Figure 6.6).

Another useful representation of top motion is to make a radial plot of the energy
as a function of the direction of **L** as viewed in the xyz body-fixed frame subject to
the constraint that $L^2 = L_x^2 + L_y^2 + L_z^2$ is a constant. This is called a *rotational energy*
(RE) surface[19]. Rotational energy surfaces for a rigid prolate symmetric top and
a rigid most asymmetric top (where the intermediate moment of inertia is halfway
between the other two) are shown in Figure 6.7. In Figure 6.7 the contour lines
correspond to angular momentum trajectories where each point on the RE surface
represents a direction **L** in the body-fixed frame. Recall that the direction of **L** is
fixed in space, but as the top rotates in the space-fixed frame, **L** moves from point
to point on the RE surface. At each point the radial height of the RE surface is
defined to be proportional to the rotational energy. Note that for **L** located on the

equator of a symmetric top RE surface the contours collapse to points, denoted by x's and the precessional frequency of $\mathbf{L}$ vanishes. An equatorial line of fixed points is the only part of the RE surface that remains for a linear rotor (diatomic molecule or linear polyatomic molecule) in a Σ state. Thus the RE surface is most useful in describing internal angular momentum trajectories and dynamics of nonlinear polyatomic molecules.

In particular, the RE surface provides a good means of visualizing why the high-J fine structure of nonrigid tops, particularly those of spherical tops with tetrahedral or octahedral symmetry, show a clustering of the rovibrational sublevels into nearly degenerate multiplets[19–23]. The total degeneracy of each cluster corresponds to the number of distinct classical trajectories with the same energy on equivalent parts of the RE surface. The classical frequency of $\mathbf{L}$ precession is roughly proportional to the fine structure spacing between neighboring clusters; the superfine structure splitting within a cluster is related to the rate at which $\mathbf{L}$ delocalizes or tunnels to other equivalent trajectories[19].

6.3 SYMMETRIC AND ASYMMETRIC TOP ENERGY LEVELS

The energy levels of a rigid rotor are found by solving the Schrödinger equation

$$\mathcal{H}_R \Psi_{JM}(\phi,\theta,\chi) = E \Psi_{JM}(\phi,\theta,\chi) \tag{6.48}$$

where, according to Eq. (6.19), the Hamiltonian for the rotational motion of a free rigid rotor is

$$\mathcal{H}_R = \frac{J_a^2}{2 I_{aa}} + \frac{J_b^2}{2 I_{bb}} + \frac{J_c^2}{2 I_{cc}} = A J_a^2 + B J_b^2 + C J_c^2 \tag{6.49}$$

In the oblate top limit $I_{aa} = I_{bb} < I_{cc}$,

$$\mathcal{H}_R = A\mathbf{J}^2 + (C - A) J_z^2 \tag{6.50}$$

where we make the identification of axes $a \leftrightarrow x$, $b \leftrightarrow y$, and $c \leftrightarrow z$ (the figure axis). The eigenfunctions $|JKM\rangle$ are given in Eq. (3.125), and the eigenvalues are

$$E(J,K) = AJ(J+1) + (C-A)K^2 \tag{6.51}$$

For each $\mathbf{J}$ level there are $2J+1$ different K levels, but those with $K \neq 0$ are doubly degenerate. For a prolate top ($I_{aa} < I_{bb} = I_{cc}$), we redefine the body-fixed coordinates $a \leftrightarrow z$ (figure axis), $b \leftrightarrow x$, and $c \leftrightarrow y$, and the rotational Hamiltonian takes the form

$$\mathcal{H}_R = C\mathbf{J}^2 + (A - C) J_z^2 \tag{6.52}$$

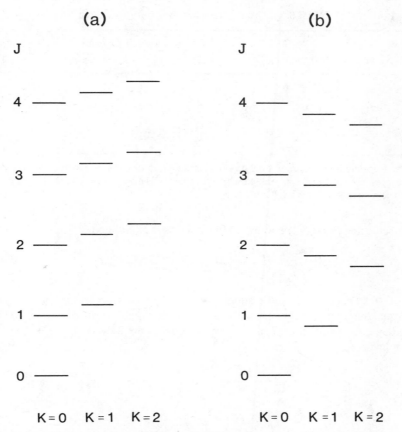

FIGURE 6.8 Energy-level structure for (a) prolate top and (b) oblate top. All levels with $K \neq 0$ are doubly degenerate.

The eigenvalues then become

$$E(J, K) = CJ(J + 1) + (A - C)K^2 \qquad (6.53)$$

which again show what is called K-type doubling for $K \neq 0$. Note that the coefficient in front of K^2 changes signs in going from an oblate to a prolate top, as it is negative in the former and positive in the latter. Figure 6.8 illustrates the energy-level structure for these two types of symmetric tops.

In the general case of an asymmetric top ($I_{aa} \neq I_{bb} \neq I_{cc}$) the eigenfunctions are linear combinations of symmetric top wave functions [see Eq. (3.124)] and we write

$$\Psi_{JM}(\phi, \theta, \chi) = \sum_K A_K |JKM\rangle \qquad (6.54)$$

TABLE 6.1   Group Characters for the D_2 (V) Group

Symmetry Designations[a]		Group Operations			
		E	$C_2(c)$	$C_2(b)$	$C_2(a)$
A	A	1	1	1	1
B_1	B_c	1	1	-1	-1
B_2	B_b	1	-1	1	-1
B_3	B_a	1	-1	-1	1

[a]The first column is the traditional group theoretic designation; the second column places a subscript on B to denote the axis about which the rotation is made. We prefer the latter because it is so easy to remember what it means.

On substituting Eq. (6.54) into (6.48), we obtain the expression

$$\sum_K A_K \mathcal{H}_R |JKM\rangle = E \sum_K A_K |JKM\rangle \qquad (6.55)$$

In Eqs. (6.54) and (6.55) summation over J and M is unnecessary since they are good quantum numbers for an asymmetric top. After multiplication by $\langle JK'M|$ and integration, we find that the energies are the roots of the $2J + 1$ by $2J + 1$ secular determinant

$$|\mathcal{H}_{K'K} - E\delta_{K'K}| = 0 \qquad (6.56)$$

where

$$\mathcal{H}_{K'K} = \langle JK'M|\mathcal{H}_R|JKM\rangle \qquad (6.57)$$

A nifty simplification can be made in the evaluation of Eq. (6.56) by using the symmetry inherent in the moment of inertia ellipsoid to factor the secular determinant into four subblocks (the Wang transformation[6]). The rotor Hamiltonian [see

TABLE 6.2   Possible Identifications of Principal Axes a, b, c with Body-Fixed Coordinates x, y, z (Both Frames Are Righthanded Coordinate Systems).

Body-Fixed Coordinates	Representation		
	I^a	II^b	III^c
x	b	c	a
y	c	a	b
z	a	b	c

[a]Most natural in the limit of a prolate top.
[b]Most natural for a top intermediate between the two limiting symmetric top cases.
[c]Most natural in the limit of an oblate top.

TABLE 6.3 Cartesian Coordinate and Euler Angle Transformations under Operation of the D_2 (V) Group

Group Operation[a]	Cartesian Coordinate Transformation	Euler Angle Transformation
E	$x \to x$ $y \to y$ $z \to z$	$\phi \to \phi$ $\theta \to \theta$ $\chi \to \chi$
$C_2(z)$	$x \to -x$ $y \to -y$ $z \to z$	$\phi \to \phi$ $\theta \to \theta$ $\chi \to \pi + \chi$
$C_2(y)$	$x \to -x$ $y \to y$ $z \to -z$	$\phi \to \phi + \pi$ $\theta \to \pi - \theta$ $\chi \to \pi - \chi$
$C_2(x)$	$x \to x$ $y \to -y$ $z \to -z$	$\phi \to \pi + \phi$ $\theta \to \pi - \theta$ $\chi \to -\chi$

[a]For identification of x, y, z with a, b, c, see Table 6.2.

Eq. (6.49)] has the property that a rotation by $180°$ about a, b, c leaves the Hamiltonian unchanged. Thus any asymmetric top belongs to the point group D_2, also known as the *Viergruppe (V)*. The group operations are twofold rotations about a, b, c, denoted by $C_2(a)$, $C_2(b)$, $C_2(c)$, and the identity E. The group characters are shown in Table 6.1.

We wish to discover how the $|JKM\rangle$ transform under the operations of the D_2 (V) group. However, the $|JKM\rangle$ are defined with respect to the xyz frame fixed in the body where z is chosen to be along the figure axis, while the group operations refer to the principal axes a, b, c. As shown in Table 6.2, there are three possible ways in which the a, b, c may be identified with the x, y, z (where both frames refer to right-handed coordinate systems). The K value in $|JKM\rangle$ depends on the representation that is used. For a near-prolate top representation I is used while for a near-oblate top representation III is most natural. For an asymmetric top intermediate between the two symmetric top limits, representation II brings out the asymmetry most clearly.

Once the top representation is selected, the effect of the operations of the D_2 (V) group may be found by determining the change in the Euler angles. For example, $C_2(z)$ causes the transformation $x \to -x, y \to -y, z \to z$. Then by inspection or by using the direction cosine matrix elements $\Phi_{Fg}(\phi, \theta, \chi)$ that relate the Cartesian vector components r_F and r_g [see Eq. (3.37)], it is readily verified that the $C_2(z)$

operation corresponds to $\phi \rightarrow \phi, \theta \rightarrow \theta, \chi \rightarrow \pi + \chi$. The results of this analysis are collected in Table 6.3. Thus we find that

$$C_2(z)\,|JKM\rangle = \left[\frac{2J+1}{8\pi^2}\right]^{\frac{1}{2}} e^{iM\phi} d^J_{MK}(\theta)\,e^{iK(\pi+\chi)}$$

$$= (-1)^K\,|JKM\rangle \tag{6.58}$$

$$C_2(y)\,|JKM\rangle = \left[\frac{2J+1}{8\pi^2}\right]^{\frac{1}{2}} e^{iM(\phi+\pi)} d^J_{MK}(\pi-\theta)\,e^{iK(\pi-\chi)}$$

$$= \left[\frac{2J+1}{8\pi^2}\right]^{\frac{1}{2}} (-1)^M e^{iM\phi}(-1)^{-J-M} d^J_{M,-K}(\theta)(-1)^K e^{-iK\chi}$$

$$= (-1)^{J-K}\,|J-KM\rangle \tag{6.59}$$

and

$$C_2(x)\,|JKM\rangle = \left[\frac{2J+1}{8\pi^2}\right]^{\frac{1}{2}} e^{iM(\phi+\pi)} d^J_{MK}(\pi-\theta)\,e^{iK(-\chi)}$$

$$= \left[\frac{2J+1}{8\pi^2}\right]^{\frac{1}{2}} (-1)^M e^{iM\phi}(-1)^{-J-M} d^J_{M,-K}(\theta)\,e^{-iK\chi}$$

$$= (-1)^{-J}\,|J-KM\rangle \tag{6.60}$$

where we use Eq. (3.67) and the identity $e^{i\pi} = -1$.

Equations (6.58)–(6.60) suggest that we use, instead of $|JKM\rangle$, a symmetry basis set constructed from linear combinations of $|J \pm KM\rangle$ that transform the same way under the operations of the D_2 (V) group. We define for $K \neq 0$

$$|JKMs\rangle = \frac{1}{\sqrt{2}}\left[|JKM\rangle + (-1)^s\,|J-KM\rangle\right] \tag{6.61}$$

where we may take $s = 0$ or 1, and for $K = 0$

$$|J0Ms\rangle = (-1)^s\,|J0M\rangle = |J0M\rangle \tag{6.62}$$

where we may take only $s = 0$. Then we have

$$E\,|JKMs\rangle = |JKMs\rangle$$

$$C_2(z)\,|JKMs\rangle = (-1)^K\,|JKMs\rangle$$

TABLE 6.4   Asymmetric Top Symmetry Species

Secular Determinant	K	s	Symmetry Designation[a] J_{even}	J_{odd}
E^+	Even	0	A	B_z
E^-	Even	1	B_z	A
O^+	Odd	0	B_x	B_y
O^-	Odd	1	B_y	B_x

[a]The identification of x, y, z with a, b, c depends on the top representation chosen (see Table 6.2).

$$C_2(y)\,|JKMs\rangle = (-1)^{J-K+s}\,|JKMs\rangle$$

$$C_2(x)\,|JKMs\rangle = (-1)^{-J+s}\,|JKMs\rangle \tag{6.63}$$

Hence the $(2J+1) \times (2J+1)$ secular determinant factors are divided into four blocks; the eigenvalues and eigenfunctions of each block belong to one of the symmetry species of the D_2 (V) group. We list in Table 6.4 the results for the various values of J, K, and s.

Equation (6.63) is more commonly written as

$$C_2(y)\,|JKMs\rangle = (-1)^{J+K+s}\,|JKMs\rangle$$

$$C_2(x)\,|JKMs\rangle = (-1)^{J+s}\,|JKMs\rangle$$

which, of course, holds true for integral J and K. However, we keep the more awkward looking forms given in Eq. (6.63) because these expressions are also correct for half-integral J and K.

We now turn to the calculation of the matrix elements of $\mathcal{H}_R$ appearing in the secular determinant subblocks. Recalling that the commutation rules for $\mathbf{J}$ referred to the molecular frame are anomalous, we apply Eq. (3.49) and obtain

$$\langle JKM|J_x|JK\pm 1M\rangle = \frac{1}{2}\,[J(J+1)-K(K\pm 1)]^{\frac{1}{2}}$$

$$\langle JKM|J_y|JK\pm 1M\rangle = \mp\frac{i}{2}\,[J(J+1)-K(K\pm 1)]^{\frac{1}{2}}$$

$$\langle JKM|J_z|JKM\rangle = K \tag{6.64}$$

We then use matrix multiplication to evaluate the desired elements, for instance

$$\langle JKM|\,J_x^2\,|JK'M\rangle = \sum_{K''}\langle JKM|J_x|JK''M\rangle\langle JK''M|J_x|JK'M\rangle \tag{6.65}$$

We find for the diagonal elements

$$\langle JKM | J_x^2 | JKM \rangle = \langle JKM | J_y^2 | JKM \rangle = \tfrac{1}{2} \left[J(J+1) - K^2 \right]$$

$$\langle JKM | J_z^2 | JKM \rangle = K^2 \qquad\qquad (6.66)$$

and for the off-diagonal elements

$$\langle JKM | J_x^2 | JK \pm 2M \rangle$$

$$= - \langle JKM | J_y^2 | JK \pm 2M \rangle$$

$$= \tfrac{1}{4} [J(J+1) - K(K \pm 1)]^{\frac{1}{2}} [J(J+1) - (K \pm 1)(K \pm 2)]^{\frac{1}{2}} \quad (6.67)$$

To be specific, let us choose to work in the prolate symmetric top basis set (representation I in Table 6.2) since most molecules are closer to the prolate top limit. We have

$$\mathcal{H}_{KK} = \tfrac{1}{2}(B + C) \left[J(J+1) - K^2 \right] + AK^2 \qquad\qquad (6.68)$$

and

$$\mathcal{H}_{K,K\pm2} = \tfrac{1}{4}(B - C) [J(J+1) - K(K \pm 1)]^{\frac{1}{2}}$$

$$\times [J(J+1) - (K \pm 1)(K \pm 2)]^{\frac{1}{2}}$$

$$(6.69)$$

Note that only even or odd K levels interact with one another.

We illustrate the foregoing by working out explicit forms for the energy levels $E(A, B, C)$ of an asymmetric rotor for some of the lowest J values. We designate the levels by $J_{K_{-1}K_1}$, where K_{-1} is the value of $|K|$ the top would approach in the limit of a prolate top ($A > B = C$) and K_1 is the value of $|K|$ in the limit of an oblate top ($A = B > C$). Figure 6.8 shows that K_{-1} runs from J to 0 and K_1 from 0 to J with decreasing energy. Hence the possible energy levels are J_{J0}, J_{J1}, $J_{J-1,1}$, $J_{J-1,2}$, and so on. It follows that we can label the energy levels instead by the index τ, where

$$\tau = K_{-1} - K_1 \qquad\qquad (6.70)$$

Then τ runs from J to $-J$ in unit steps in order of decreasing energy. However, for symmetry purposes it is more convenient to use the $K_{-1}K_1$ label. Note that K_{-1} is associated with a rotation about the a axis and K_1, the c axis. Then the D_2 (V)

group designations may be identified with the (even, odd) parity of the $K_{-1}K_1$ pair as follows (see Table 6.1):

$$A \longleftrightarrow ee$$

$$B_c \longleftrightarrow oe$$

$$B_b \longleftrightarrow oo$$

$$B_a \longleftrightarrow eo \tag{6.71}$$

The symmetry about the b axis may be obtained from the parity of the sum $K_{-1} + K_1$.

For $J = 0$ there is only one level, 0_{00}, and its energy is seen at once to be 0. For $J = 1$ there are three levels 1_{10}, 1_{11}, and 1_{01}. The hypothetical 1_{00} level of A symmetry comes from the E^- secular determinant subblock, which can have no entries for $K = 0$ and $s = 1$. The energy of the 1_{10} level of B_c symmetry is calculated from the 1×1 O^+ secular determinant

$$|\langle 11M0|\mathcal{H}_R|11M0\rangle - E| = 0 \tag{6.72}$$

from which we find

$$E(A,B,C) = \frac{1}{\sqrt{2}}\left[\langle 1\,1\,M| + \langle 1\,-1\,M|\right]\mathcal{H}_R\frac{1}{\sqrt{2}}\left[|1\,1\,M\rangle + |1\,-1\,M\rangle\right]$$

$$= \tfrac{1}{2}\left\{\left[\tfrac{1}{2}(B+C)+A\right] + \left[\tfrac{1}{2}(B+C)+A\right]\right.$$

$$\left. + \tfrac{1}{2}(B-C) + \tfrac{1}{2}(B-C)\right\}$$

$$= A + B \tag{6.73}$$

Similarly, the energy of the 1_{11} level (B_b symmetry) is calculated from the 1×1 O^- secular determinant

$$|\langle 11M1|\mathcal{H}_R|11M1\rangle - E| = 0 \tag{6.74}$$

which has the solution

$$E(A,B,C) = \frac{1}{\sqrt{2}}\left[\langle 1\,1\,M| - \langle 1\,-1\,M|\right]\mathcal{H}_R\frac{1}{\sqrt{2}}\left[|1\,1\,M\rangle - |1\,-1\,M\rangle\right]$$

$$= \tfrac{1}{2}\left\{\left[\tfrac{1}{2}(B+C)+A\right] + \left[\tfrac{1}{2}(B+C)+A\right]\right.$$

$$\left. - \tfrac{1}{2}(B-C) - \tfrac{1}{2}(B-C)\right\}$$

$$= A + C \tag{6.75}$$

and the 1_{01} level (B_a symmetry) from the 1×1 E^+ secular determinant

$$|\langle 10\,M0\,|\,\mathcal{H}_R\,|10\,M0\rangle - E| = 0 \qquad (6.76)$$

whose solution is

$$E(A,B,C) = B + C \qquad (6.77)$$

For $J = 2$ there are five levels, denoted as 2_{20}, 2_{21}, 2_{11}, 2_{12}, and 2_{02}. The levels 2_{21}, 2_{11}, and 2_{12} each belong to a different symmetry designation of the D_2 (V) group, namely, B_a, B_b, and B_c, respectively. Hence the energy for each of these levels involves only the evaluation of a 1×1 secular determinant, as in the $J = 1$ case. It is easily shown that the energy is $4A + B + C$ for the 2_{21} level, $A + 4B + C$ for the 2_{11} level, and $A + B + 4C$ for the 2_{12} level. However, the 2_{20} and 2_{02} levels both are of A symmetry, and their energies are the roots of the 2×2 E^+ secular determinant:

$$\begin{vmatrix} \langle 20\,M0\,|\,\mathcal{H}_R\,|20\,M0\rangle - E & \langle 20\,M0\,|\,\mathcal{H}_R\,|22\,M0\rangle \\ \langle 22\,M0\,|\,\mathcal{H}_R\,|20\,M0\rangle & \langle 22\,M0\,|\,\mathcal{H}_R\,|22\,M0\rangle - E \end{vmatrix} = 0 \qquad (6.78)$$

which has the explicit form

$$\begin{vmatrix} 3(B+C) - E & (B-C)\sqrt{3} \\ (B-C)\sqrt{3} & B + C + 4A - E \end{vmatrix} = 0 \qquad (6.79)$$

The roots are $2A + 2B + 2C + 2[(B-C)^2 + (A-C)(A-B)]^{1/2}$ for the 2_{20} level and $2A + 2B + 2C - 2[(B-C)^2 + (A-C)(A-B)]^{1/2}$ for the 2_{02} level. The energy assignments are made by identifying 2_{20} with the higher energy and 2_{02} with the lower energy in the prolate top limit $A > B = C$ (see Figure 6.8). Explicit algebraic expressions up to $J = 3$ are given in Table 6.5, and we sketch in Figure 6.9 how the energies of an asymmetric rotor vary between its prolate and oblate top limits for these levels. We note that no two levels cross, a result that may be proved to hold for all J[5, 10].

As first pointed out by Ray[10], it is advantageous to introduce an asymmetry parameter κ defined by

$$\kappa = \frac{2B - A - C}{A - C} \qquad (6.80)$$

in calculating the energy level structure of an asymmetric top. Note that κ ranges between -1 for a prolate top ($B = C$) and $+1$ for an oblate top ($A = B$). When $\kappa = 0$, $B = \frac{1}{2}(A + C)$, and as Figure 6.9 shows, the energy levels are symmetrically distributed about their mean value (middle level).

The motivation behind this choice for κ is that the energy levels of any asymmetric top can be parameterized in terms of this single variable. This may be shown as

TABLE 6.5   Asymmetric Top Energy Levels for $J \leq 3$

Level[a]	Energy $E(A, B, C)$	Symmetry
0_{00}	0	A
1_{10}	$A + B$	B_c
1_{11}	$A + C$	B_b
1_{01}	$B + C$	B_a
2_{20}	$2A + 2B + 2C + 2[(B - C)^2 + (A - C)(A - B)]^{\frac{1}{2}}$	A
2_{21}	$4A + B + C$	B_a
2_{11}	$A + 4B + C$	B_b
2_{12}	$A + B + 4C$	B_c
2_{02}	$2A + 2B + 2C - 2[(B - C)^2 + (A - C)(A - B)]^{\frac{1}{2}}$	A
3_{30}	$5A + 5B + 2C + 2[4(A - B)^2 + (A - C)(B - C)]^{\frac{1}{2}}$	B_c
3_{31}	$5A + 2B + 5C + 2[4(A - C)^2 - (A - B)(B - C)]^{\frac{1}{2}}$	B_b
3_{21}	$2A + 5B + 5C + 2[4(B - C)^2 + (A - B)(A - C)]^{\frac{1}{2}}$	B_a
3_{22}	$4A + 4B + 4C$	A
3_{12}	$5A + 5B + 2C - 2[4(A - B)^2 + (A - C)(B - C)]^{\frac{1}{2}}$	B_c
3_{13}	$5A + 2B + 5C - 2[4(A - C)^2 - (A - B)(B - C)]^{\frac{1}{2}}$	B_b
3_{03}	$2A + 5B + 5C - 2[4(B - C)^2 + (A - B)(A - C)]^{\frac{1}{2}}$	B_a

[a] $J_{K\text{(prolate)} K\text{(oblate)}}$

follows. Let σ and ρ be scalar quantities. With a simple change of variables the Hamiltonian becomes

$$\mathcal{H}_R(\sigma A + \rho, \sigma B + \rho, \sigma C + \rho)$$

$$= (\sigma A + \rho) J_a^2 + (\sigma B + \rho) J_b^2 + (\sigma C + \rho) J_c^2$$

$$= \sigma \left(A J_a^2 + B J_b^2 + C J_c^2 \right) + \rho \left(J_a^2 + J_b^2 + J_c^2 \right)$$

$$= \sigma \mathcal{H}_R(A, B, C) + \rho \mathbf{J}^2 \tag{6.81}$$

and the corresponding eigenenergies are given by

$$E(\sigma A + \rho, \sigma B + \rho, \sigma C + \rho) = \sigma E(A, B, C) + \rho J(J + 1) \tag{6.82}$$

In particular, let us choose

$$\sigma = \frac{2}{A - C} \tag{6.83}$$

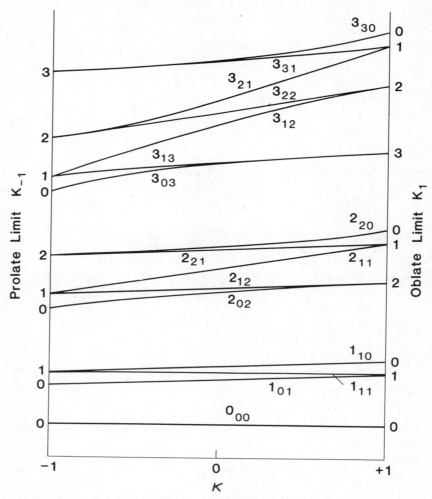

FIGURE 6.9 Energy levels of an asymmetric top as a function of the asymmetry parameter κ, calculated for $A = 5$ and $C = 3$.

and

$$\rho = -(A + C)/(A - C) \tag{6.84}$$

so that

$$\sigma A + \rho = 1$$

$$\sigma B + \rho = (2B - A - C)/(A - C) = \kappa$$

$$\sigma C + \rho = -1 \tag{6.85}$$

Substitution of Eqs. (6.83)–(6.85) into Eq. (6.82) yields the result

$$E(1, \kappa, -1) = E_J(\kappa) = \frac{2}{A - C} E(A, B, C) - \frac{A + C}{A - C} J(J + 1) \qquad (6.86)$$

which may be rearranged to read

$$E(A, B, C) = \tfrac{1}{2}(A + C)J(J + 1) + \tfrac{1}{2}(A - C)E_{J\tau}(\kappa) \qquad (6.87)$$

Hence the energy levels of an asymmetric top, labeled by the index τ [see Eq. (6.70)], may be determined from knowledge of the universal energy function $E_J(\kappa)$ of a rigid rotor with angular momentum $\mathbf{J}$ having the three rotational constants $A = 1$, $B = \kappa$, $C = -1$.

Values of $E_{J\tau}(\kappa)$ for all levels with $J \leq 12$ and for values of κ from 0 to 1 in steps of 0.01 have been computed by Turner, Hicks, and Reitwiesner[24] and are listed in Appendix IV of the monograph *Microwave Spectroscopy* by Townes and Schawlow[25]. Energy levels tables with $J \leq 40$ and for values of κ from -1 to 0 in steps of 0.1 have been compiled by Erlandsson[26] and are reproduced in Appendix IV of the monograph *Molecular Vib-Rotors* by Allen and Cross[27]. Only positive or negative values of κ need be tabulated since it may be shown[10] that (see Figure 6.9)

$$E_{J\tau}(\kappa) = -E_{J,-\tau}(-\kappa) \qquad (6.88)$$

The energy levels of a rigid asymmetric rotor have also been calculated by semiclassical methods with the use of a uniform semiclassical quantum condition[19, 28]. The results give the correct number of rotational levels for a given J and are remarkably accurate.

6.4 NONRIGID BEHAVIOR: THE VAN VLECK TRANSFORMATION

Although the rotational motion of a molecule closely resembles that of a rigid body, real molecules vibrate (many times) as they rotate, causing what is called *vibration–rotation* interaction. This interaction has the effect of causing the intramolecular bond angles and bond distances to change as the molecule rotates, thus coupling nearby vibrational levels within the same electronic state. Also, as the molecule vibrates and rotates there is the possibility of interaction between different electronic states as the electronic and nuclear motions are not exactly separable (failure of the Born–Oppenheimer approximation). All these effects cause the rigid rotor model to break down. We seek in this section a way of accounting for nonrigid behavior.

As is well known, a polyatomic molecule with N nuclei has $3N - 6$ vibrational degrees of freedom ($3N - 5$ if all the nuclei lie on a straight line). For small vibrations about the equilibrium configuration, it is always possible to choose an orthogonal set of internal coordinates $\{Q_k\} = Q_1, Q_2, \ldots, Q_{3N-6}$, so that the system undergoes simple harmonic motion at some vibrational frequency ν_k along the vibrational coordinate Q_k, which is called a *normal mode*[29]. Under the conditions where we ignore higher-order anharmonic effects, the vibrational motion of the system along each of its normal modes is independent. Then the vibrational energy is a sum of the energies $h\nu_k(v_k + \frac{1}{2})$ of each normal mode with vibrational quantum number v_k, and the vibrational wave function is a product of the individual wave functions $\psi_{v_k}(Q_k)$ of each normal mode (which are harmonic oscillator wave functions), that is,

$$E_V = h \sum_k \nu_k \left(v_k + \tfrac{1}{2}\right) \tag{6.89}$$

and

$$\psi_V = \prod_k \psi_{v_k}(Q_k) \tag{6.90}$$

Note that we let the index V denote the collection of vibrational quantum numbers $\{v_k\}$.

We suppose as a good starting point that the Hamiltonian for our nonrigid molecule can be separated into two parts:

$$\mathcal{H} = \mathcal{H}_V + \mathcal{H}_R \tag{6.91}$$

where $\mathcal{H}_V$ describes a vibrating oscillator and $\mathcal{H}_R$ a rigid rotor. Then we introduce the wave functions

$$\psi_{VR} = \psi_V \psi_R, \qquad |VR\rangle = |V\rangle |R\rangle \tag{6.92}$$

as a convenient basis set where

$$\mathcal{H} |VR\rangle = (E_V + E_R) |VR\rangle \tag{6.93}$$

Here R denotes the collection of rotational quantum numbers τ, J, K, M necessary to specify the rotational wave function with rotational energy E_R. We could also denote the different electronic states, but for simplicity we will consider only the interactions between the levels of one electronic state and, therefore, drop this designation in what follows.

Let us concentrate our attention on the rotational level structure within one specific nondegenerate[30] vibrational state of a nondegenerate[31] electronic state by considering only those levels with the same total angular momentum **J**. There are in general $2J + 1$ such levels with the same space-fixed projection of **J**, and we refer to these levels as the $|V; JKM\rangle$ block. Once again it is convenient to use the symmetric top wave functions $|JKM\rangle$ as a basis set because they span the $(2J + 1)$ by $(2J + 1)$ space of R.

Nonrigid behavior makes $\mathcal{H}_R$ vary with nuclear motion, causing the $|V; JKM\rangle$ block to be perturbed. However, a perturbation, like beauty, is in the eye of the beholder! Occasionally, the perturbation can be ascribed to a small number of interacting levels belonging to different $|V; JKM\rangle$ blocks. Then the effect of the perturbation can be treated by diagonalizing the matrix of interacting levels (all having the same total angular momentum). This procedure, which amounts to all orders of perturbation theory, requires that the location and symmetry designations of the perturbing levels be known. The result is that the energy-level scheme deviates from the regular pattern of the other unperturbed levels. Some choose to regard this phenomenon as a nuisance; others regard this as an opportunity to have a "window" on often otherwise hidden levels.

More generally, the perturbations are small in magnitude but large in number. Moreover, they involve many interacting levels from distant blocks whose energy positions are often poorly known (if at all). In this latter case, an exact treatment (diagonalization of a semi-infinite number of interacting levels) is impractical, if not impossible, and resort must be made to some approximate procedure.

The Van Vleck transformation, also called a *contact transformation*, is such an approximation that treats the interaction between the $(2J + 1)$ by $(2J + 1)$ blocks by incorporating their effects on the $|V; JKM\rangle$ block of interest[32–35]. When this procedure is applicable, there is no longer the need to diagonalize the supermatrix of interacting blocks. In what follows, we outline the Van Vleck transformation, which bears a strong resemblance to second-order perturbation theory, and illustrate its use by considering the effects of centrifugal distortion on the energy levels of a symmetric top.

Let the molecular Hamiltonian be written as

$$\mathcal{H} = \mathcal{H}_0 + \lambda \mathcal{H}_1 \tag{6.94}$$

where λ denotes the order of the perturbation. The parameter λ is a bookkeeping device; when we are finished, we will set $\lambda = 1$. We assume that the matrix elements of $\mathcal{H}_0$ lie entirely within the diagonal $|V; JKM\rangle$ blocks while the matrix elements of $\mathcal{H}_1$ may lie inside the outside diagonal blocks. The idea behind the Van Vleck transformation is to apply a unitary transformation such that it removes to first order in λ the matrix elements of $\mathcal{H}_1$ that lie outside the diagonal blocks while at the same time preserves to first order in λ the matrix elements of $\mathcal{H}_1$ that lie inside the diagonal blocks. Then the remaining matrix elements of $\mathcal{H}_1$ that lie outside the diagonal blocks are second order in λ and contribute to the energy only to fourth order in λ. Thus through third order in λ, the transformed matrix of $\mathcal{H}$, which we call $\mathcal{G}$, consists of diagonal blocks of the same dimensions as found in $\mathcal{H}_0$. If desired, $\mathcal{G}$ may be subject to a second Van Vleck transformation, and so on.

We turn to the steps necessary to carry out this procedure. Any unitary transformation $\mathbf{T}$ can be expressed as

$$\mathbf{T} = \exp(i\lambda \mathbf{S}) \tag{6.95}$$

where $\mathbf{S}$ is a Hermitian matrix. Moreover, we can expand $\mathbf{T}$ in powers of λ and obtain the series

$$\mathbf{T} = \mathbf{I} + i\lambda \mathbf{S} - \lambda^2 \mathbf{S}^2/2! - i\lambda^3 \mathbf{S}^3/3! + \cdots \tag{6.96}$$

where $\mathbf{I}$ is the identity matrix. Hence the transformed matrix $\mathcal{G}$ can be written as

$$\mathcal{G} = \mathbf{T}^{-1}\mathcal{H}\mathbf{T}$$

$$= [\mathbf{I} - i\lambda\mathbf{S} - \lambda^2\mathbf{S}^2/2 + i\lambda^3\mathbf{S}^3/6 + \cdots][\mathcal{H}_0 + \lambda\mathcal{H}_1]$$

$$\times[\mathbf{I} + i\lambda\mathbf{S} - \lambda^2\mathbf{S}^2/2 - i\lambda^3\mathbf{S}^3/6 + \cdots]$$

$$= \mathcal{G}_0 + \lambda\mathcal{G}_1 + \lambda^2\mathcal{G}_2 + \lambda^3\mathcal{G}_3 + \cdots \tag{6.97}$$

Equating like powers of λ, we find

$$\mathcal{G}_0 = \mathcal{H}_0$$

$$\mathcal{G}_1 = \mathcal{H}_1 + i(\mathcal{H}_0\mathbf{S} - \mathbf{S}\mathcal{H}_0)$$

$$\mathcal{G}_2 = i(\mathcal{H}_1\mathbf{S} - \mathbf{S}\mathcal{H}_1) + \mathbf{S}\mathcal{H}_0\mathbf{S} - \frac{1}{2}(\mathcal{H}_0\mathbf{S}^2 + \mathbf{S}^2\mathcal{H}_0)$$

$$\mathcal{G}_3 = \mathbf{S}\mathcal{H}_1\mathbf{S} - \frac{1}{2}(\mathcal{H}_1\mathbf{S}^2 + \mathbf{S}^2\mathcal{H}_1) + \frac{1}{2}i(\mathbf{S}\mathcal{H}_0\mathbf{S}^2 - \mathbf{S}^2\mathcal{H}_0\mathbf{S})$$

$$- \frac{1}{6}i(\mathcal{H}_0\mathbf{S}^3 - \mathbf{S}^3\mathcal{H}_0) \tag{6.98}$$

Let the levels of the $|V; JKM\rangle$ block of interest be indexed by j, k, ℓ, and so on, and the levels of other blocks by α, β, γ, and so forth. The Van Vleck transformation must meet two requirements. The first requirement is that $\langle j|\mathcal{G}_1|k\rangle$ equals $\langle j|\mathcal{H}_1|k\rangle$. Hence we demand that

$$0 = \langle j|\mathcal{G}_1|k\rangle - \langle j|\mathcal{H}_1|k\rangle = \langle j|i(\mathcal{H}_0\mathbf{S} - \mathbf{S}\mathcal{H}_0)|k\rangle$$

$$= (E_j - E_k)\langle j|\mathbf{S}|k\rangle \tag{6.99}$$

which is satisfied in general by setting

$$\langle j|\mathbf{S}|k\rangle = 0 \tag{6.100}$$

The second requirement is that $\langle j|\mathcal{G}_1|\alpha\rangle$ vanishes. We have

$$0 = \langle j|\mathcal{H}_1|\alpha\rangle + i\langle j|\mathcal{H}_0\mathbf{S} - \mathbf{S}\mathcal{H}_0|\alpha\rangle$$

$$= \langle j|\mathcal{H}_1|\alpha\rangle + i(E_j - E_\alpha)\langle j|\mathbf{S}|\alpha\rangle \tag{6.101}$$

which is satisfied in general by setting

$$\langle j|\mathbf{S}|\alpha\rangle = \frac{i\langle j|\mathcal{H}_1|\alpha\rangle}{E_j - E_\alpha} \tag{6.102}$$

Equations (6.100) and (6.102) define the nature of **S** that generates the unitary (contact) transformation **T**. Applying these relations to Eq. (6.97), we obtain the matrix elements of the transformed Hamiltonian within the same $(2J+1)$ by $(2J+1)$ block $|V; JKM\rangle$:

$$\langle j| \mathcal{G}_0 |k\rangle = \langle j| \mathcal{H}_0 |k\rangle = E_j \delta_{jk}$$

$$\langle j| \mathcal{G}_1 |k\rangle = \langle j| \mathcal{H}_1 |k\rangle$$

$$\langle j| \mathcal{G}_2 |k\rangle = \sum_\alpha \frac{[\frac{1}{2}(E_j + E_k) - E_\alpha] \langle j| \mathcal{H}_1 |\alpha\rangle \langle \alpha| \mathcal{H}_1 |k\rangle}{(E_j - E_\alpha)(E_k - E_\alpha)}$$

$$\langle j| \mathcal{G}_3 |k\rangle = \sum_\alpha \left[\sum_\beta \frac{\langle j| \mathcal{H}_1 |\alpha\rangle \langle \alpha| \mathcal{H}_1 |\beta\rangle \langle \beta| \mathcal{H}_1 |k\rangle}{(E_j - E_\alpha)(E_k - E_\alpha)} \right.$$

$$-\frac{1}{2} \sum_\ell \frac{\langle j| \mathcal{H}_1 |\ell\rangle \langle \ell| \mathcal{H}_1 |\alpha\rangle \langle \alpha| \mathcal{H}_1 |k\rangle}{(E_\ell - E_\alpha)(E_k - E_\alpha)}$$

$$\left. -\frac{1}{2} \sum_\ell \frac{\langle j| \mathcal{H}_1 |\alpha\rangle \langle \alpha| \mathcal{H}_1 |\ell\rangle \langle \ell| \mathcal{H}_1 |k\rangle}{(E_\ell - E_\alpha)(E_j - E_\alpha)} \right] \qquad (6.103)$$

In most applications it suffices to retain only the second-order correction. A further simplification results if the energy separation between interacting blocks is much larger than the energy separation between levels in the same block so that the levels within the same block are quasi-degenerate compared to level separations between blocks. Then to good approximation the correction matrix $\mathcal{H}_{jk}^{(2)} = \langle j| \mathcal{G}_2 |k\rangle$ may be written as

$$\mathcal{H}_{jk}^{(2)} = \sum_\alpha \frac{\langle j| \mathcal{H}_1 |\alpha\rangle \langle \alpha| \mathcal{H}_1 |k\rangle}{E_{VJ} - E_{V'J}} \qquad (6.104)$$

where $E_{VJ} - E_{V'J}$ is the energy difference between the centers of the unperturbed blocks $|V; JKM\rangle$ and $|V'; JKM\rangle$. Equation (6.104) closely resembles the form of second-order perturbation theory. We can factor $\mathcal{H}_{jk}^{(2)}$ as $f_{jk}(J, K)A_{jk}$, where $f_{jk}(J, K)$ contains the dependence of $\mathcal{H}_{jk}^{(2)}$ on the quantum numbers J and K and A_{jk} is a term independent of these quantum numbers. As a result, the Van Vleck transformation introduces new terms into the effective molecular Hamiltonian to account for the perturbation $\mathcal{H}_1$ on the levels of the $|V; JKM\rangle$ block. The "constants" A_{jk} are then found empirically by comparison with spectroscopic data.

An example will help to make the above less abstract. Consider how centrifugal distortion affects the energy levels of a prolate symmetric top. The rotational Hamiltonian is, according to Eq. (6.52), given by $\mathcal{H}_R = C\mathbf{J}^2 + (A - C)J_z^2$, where for a rigid body A and C are rotational constants inversely proportional to twice the moment of inertia about the a and c axes, respectively, where a is the figure axis (the z axis in the molecule-fixed frame). When the molecule vibrates, $\mathcal{H}_R$ can be written in the same form; however, A and C are no longer constants but depend on the positions and conjugate momenta of the nuclei[36]. Hence we must regard A and C as operators in the space of molecular vibrations. We then identify the expectation value $\langle V | \mathbf{A} | V \rangle$ with A_V and $\langle V | \mathbf{C} | V \rangle$ with C_V. Application of the Van Vleck transformation via Eq. (6.104) is particularly simple in this case because for the prolate symmetric top rotational Hamiltonian there are no off-diagonal matrix elements in K. We find for the second-order centrifugal distortion correction

$$\mathcal{H}^{(2)}_{VJK;VJK} = \sum_{\substack{V' \\ J'K'M'}} \frac{\langle V; JKM | \mathcal{H}_R | V'; J'K'M' \rangle \langle V'; J'K'M' | \mathcal{H}_R | V; JKM \rangle}{E_{VJ} - E_{V'J}}$$

$$= \sum_{V'} \frac{\langle V | \mathbf{C} | V' \rangle \langle V' | \mathbf{C} | V \rangle}{E_{VJ} - E_{V'J}}$$

$$\times \sum_{J'K'M'} \langle JKM | \mathbf{J}^2 | J'K'M' \rangle \langle J'K'M' | \mathbf{J}^2 | JKM \rangle$$

$$+ \sum_{V'} \frac{\langle V | \mathbf{C} | V' \rangle \langle V' | \mathbf{A} - \mathbf{C} | V \rangle + \langle V | \mathbf{A} - \mathbf{C} | V' \rangle \langle V' | \mathbf{C} | V \rangle}{E_{VJ} - E_{V'J}}$$

$$\times \sum_{J'K'M'} \langle JKM | J_z^2 | J'K'M' \rangle \langle J'K'M' | \mathbf{J}^2 | JKM \rangle$$

$$+ \sum_{V'} \frac{\langle V | \mathbf{A} - \mathbf{C} | V' \rangle \langle V' | \mathbf{A} - \mathbf{C} | V \rangle}{E_{VJ} - E_{V'J}}$$

$$\times \sum_{J'K'M'} \langle JKM | J_z^2 | J'K'M' \rangle \langle J'K'M' | J_z^2 | JKM \rangle \tag{6.105}$$

This is essentially equivalent to introducing a centrifugal distortion Hamiltonian for the level V of the form

$$\mathcal{H}_{\text{cent}} = -D_J \mathbf{J}^4 - D_{JK} J_z^2 \mathbf{J}^2 - D_K J_z^4 \tag{6.106}$$

where we identify[37]

$$D_J = -\sum_{V'} \frac{\langle V | \mathbf{C} | V' \rangle \langle V' | \mathbf{C} | V \rangle}{E_V - E_{V'}} \tag{6.107}$$

$$D_{JK} = -\sum_{V'} \frac{\langle V | \mathbf{C} | V' \rangle \langle V' | \mathbf{A} - \mathbf{C} | V \rangle + \langle V | \mathbf{A} - \mathbf{C} | V' \rangle \langle V' | \mathbf{C} | V \rangle}{E_V - E_{V'}} \tag{6.108}$$

$$D_K = -\sum_{V'} \frac{\langle V | \mathbf{A} - \mathbf{C} | V' \rangle \langle V' | \mathbf{A} - \mathbf{C} | V \rangle}{E_V - E_{V'}} \qquad (6.109)$$

as three phenomenological centrifugal distortion constants.

The rotational energy level structure of a vibrating prolate symmetric top then becomes

$$E_R = C_V \, J(J+1) + (A_V - C_V) K^2$$

$$- D_J [J(J+1)]^2 - D_{JK} J(J+1) K^2 - D_K K^4 \qquad (6.110)$$

which is a result first obtained by Slawsky and Dennison[38]. The correction terms for centrifugal distortion depend on only even powers of the angular momentum. This is so because the distortion effects do not depend on the direction of rotation about any axis. The term $D_{JK} J(J+1) K^2$ has the effect that the K stack levels (levels of different J but the same K) will no longer coincide exactly when shifted by an appropriate amount.

Centrifugal distortion corrections for asymmetric rotors have been worked out in a similar manner, only here the centrifugal distortion Hamiltonian is of the form[39]

$$\mathcal{H}_{\text{cent}} = \frac{1}{4} \sum_{\alpha,\beta,\gamma,\delta} \tau_{\alpha\beta\gamma\delta} J_\alpha J_\beta J_\gamma J_\delta$$

where α, β, γ, δ may each represent either x, y, or z. The full vibration–rotation problem is very complex and treated in detail elsewhere[36,40–42]. By use of the Van Vleck transformation it is possible to describe quite complicated molecular interactions and reduce them to a small number of phenomenological molecular constants.

6.5 LINE STRENGTH FACTORS

The line strength $S\left(J'; J''\right)$ of the electric-dipole-allowed molecular transition connecting the upper level characterized by the total angular momentum J' to the lower level J'' is defined in the same manner as in atomic spectroscopy[43], namely,

$$S\left(J'; J''\right) = \sum_{M',M''} |\langle J' M' | \boldsymbol{\mu} | J'' M'' \rangle|^2$$

$$= \sum_{M',M''} |\langle J'' M'' | \boldsymbol{\mu} | J' M' \rangle|^2$$

$$= \sum_{M',M''} S\left(J'M'; J''M''\right) \qquad (6.111)$$

where the summation is over all magnetic sublevels of both levels. It may be shown[43] that the line strength factor may be related to the Einstein coefficient for spontaneous emission by the relation

$$A\left(J' \to J''\right) = \frac{64\,\pi\nu^3}{3\,hc^3} \frac{S\left(J';J''\right)}{2\,J'+1} \tag{6.112}$$

where ν is the frequency of the $J' \to J''$ transition, h is Planck's constant, and c is the speed of light. Thus the central issue in comparing molecular line intensities is the calculation of the line strength factor $S\left(J';J''\right)$, which involves the evaluation of the matrix elements of the electric dipole operator μ between the wave functions $\psi_{J'M'}$ and $\psi_{J''M''}$.

The operator μ consists of a sum over all charges, nuclear and electronic, weighted by their position vectors measured from a common origin. For an electric dipole transition the fundamental interaction is $\mu \cdot \mathbf{E}$, where $\mathbf{E}$ is the electric vector of the radiation field. The line strength refers to what is called "natural excitation" whereby the excitation is isotropic as opposed to directional[43]. Then we may rewrite Eq. (6.111) as

$$S\left(J';J''\right) = 3 \sum_{M',M''} |\langle J'M'|\mu_Z|J''M''\rangle|^2 \tag{6.113}$$

because each space-fixed component of μ contributes equally to the line strength factor. It is more convenient to refer μ in the space-fixed frame ($F = X, Y, Z$) to the molecule-fixed frame ($g = x, y, z$). According to Eq. (5.9)

$$\mu\left(1,p\right) = \sum_q D_{pq}^{1*}\left(\phi, \theta, \chi\right)\mu\left(1,q\right) \tag{6.114}$$

and for μ_Z

$$\mu\left(1,0\right) = \sum_q D_{0q}^{1*}\mu\left(1,q\right) \tag{6.115}$$

Hence the line strength factor is given by

$$S\left(J';J''\right) = 3 \sum_{M',M''} \left|\langle J'M'|\sum_q D_{0q}^{1*}\mu\left(1,q\right)|J''M''\rangle\right|^2 \tag{6.116}$$

We begin by specializing our consideration to a symmetric top. For the transition $J'K' \longleftrightarrow J''K''$ there is only one component of μ in the molecule-fixed frame which can contribute, namely, the component q for which $K' - K'' = q$. We set the magnitude of the square of the dipole moment to unity so that our factors refer simply to relative rotational line strengths independent of the nature of the molecule. We are then able to carry out the evaluation of Eq. (6.116) in a straightforward manner:

$$S\left(J'K'; J''K''\right)$$

$$= 3 \sum_{M',M''} \left| \left[\left(\frac{2J'+1}{8\pi^2} \right) \left(\frac{2J''+1}{8\pi^2} \right) \right]^{\frac{1}{2}} \int D_{M'K'}^{J'} D_{0q}^{1*} D_{M''K''}^{J''*} d\Omega \right|^2$$

$$= 3 \sum_{M',M''} \left| \frac{\left[(2J'+1)(2J''+1) \right]^{\frac{1}{2}}}{8\pi^2} \left[\int D_{M'K'}^{J'*} D_{0q}^{1} D_{M''K''}^{J''} d\Omega \right]^* \right|^2$$

$$= 3 \sum_{M',M''} \left| \left(\frac{2J''+1}{2J'+1} \right)^{\frac{1}{2}} \langle J'M', 10 | J''M'' \rangle \langle J'K', 1q | J''K'' \rangle \right|^2$$

$$= 3 \frac{2J''+1}{2J'+1} \langle J'K', 1q | J''K'' \rangle^2 \sum_{M',M''} \langle J'M', 10 | J''M'' \rangle^2$$

$$= 3 \frac{2J''+1}{2J'+1} \langle J'K', 1q | J''K'' \rangle^2 \sum_{M} \left(\frac{2J'+1}{3} \right) \langle JM, J' -M | 10 \rangle^2$$

$$= (2J''+1) \langle J'K', 1q | J''K'' \rangle^2$$

$$= (2J'+1)(2J''+1) \begin{pmatrix} J' & 1 & J'' \\ -K' & K'-K'' & K'' \end{pmatrix}^2 \tag{6.117}$$

Using Eq. (2.33), we obtain at once the sum rules:

$$\sum_{J',K'} S\left(J'K'; J''K''\right) = 2J'' + 1 \tag{6.118}$$

and

$$\sum_{J'',K''} S\left(J'K'; J''K''\right) = 2J' + 1 \tag{6.119}$$

The $S\left(J'K'; J''K''\right)$ are called *Hönl–London factors*, named after the two workers who first derived their value but by quite different means involving "old" quantum theory[44]. We introduce the standard spectroscopic notation:

$$\Delta J = J' - J'' = -1 \qquad P \text{ branch}$$

$$\Delta J = J' - J'' = 0 \qquad Q \text{ branch}$$

$$\Delta J = J' - J'' = +1 \qquad R \text{ branch}$$

TABLE 6.6   The Hönl–London Factors[a]

Transition	$S\left(J'K'; J''K''\right)$

$\Delta K = +1$

$P\left(J''\right)$	$\dfrac{\left(J'' - 1 - K''\right)\left(J'' - K''\right)}{4J''}$
$Q\left(J''\right)$	$\dfrac{\left(J'' + 1 + K''\right)\left(J'' - K''\right)\left(2J'' + 1\right)}{4J''\left(J'' + 1\right)}$
$R\left(J''\right)$	$\dfrac{\left(J'' + 2 + K''\right)\left(J'' + 1 + K''\right)}{4\left(J'' + 1\right)}$

$\Delta K = 0$

$P\left(J''\right)$	$\dfrac{\left(J'' + K''\right)\left(J'' - K''\right)}{J''}$
$Q\left(J''\right)$	$\dfrac{\left(2J'' + 1\right)K''^{2}}{J''\left(J'' + 1\right)}$
$R\left(J''\right)$	$\dfrac{\left(J'' + 1 + K''\right)\left(J'' + 1 - K''\right)}{J'' + 1}$

$\Delta K = -1$

$P\left(J''\right)$	$\dfrac{\left(J'' - 1 + K''\right)\left(J'' + K''\right)}{4J''}$
$Q\left(J''\right)$	$\dfrac{\left(J'' + 1 - K''\right)\left(J'' + K''\right)\left(2J'' + 1\right)}{4J''\left(J'' + 1\right)}$
$R\left(J''\right)$	$\dfrac{\left(J'' + 2 - K''\right)\left(J'' + 1 - K''\right)}{4\left(J'' + 1\right)}$

[a]Note that the $P\left(0\right)$ and $Q\left(0\right)$ transitions do not exist.

Table 6.6 lists the Hönl–London factors for the nine possible types of symmetric top transitions.

Often we are interested in the form of the rotational line strengths when $J'' \gg K''$. Table 6.6 shows that for a $\Delta K = 0$ transition (with the transition dipole parallel to the figure axis of the top) $P\left(J''\right) \simeq J''$, $Q\left(J''\right) \simeq 0$, and $R\left(J''\right) \simeq \left(J'' + 1\right)$, that is, the P and R branch lines are of comparable strength while the lines of the Q branch have appreciable strength only for the lowest J'' values. For a $\Delta K = \pm 1$ transition (with the transition dipole perpendicular to the figure axis of the top) $P\left(J''\right) \simeq \left(J'' - 1\right)/4$, $Q\left(J''\right) \simeq \left(2J'' + 1\right)/4$, and $R\left(J''\right) \simeq \left(J'' + 2\right)/4$, that is, the lines of the P and R branches for a given J'' value have comparable strength but the lines of the Q branch for the same J'' value are about twice as strong.

For an asymmetric top we may write the rotational wave function belonging to the energy level $E_{J\tau}(\kappa)$ as a linear combination of symmetric top wave functions:

$$|\tau'' J'' M''\rangle = \sum_{K''} a_{\tau'' K''} |J'' K'' M''\rangle \qquad (6.120)$$

and

$$|\tau' J' M'\rangle = \sum_{K'} a_{\tau' K'} |J' K' M'\rangle \qquad (6.121)$$

where the coefficients $a_{\tau K}$ are real and satisfy the condition

$$\sum_{K} a_{\tau K}^2 = 1 \qquad (6.122)$$

On subsitution of Eqs. (6.120) and (6.121) into (6.116), we obtain for the rotational line strength of an asymmetric top transition $\tau' J' \longleftrightarrow \tau'' J''$

$$S\left(\tau' J'; \tau'' J''\right) = 3 \sum_{M} \left| \sum_{K'} \sum_{K''} a_{\tau' K'} a_{\tau'' K''} \left\langle J' K' M \left| \mu_Z \right| J'' K'' M \right\rangle \right|^2$$

$$= \left(2 J' + 1\right) \left(2 J'' + 1\right) \left| \sum_{K'} \sum_{K''} a_{\tau' K'} a_{\tau'' K''} \right.$$

$$\left. \times \begin{pmatrix} J'' & 1 & J' \\ K'' & K' - K'' & -K' \end{pmatrix} \right|^2 \qquad (6.123)$$

Once again the $S\left(\tau' J'; \tau'' J''\right)$ factors may be shown to satisfy the sum rules:

$$\sum_{\tau', J'} S\left(\tau' J'; \tau'' J''\right) = 2 J'' + 1 \qquad (6.124)$$

and

$$\sum_{\tau'', J''} S\left(\tau' J'; \tau'' J''\right) = 2 J' + 1 \qquad (6.125)$$

Equations (6.124) and (6.125) are a consequence of Eqs. (6.118), (6.119), and (6.122). Thus the calculation of rotational line strengths for asymmetric top transitions reduces to the problem of finding the values of the coefficients $a_{\tau'' K''}$ and $a_{\tau' K'}$. As shown in Section 6.3, this is equivalent to finding the elements of the columns (vectors) of the unitary matrix that diagonalizes the asymmetric top Hamiltonian.

To the P, Q, R branch notation used to designate $\Delta J = -1, 0, +1$ it is convenient to add as subscripts on the right the values of ΔK_{-1} and ΔK_{+1} to distinguish different subbranches[45]. For example, $P_{3,-2}$ indicates $\Delta J = -1$, $\Delta K_{-1} = 3$, and $\Delta K_{+1} = -2$. In pure rotational or vibrational–rotational spectra where the top geometry almost does not change between the initial and final states, the principal subbranches are those for which ΔK_{-1} and ΔK_{+1} change by 0 or ± 1 and are thus allowed in both the prolate and oblate top limits. The next most important subbranches are those for which one of the indices changes by 0 or ± 1 but the change in the other index is greater than one. Such transitions are allowed in one of the two top limits. The least important are those transitions forbidden in both top limits.

For an asymmetric top there are in general three nonvanishing components of the transition dipole, μ_a, μ_b, and μ_c, which transform as B_a, B_b, and B_c, respectively, under the symmetry operations of the D_2 (V) group. For the matrix element $\langle \tau' J' | \, \boldsymbol{\mu} \, | \tau'' J'' \rangle$ not to vanish it is necessary that the product of the three factors be symmetric, that is, belong to A, under all the group operations. Making use of this fact, we obtain the following selection rules for asymmetric top transitions:

$$\mu_a \neq 0 \qquad A \longleftrightarrow B_a \qquad B_b \longleftrightarrow B_c \qquad \Delta K_a = 0, \pm 2, \ldots$$
$$ee \longleftrightarrow eo \qquad oo \longleftrightarrow oe \qquad \Delta K_c = \pm 1, \pm 3, \ldots$$

$$\mu_b \neq 0 \qquad A \longleftrightarrow B_b \qquad B_a \longleftrightarrow B_c \qquad \Delta K_a = \pm 1, \pm 3, \ldots$$
$$ee \longleftrightarrow oo \qquad eo \longleftrightarrow oe \qquad \Delta K_c = \pm 1, \pm 3, \ldots$$

$$\mu_c \neq 0 \qquad A \longleftrightarrow B_c \qquad B_a \longleftrightarrow B_b \qquad \Delta K_a = \pm 1, \pm 3, \ldots$$
$$ee \longleftrightarrow oe \qquad eo \longleftrightarrow oo \qquad \Delta K_c = 0, \pm 2, \ldots$$

We have completed the calculation of the rotational line strength factors for a rigid body undergoing an electric dipole transition. We take up the more general problem of finding the line strengths of molecular transitions. A full explication of this topic is beyond our scope but the following overview is intended as an outline of the most important features. Once again we begin by invoking the Born–Oppenheimer approximation[46] and writing the total wave function ψ_{JM} as a product of an electronic, vibrational, and rotational part:

$$\psi_{JM} = \psi_E \psi_V \psi_R \qquad (6.126)$$

Here E, V, and R stand for the complete set of electronic, vibrational, and rotational quantum numbers, respectively. Each ψ_E, ψ_V, ψ_R form a complete orthonormal basis set spanning their respective spaces. Hence when more accurate calculations are required, we can expand ψ_{JM} in terms of these functions

$$\psi_{JM} = \sum_{E,V,R} C(E, V, R) \, \psi_E \psi_V \psi_R \qquad (6.127)$$

For atomic transitions one chooses the origin to be the atomic nucleus; then μ depends almost always on the electronic coordinates alone (disregarding γ-ray transitions). For molecular transitions, μ is always a sum of an electronic part μ^e and a nuclear part μ^n:

$$\langle \psi_{J'M'} | \mu | \psi_{J''M''} \rangle = \langle \psi_{E'} \psi_{V'} \psi_{R'} | \mu^e | \psi_{E''} \psi_{V''} \psi_{R''} \rangle + \langle \psi_{E'} \psi_{V'} \psi_{R'} | \mu^n | \psi_{E''} \psi_{V''} \psi_{R''} \rangle$$

(6.128)

Because μ^n is independent of the electronic coordinates, the second integral in Eq. (6.128) vanishes for $E' \neq E''$ since $\langle \psi_{E'} | \psi_{E''} \rangle = \delta_{E'E''}$. Hence for electronic transitions in molecules, only the electronic part of the electric dipole operator needs to be considered, while for transitions within the same electronic state, the full dipole moment operator contributes. Indeed, in the latter case $\left(E' = E'' \right)$, the operator is the expectation value of the dipole moment in the electronic state, $\langle \mu \rangle_E$.

Equation (6.128) readily factors into the product of two terms so that

$$S \left(J'; J'' \right) = S_{J'J''}^{EV} S_{J'J''}^{R}$$

(6.129)

where $S_{J'J''}^{R}$ is the rotational line strength factor we discussed previously, and $S_{J'J''}^{EV}$ is the strength of the $E'V' \longleftrightarrow E''V''$ vibronic band.

First let us discuss the form of $S_{J'J''}^{EV}$ for $E' \neq E''$. Here we have

$$S_{J'J''}^{EV} = \left| \langle V' | D_{E'E''} | V'' \rangle \right|^2$$

(6.130)

where

$$D_{E'E''} = \langle E' | \mu^e | E'' \rangle$$

(6.131)

is the transition dipole moment connecting the two electronic states and is a weak function of the nuclear coordinates in that the electronic wave functions $\psi_{E'}$ and $\psi_{E''}$ depend only parametrically on them. Consequently, $D_{E'E''}$ may be treated as a constant (Condon approximation[47]) and Eq. (6.130) then becomes [see Eq. (6.90)]:

$$S_{J'J''}^{EV} = D_{E'E''}^2 \left| \langle V' | V'' \rangle \right|^2 = D_{E'E''}^2 \prod_k q_{v_k' v_k''}$$

(6.132)

where the square of the vibrational overlap integral

$$q_{v_k' v_k''} = \left| \left\langle \psi_{v_k'} (Q_k) \middle| \psi_{v_k''} (Q_k) \right\rangle \right|^2$$

(6.133)

is called the Franck–Condon factor for the normal mode k. Because the vibrational wave functions are oscillatory functions of nuclear displacement, the Franck–Condon factors vary markedly with vibrational quantum number depending on whether the overlap of the vibrational wave function is in or out of phase. This quantum mechanical effect has been termed "internal diffraction" by Condon, who remarked:

"internal diffraction is just as real and forceful a proof for the wave nature of nuclear motion as any of the basic external diffraction experiments ... which also are determined by a phase relation between initial and final wave functions[47]." For more accurate work the variation of $D_{E'E''}$ with nuclear displacement must also be taken into account[48, 49].

Consider next the electric dipole transitions connecting $V'R'$ and $V''R''$ within the same electronic state E. The electronic dipole moment operator $\langle \mu \rangle_E$ depends both on electronic and nuclear coordinates suggesting that we consider an expansion of dipole moment relative to one of the molecule-fixed axes $g = x, y, z$ about the equilibrium configuration

$$\langle \mu_g \rangle_E = (\langle \mu_g \rangle_E)_0 + \sum_k \left(\frac{\partial \langle \mu_g \rangle_E}{\partial Q_k} \right)_0 Q_k + \dots \tag{6.134}$$

Then the $V' - V''$ band strength factor becomes

$$S^{EV}_{J'J''} = \left| \langle V' | \langle \mu_g \rangle_E | V'' \rangle \right|^2$$

$$= \left| \langle V' | (\langle \mu_g \rangle_E)_0 + \sum_k \left(\frac{\partial \langle \mu_g \rangle_E}{\partial Q_k} \right)_0 Q_k + \dots | V'' \rangle \right|^2 \tag{6.135}$$

Note that terms involving $\langle V' | V'' \rangle$ vanish unless $V' = V''$, whereas terms involving $\langle V' | Q_k | V'' \rangle$ vanish unless $V' \neq V''$ since $\langle V | Q_k | V \rangle$ is an odd function in the normal coordinate Q_k. Then to lowest order, the term $\left| \langle V' | (\langle \mu_g \rangle_E)_0 | V'' \rangle \right|^2$ gives rise to a pure rotation spectrum ($V' = V''$) that is seen to require a nonvanishing permanent dipole moment along at least one of the principal axes $g = a, b, c$. And to lowest order, the term $\left| \sum_k (\partial \langle \mu_g \rangle_E / \partial Q_k)_0 \langle V' | Q_k | V'' \rangle \right|^2$ gives rise to a vibration–rotation spectrum ($V' \neq V$) that is seen to require a changing dipole moment along some normal coordinate Q_k. If the vibrational wave functions are represented as products of harmonic oscillator wave functions, as in Eq. (6.90), then only $\Delta v_k = \pm 1$ transitions are allowed in which all the vibrational quantum numbers but one stay fixed. This result is a consequence of the fact that

$$\left\langle \psi_{v'_k} \middle| Q_k \middle| \psi_{v''_k} \right\rangle$$

has nonzero values for only $v'_k - v''_k = \pm 1$. Such transitions are called *fundamental transitions* because they are the most intense features of the vibration–rotation spectrum. To obtain *overtone transitions* in which $\Delta v_k > 1$ or *combination bands* in which the quantum numbers associated with more than one normal mode change, it is necessary to consider both higher order terms in the normal mode expansion of $\langle \mu_g \rangle_E$ (electrical anharmonicity) and coupling between normal modes caused by mechanical anharmonicity.

NOTES AND REFERENCES

1. F. Reiche and H. Rademacher, Z. Phys., **39**, 444 (1926); H. Rademacher and F. Reiche, Z. Phys., **41**, 453 (1927).

2. D. M. Dennison, Phys. Rev., **28**, 318 (1926); Rev. Mod. Phys., **3**, 280 (1931).

3. R. L. Kronig and I. I. Rabi, Phys. Rev., **29**, 262 (1927).

4. E. E. Witmer, Proc. Natl. Acad. Sci. (USA), **13**, 60 (1927); Phys. Rev., **74**, 1247, 1250 (1948).

5. H. A. Kramers and G. P. Ittman, Z. Phys., **53**, 553 (1929); **58**, 217 (1929); **60**, 663 (1930).

6. S. C. Wang, Phys. Rev., **34**, 243 (1929).

7. O. Klein, Z. Phys., **58**, 730 (1929).

8. H. B. G. Casimir, Rotation of a Rigid Body in Quantum Mechanics (J. B. Wolter's, The Hague, 1931).

9. H. H. Nielsen, Phys. Rev., **38**, 1432 (1931).

10. B. S. Ray, Z. Phys., **78**, 74 (1932).

11. G. W. King, R. M. Hainer, and P. C. Cross, J. Chem. Phys., **11**, 27 (1943).

12. P. C. Cross, R. M. Hainer, and G. W. King, J. Chem. Phys., **12**, 210 (1944).

13. G. W. King, J. Chem. Phys., **15**, 820 (1947).

14. R. M. Hainer, P. C. Cross, and G. W. King, J. Chem. Phys., **17**, 826 (1949).

15. C. Van Winter, Physica, **20**, 274 (1954).

16. My favorite reference is J. L. Synge and B. A. Griffith, Principles of Mechanics (McGraw-Hill, New York, 1959), pp. 374–383.

17. An asymmetric top set in motion with angular momentum pointing near the b axis will spin for awhile and then precipitously flop almost completely over spinning about the opposite axis for awhile before repeating this cycle. See W. G. Harter and C. C. Kim, Am. J. Phys., **44**, 1080 (1976), who describe an apparatus for demonstrating this behavior as well as show how to calculate the time between reversals (top flops). If the top system has dissipative energy losses (as it would if filled with a fluid), then the only stable rotation is about the axis associated with the largest moment of inertia. On a molecular scale, such losses may be caused by extensive vibration–rotation coupling in a nonrigid molecule for which the time scale for recurrence exceeds the relaxation time (caused by collision or radiative emission, for example) so that the top flop is irreversible. For more information on the treatment of radiationless transitions in isolated molecules, see M. Bixon and J. Jortner, J. Chem. Phys., **48**, 715 (1968); D. P. Chock, J. Jortner, and S. A. Rice, J. Chem. Phys., **49**, 610 (1968); P. Avouris, W. M. Gelbart, and M. A. El-Sayed, Chem. Rev., **77**, 793 (1977); and K. F. Freed, Acc. Chem. Res., **11**, 74 (1978).

18. G. M. Nathanson and G. M. McClelland, J. Chem. Phys., **81**, 629 (1984); Chem. Phys. Lett., **114**, 441 (1985)

19. W. G. Harter and C. W. Patterson, J. Chem. Phys., **80**, 4241 (1984).

20. A. J. Dorney and J. K. G. Watson, J. Molec. Spectrosc., **42**, 135 (1972).

21. K. Fox, H. W. Galbraith, B. J. Krohn, and J. D. Louck, *Phys. Rev. A*, **15**, 1363 (1977).

22. W. G. Harter and C. W. Patterson, *J. Chem. Phys.*, **66**, 4872, 4886 (1977).

23. W. G. Harter, C. W. Patterson, and F. J. da Paixao, *Rev. Mod. Phys.*, **50**, 37 (1978).

24. T. E. Turner, B. L. Hicks, and G. Reitwiesner, Ballistics Research Laboratories Report No. 878, Ballistics Research Laboratories, Aberdeen, Maryland (1953).

25. C. H. Townes and A. L. Schawlow, *Microwave Spectroscopy* (McGraw-Hill, New York, 1955).

26. G. Erlandsson, *Arkiv Fysik*, **10**, 65 (1956).

27. H. C. Allen, Jr. and P. C. Cross, *Molecular Vib-Rotors* (Wiley, New York, 1963).

28. S. M. Colwell, N. C. Handy, and W. H. Miller, *J. Chem. Phys.*, **68**, 745 (1978).

29. G. Herzberg, *Infrared and Raman Spectra of Polyatomic Molecules* (Van Nostrand, Princeton, NJ, 1945).

30. Polyatomic molecules may have vibrations whose motions behave as internal angular momenta. An example is the doubly degenerate bending mode of a linear triatomic molecule. Here two mutually perpendicular bends with a phase difference of $\pm\pi/2$ are equivalent to a pure rotation of the bent molecule about the longitudinal axis. It may be shown that vibrational angular momenta occur only for degenerate vibrational modes. Furthermore, the degenerate vibrational modes are easily identified from the point group symmetry of the molecule (see reference [29]). In particular, if a molecule has a threefold or higher axis of rotation as a symmetry element, there are degenerate vibrational modes present.

31. Degenerate electronic states in polyatomic molecules introduce a number of interesting complications. See G. Herzberg, *Electronic Spectra and Electronic Structure of Polyatomic Molecules* (Van Nostrand, Princeton, NJ, 1966) for more information. In brief, for a degenerate electronic state of a *linear* polyatomic molecule, the interaction of the electronic and vibrational angular momenta (associated with degenerate bending motion) causes the potential energy function to split into two when the molecule is bent. This is called the *Renner–Teller* effect. See Ch. Jungen and A. J. Merer, *Molec. Phys.*, **40**, 1 (1980). For a degenerate electronic state of a *nonlinear* polyatomic molecule, there are nontotally symmetric displacements of the nuclei that cause a splitting of the potential energy function such that the potential minima are no longer located in the symmetrical position. This vibrational–electronic interaction is called the Jahn–Teller effect. See T. A. Miller and V. E. Bondybey, in *Molecular Ions: Spectroscopy, Structure and Chemistry*, T. A. Miller and V. E. Bondybey, eds. (North-Holland, Amsterdam, 1983), pp. 201–229.

32. J. H. Van Vleck, *Phys. Rev.*, **33**, 467 (1929).

33. O. M. Jordahl, *Phys. Rev.*, **45**, 87 (1934).

34. See E. C. Kemble, *Fundamental Principles of Quantum Mechanics* (McGraw-Hill, New York, 1937), pp. 394–396.

35. D. R. Herschbach, unpublished notes; see J. E. Wollrab, *Rotational Spectra and Molecular Structure* (Academic Press, New York, 1967), Appendix 7. In the treatment we present, it is made explicit that the expression for $\mathcal{H}_{ij}$ includes more than one interacting block, but we have omitted a term in $\langle i| \mathcal{G}_3 |j\rangle$ that is $(E_i - E_j)/(E_\alpha - E_\beta)$ times smaller than the other terms, where α and β belong to different blocks.

36. The vibration–rotation Hamiltonian is derived in E. B. Wilson, Jr., J. C. Decius, and P. C. Cross, *Molecular Vibrations* (McGraw-Hill, New York, 1955) and in D. Papoušek and M. R. Aliev, *Molecular Vibrational–Rotational Spectra* (Elsevier, Amsterdam, 1982).

37. We drop the J dependence on the energy difference (which would show up as a higher order correction).

38. Z. I. Slawsky and D. M. Dennison, *J. Chem. Phys.*, **7**, 509 (1939).

39. J. B. Howard and E. B. Wilson, Jr., *J. Chem. Phys.*, **4**, 260 (1936); E. B. Wilson, Jr., *J. Chem. Phys.*, **5**, 617 (1937).

40. J. K. G. Watson, in *Vibrational Spectra and Structure*, Vol. 6, J. R. Durig, ed. (Elsevier, Amsterdam, 1977), pp. 1–89.

41. M. R. Aliev and J. K. G. Watson, in *Molecular Spectroscopy: Modern Research*, Vol. III, K. Narahari Rao, ed. (Academic Press, New York, 1985), pp. 1–67.

42. E. Hirota, *High-Resolution Spectroscopy of Transient Molecules* (Springer-Verlag, Berlin, 1985).

43. E. U. Condon and G. H. Shortley, *Theory of Atomic Spectra* (Cambridge University Press, Cambridge, England, 1935).

44. H. Hönl and F. London, *Z. Phys.*, **33**, 803 (1925).

45. Reference [27], pp. 86-93.

46. See W. H. Flygare, *Molecular Structure and Dynamics* (Prentice-Hall, Englewood Cliffs, NJ, 1978).

47. E. U. Condon, *Am. J. Phys.*, **15**, 365 (1947); *Phys. Rev.*, **32**, 858 (1928).

48. C. Noda and R. N. Zare, *J. Molec. Spectrosc.*, **95**, 254 (1982).

49. For example, an electronic transition is said to be pure electric dipole forbidden if $\langle E'V'R' | \sum_g (\mu_g)_0 | E''V''R'' \rangle$ vanishes. However, if by retaining additional terms, the electric dipole moment matrix element is nonzero, the transition is said to be vibronically induced. See G. Herzberg, *Electronic Spectra of Polyatomic Molecules* (Van Nostrand, New York, 1966).

PROBLEM SET 4

1. Let $\hat{n}$ be a unit vector pointing along the angular velocity vector ω so that

$$\omega = \omega \hat{n} \qquad (6.136)$$

The kinetic energy of the rigid body in the center-of-mass frame is given by [see Eq. (6.18)]

$$T = \frac{1}{2}\omega \cdot \mathbf{J} \qquad (6.137)$$

but [see Eq. (6.11)]

$$\mathbf{J} = \underset{\sim}{\mathbf{I}} \cdot \omega \qquad (6.138)$$

Hence

$$T = \frac{1}{2}\omega \cdot \underset{\sim}{\mathbf{I}} \cdot \omega = \frac{1}{2}\omega^2 \hat{n} \cdot \underset{\sim}{\mathbf{I}} \cdot \hat{n} = \frac{1}{2}I\omega^2 \qquad (6.139)$$

where I is a scalar quantity given by

$$I = \sum_i m_i \left[r_i^2 - (\mathbf{r}_i \cdot \hat{n})^2 \right] \qquad (6.140)$$

It is traditional to call I the moment of inertia about the axis of rotation.

A. Evaluate $\hat{n} \cdot \underset{\sim}{\mathbf{I}} \cdot \hat{n}$ and show that Eq. (6.140) results.

B. The distance of the particle i from the axis of rotation $\hat{n}$ is $r_i \sin \theta_i$, where θ_i is the angle included between $\mathbf{r}_i$ and $\hat{n}$. Thus this distance is the same as the magnitude of the vector $\mathbf{r}_i \times \hat{n}$. The customary definition of the moment of inertia about an axis of rotation is the mass times the square of the distance of the object from the rotation axis. Hence for a rigid body

$$I = \sum_i m_i(\mathbf{r}_i \times \hat{n}) \cdot (\mathbf{r}_i \times \hat{n}) \qquad (6.141)$$

Show that Eq. (6.141) is equivalent to Eq. (6.140).

The value of I depends on the direction of ω. In general, ω changes its direction with respect to the center-of-mass frame as a function of time, causing $\mathbf{J}$ also to change its direction. At any given moment

$$\hat{n} \cdot \mathbf{J} = \hat{n} \cdot \underset{\sim}{\mathbf{I}} \cdot \hat{n}\omega = \omega I \qquad (6.142)$$

2. Microwave spectrum of methylene fluoride.

A. Calculate the principal moments of inertia (expressed in amu Å^2) and rotation constants A, B, C (expressed in MHz and in cm^{-1}) for the methylene fluoride molecule, $^{12}\text{CH}_2\text{F}_2$. Use the following parameters:

$$r_{\text{CH}} = 1.092 \text{ Å}, \qquad r_{\text{CF}} = 1.358 \text{ Å}$$
$$\angle\text{HCH} = 111°52', \qquad \angle\text{FCF} = 108°17'$$
$$m_{\text{H}} = 1.007825 \text{ amu}, \quad m_{\text{C}} = 12.000000 \text{ amu}, \quad m_{\text{F}} = 18.99840 \text{ amu}$$

A handy conversion factor is

$$\hbar/4\pi = 5.05377 \times 10^5 \text{ amu Å}^2 \text{ MHz}$$

B. List the frequencies (in MHz) of the *allowed* rotational transitions for methylene fluoride with $J \leq 2$.

C. Discuss whether the observation of the above transitions could be used to determine the bond angles and bond lengths in methylene fluoride. Are any assumptions or additional information needed?

3. In terms of the rotational constants A, B, C, show that the asymmetric top energy levels for $J = 3$ have the form:

Level[a]	General Formula	Prolate Limit $A > B = C$
3_{03}	$2A + 5B + 5C - 2[4(B-C)^2 + (A-B)(A-C)]^{\frac{1}{2}}$	$12B$
3_{13}	$5A + 2B + 5C - 2[4(A-C)^2 - (A-B)(B-C)]^{\frac{1}{2}}$	$11B + A$
3_{12}	$5A + 5B + 2C - 2[4(A-B)^2 + (A-C)(B-C)]^{\frac{1}{2}}$	$11B + A$
3_{22}	$4A + 4B + 4C$	$8B + 4A$
3_{21}	$2A + 5B + 5C + 2[4(B-C)^2 + (A-B)(A-C)]^{\frac{1}{2}}$	$8B + 4A$
3_{31}	$5A + 2B + 5C + 2[4(A-C)^2 - (A-B)(B-C)]^{\frac{1}{2}}$	$3B + 9A$
3_{30}	$5A + 5B + 2C + 2[4(A-B)^2 + (A-C)(B-C)]^{\frac{1}{2}}$	$3B + 9A$

[a] $J_{K(\text{prolate}) K(\text{oblate})}$

Hint: An inspection of the symmetric form of the solutions should save you some time; it is not necessary to find the roots of each secular determinant.

4. Consider a symmetric triatomic molecule AB_2 that belongs to the point group C_{2v}. Let $r = r_{\text{AB}}$ denote the length of the AB bond and $\theta = \theta_{\text{BAB}}$ denote the included angle. The masses are designated by M_A and M_B.

A. Show that the three moments of inertia are given by

$$I_1 = 2\, M_B\, r_{AB}^2 \sin^2 \left(\frac{\theta_{BAB}}{2} \right) \tag{6.143}$$

$$I_2 = \frac{2\, M_A\, M_B}{M_A + 2\, M_B}\, r_{AB}^2 \cos^2 \left(\frac{\theta_{BAB}}{2} \right) \tag{6.144}$$

$$I_3 = 2\, M_B\, r_{AB}^2 \left[\frac{M_A}{M_A + 2\, M_B} \cos^2 \left(\frac{\theta_{BAB}}{2} \right) + \sin^2 \left(\frac{\theta_{BAB}}{2} \right) \right] \tag{6.145}$$

where 1 is the principal axis in the plane of the molecule that bisects the θ_{BAB} angle, 2 is the principal axis in the plane of the molecule that is perpendicular to 1, and 3 is the principal axis perpendicular to the plane of the molecule (i.e., perpendicular to axes 1 and 2). We can identify I_{cc} with I_3. In order for I_1 to equal I_2, it is necessary that

$$\theta_{BAB} = 2 \arctan \left[\frac{M_A}{M_A + 2\, M_B} \right]^{\frac{1}{2}} = \theta_{BAB}^0 \tag{6.146}$$

For $\theta > \theta_{BAB}^0$, $I_{aa} = I_2$ and $I_{bb} = I_1$ while for $\theta < \theta_{BAB}^0$, $I_{aa} = I_1$ and $I_{bb} = I_2$.

B. Show that the asymmetry parameter κ depends only on the mass ratio

$$r = \frac{M_A}{M_A + 2\, M_B} \tag{6.147}$$

C. Make a plot of κ versus θ_{BAB} for $r = \frac{1}{3}$ (corresponding to $M_A = M_B$), $r = \frac{1}{2}$ (corresponding to $M_A = 2\, M_B$), and $r = \frac{1}{5}$ (corresponding to $M_A = 0.5\, M_B$). The only case for which the AB_2 molecule is exactly a symmetric top is when the AB_2 molecule is linear ($\theta_{BAB} = 180°$). Examination of the graph of κ versus θ_{BAB} will explain why most AB_2 triatomic molecules are near prolate symmetric tops, although θ_{BAB} often differs markedly from 180°.

D. We define the inertia defect Δ as

$$\Delta = I_{cc} - I_{bb} - I_{aa} \tag{6.148}$$

Note that $\Delta = 0$ for a symmetric triatomic molecule. Prove that $\Delta = 0$ for any planar rigid rotor.

E. Suggest how a real "planar" molecule, such as benzene or ozone, can have a nonvanishing inertia defect.

For a general discussion of the inertia defect, see J. E. Wollrab, *Rotational Spectra and Molecular Structure* (Academic Press, New York, 1967).

APPLICATION 15
INTRODUCTION TO DIATOMIC MOLECULES

The rotational energy levels, wave functions, and line strengths of diatomic molecules may be worked out in a manner similar to those presented for asymmetric tops. The following treatment is intended only as a brief introduction to this rather extensive topic[1–21]. We concentrate our attention on $^2\Sigma$ and $^2\Pi$ electronic states because these are the simplest and most commonly encountered examples of open-shell diatomics.

We begin by presenting our cast of angular momentum characters. We ignore nuclear spin, a tiny shy fellow, because hyperfine structure is seldom resolved in optical spectroscopy, and the effects of nuclear spin can be readily taken into account, if needed[22]. We define

L — electronic orbital angular momentum

S — electronic spin angular momentum

J — total angular momentum

N — total angular momentum excluding electronic spin (i.e., **N** = **J** − **S**)

R — nuclear rotational angular momentum (i.e., **R** = **N** − **L**).

We also introduce the Hund's case (a) coupling scheme (see Figure 1). In Figure 1 **L** and **S** are both "tied" to the internuclear axis (the z axis in the molecule-fixed frame) making the signed projections Λ and Σ, respectively, while **R** is at right angles to the internuclear axis. Thus the total angular momentum **J** makes the projection M on the space-fixed Z axis and the projection $\Omega = \Lambda + \Sigma$ on the molecule-fixed z axis. Consequently, in case (a) coupling the rotational wave function is the symmetric top wave function $|J\Omega M\rangle$.

A linear molecule has only two rotational degrees of freedom. Hence the Euler angle χ is redundant and its value can be fixed arbitrarily. We follow the convention[14] that $\chi = 0$ and write for the rotational wave function

$$|J\Omega M\rangle = \left[\frac{2J+1}{4\pi}\right]^{\frac{1}{2}} D_{M\Omega}^{J*}(\phi, \theta, 0) \tag{1}$$

where the normalization factor has been altered to reflect the fact that integration over the solid angle element no longer includes χ. The choice of $\chi = 0$ implies that the molecule-fixed y axis coincides with the line of nodes, that is, lies in the plane of rotation, while the molecule-fixed x axis is perpendicular to the plane of rotation,

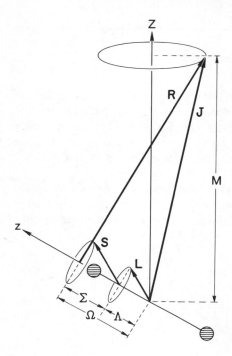

FIGURE 1   Vector diagram for Hund's case (a) coupling.

which is the direction $\mathbf{J}$ points in the high J limit. The alternative choice[11, 20], $\chi = \pi/2$, places the molecule-fixed y axis perpendicular to the plane of rotation.

The case (a) wave functions may be written as a product of an electronic orbital part $|n\Lambda\rangle$, an electronic spin part $|S\Sigma\rangle$, a vibrational part $|v\rangle$, and a rotational part $|J\Omega M\rangle$:

$$|\psi\left(n\,^{2S+1}\Lambda_{\Omega};vJM\right)\rangle = |n\Lambda\rangle\,|S\Sigma\rangle\,|v\rangle\,|J\Omega M\rangle \qquad (2)$$

The separation $|n\Lambda\,S\Sigma\rangle = |n\Lambda\rangle\,|S\Sigma\rangle$ is strictly valid for only one- and two-valence electron molecules[13]. It is retained here because we are interested primarily in the symmetry nature of the electronic part of the wave function, which we shall also write as $|n\,^{2S+1}\Lambda_{\Omega}\rangle$. These basis set functions prove to be a convenient starting point because the operators L_z, $\mathbf{J}^2$, J_z, $\mathbf{S}^2$, and S_z are all diagonal in this coupling scheme.

$^1\Sigma$ States

Therotational Hamiltonian is simply

$$\mathcal{H}_{\text{rot}} = B(r)\mathbf{R}^2 \qquad (3)$$

where

$$B(r) = \frac{1}{2I} = \frac{1}{2\mu r^2} \tag{4}$$

Thus in the vibrational level v the effective Hamiltonian becomes

$$\mathcal{H}_{\text{rot}}^{(v)} = B_v \mathbf{J}^2 \tag{5}$$

and the energy levels are given by

$$E(^1\Sigma; vJ) = B_v J(J + 1) = B_v N(N + 1) \tag{6}$$

since $\mathbf{J} = \mathbf{N} = \mathbf{R}$ when $\mathbf{L} = \mathbf{S} = 0$. Equation (6) is recognized as the familiar expression for the rotational energy levels of a linear rigid rotor. The mechanical interpretation of B_v as proportional to $\langle v| 1/r^2 |v\rangle$ assumes that mixing with other electronic or vibrational levels is negligible.

A. Use the Van Vleck transformation to calculate the first centrifugal distortion correction to the rotational energy levels of a $^1\Sigma$ state. Show that

$$E(^1\Sigma; vJ) = B_v J(J + 1) - D_v [J(J + 1)]^2 \tag{7}$$

where

$$D_v = -\sum_{v' \neq v} \frac{\langle v| B(r) |v'\rangle \langle v'| B(r) |v\rangle}{E_v - E_{v'}} \tag{8}$$

$^1\Pi$ States

In $^1\Pi$ electronic states $|\Lambda| = 1$, and we must introduce the effects of $\mathbf{L}$. The rotational Hamiltonian is

$$\mathcal{H}_{\text{rot}} = B(r)\mathbf{R}^2 = B(r)(\mathbf{J} - \mathbf{L})^2 \tag{9}$$

that is,

$$\mathcal{H}_{\text{rot}} = B(r)\left[\mathbf{J}^2 - 2J_z L_z + L_z^2\right] + B(r)\frac{1}{2}[L_+ L_- + L_- L_+]$$

$$-B(r)[J_+ L_- + J_- L_+] \tag{10}$$

The second term in Eq. (10) has no J dependence, and its expectation value is incorporated into the band origin (which is a polite way of saying that it is neglected).

The third term in Eq. (10) has no nonvanishing expectation value; instead, as we shall see, it couples states that differ by one unit in Λ, that is, couples Π states to Σ and Δ states. The first term in Eq. (10) does have an expectation value, and we readily find for the rotational energy of a $^1\Pi$ state ($|\Lambda| = 1$):

$$E(^1\Pi; vJ) = B_v \left[J(J+1) - 2\Omega\Lambda + \Lambda^2 \right]$$

$$= B_v [J(J+1) - 1] \tag{11}$$

where for $^1\Pi$ states the lowest J value is $J = 1$. Again because $S = 0$ we could also replace J by N in Eq. (11), if desired. To this approximation, the rotational energy level structure of a $^1\Pi$ electronic state looks just like a $^1\Sigma$ state, except the $J = 0$ level is missing. Note that each level is doubly degenerate as $\Lambda = +1$ and $\Lambda = -1$ give the same energy; that is, it does not matter whether $\mathbf{L}$ precesses clockwise or counterclockwise about the internuclear axis. This is called Λ doubling or Λ-type doubling. It is the diatomic molecule analog of K-type doubling in a symmetric top.

$^2\Sigma$ States

In a $^2\Sigma$ state, $\Lambda = 0$ but $S = \frac{1}{2}$ so that $\Omega = +\frac{1}{2}$ or $-\frac{1}{2}$ according to whether $\Sigma = +\frac{1}{2}$ or $-\frac{1}{2}$. Rather than evaluate a 2×2 secular determinant, we rewrite the molecular wave function as an appropriate linear combination of the case (a) wave functions in the same spirit as the Wang transformation

$$\left| n \, ^{2S+1}\Lambda_\Omega v J M p^\pm \right\rangle$$

$$= \frac{1}{\sqrt{2}} \left[|n\Lambda\rangle \, |S\Sigma\rangle \, |J\Omega M\rangle \pm |n-\Lambda\rangle \, |S-\Sigma\rangle \, |J-\Omega M\rangle \right] |v\rangle \tag{12}$$

and for the special case $\Lambda = 0, \Sigma = 0$

$$\left| n \, ^{2S+1}\Sigma_0 \, v J M p^+ \right\rangle = |n0\rangle \, |S0\rangle \, |J0 M\rangle \, |v\rangle \tag{13}$$

where $p^\pm$ denotes the sign of the linear combination. As we shall see, $p^\pm$ is related to the parity of the wave function. The rotational part of the Hamiltonian is once again $B(r)\mathbf{R}^2 = B(r)(\mathbf{J} - \mathbf{S})^2$, but we must also include the phenomenological spin-rotation coupling term $\gamma(r)\mathbf{N} \cdot \mathbf{S}$ (which is usually caused almost exclusively by spin–orbit interaction that couples $^2\Sigma$ and $^2\Pi$ states[23]). Hence for the vibrational level v the rotational Hamiltonian may be written for $^2\Sigma$ states as

$$\mathcal{H}_{\text{rot}}^{(v)} = B_v(\mathbf{J} - \mathbf{S})^2 + \gamma_v(\mathbf{J} - \mathbf{S}) \cdot \mathbf{S}$$

$$= B_v \left[\mathbf{J}^2 - 2 J_z S_z + \mathbf{S}^2 \right] + \gamma_v \left[J_z S_z - \mathbf{S}^2 \right]$$

$$- (B_v - \gamma_v/2) \left[J_+ S_- + J_- S_+ \right] \tag{14}$$

where B_v is the rotational constant and γ_v the spin–rotation constant.

B. Show that

$$E\left(^2\Sigma; vJp^+\right) = B_v\left(J - \tfrac{1}{2}\right)\left(J + \tfrac{1}{2}\right) + \tfrac{1}{2}\gamma_v\left(J - \tfrac{1}{2}\right)$$

$$= B_v N\left(N + 1\right) + \tfrac{1}{2}\gamma_v N \tag{15}$$

where $J = N + \tfrac{1}{2}$, and

$$E\left(^2\Sigma; vJp^-\right) = B_v\left(J + \tfrac{1}{2}\right)\left(J + \tfrac{3}{2}\right) - \frac{1}{2}\gamma_v\left(J + \tfrac{3}{2}\right)$$

$$= B_v N\left(N + 1\right) - \tfrac{1}{2}\gamma_v\left(N + 1\right) \tag{16}$$

where $J = N - \tfrac{1}{2}$. *Hint*: Evaluate $\langle\psi(p^\pm)|\mathcal{H}_{\text{rot}}^{(v)}|\psi(p^\pm)\rangle$. Use the phase conventions (see Section 3.4):

$$\langle J\Omega M|\mathbf{J}^2|J\Omega M\rangle = J(J + 1) \tag{17}$$

$$\langle J\Omega M|J_z|J\Omega M\rangle = \Omega \tag{18}$$

$$\langle J\Omega \mp 1M|J_\pm|J\Omega M\rangle = [J(J + 1) - \Omega(\Omega \mp 1)]^{\frac{1}{2}} \tag{19}$$

$$\langle S\Sigma|\mathbf{S}^2|S\Sigma\rangle = S(S + 1) \tag{20}$$

$$\langle S\Sigma|S_z|S\Sigma\rangle = \Sigma \tag{21}$$

$$\langle S\Sigma \pm 1|S_\pm|S\Sigma\rangle = [S(S + 1) - \Sigma(\Sigma \pm 1)]^{\frac{1}{2}} \tag{22}$$

Note carefully the similarities and differences between Eqs. (19) and (22).

Equations (15) and (16) show that (even in the absence of perturbations[20]) the rotational levels of a $^2\Sigma$ state are split by an amount

$$E\left(^2\Sigma^+; vJp^+\right) - E\left(^2\Sigma^+; vJ - 1p^-\right) = \gamma_v J \tag{23}$$

where the magnitude of the spin–rotation splitting increases linearly with J (in the absence of centrifugal distortion corrections to γ_v). Thus each rotational level except the $J = \tfrac{1}{2}$ level of p^+ parity occurs in closely spaced pairs. The levels with $J = N + \tfrac{1}{2}$ are called F_1, and those with $J = N - \tfrac{1}{2}$ are called F_2. Hence

$F_1(N) - F_2(N) = \frac{1}{2}\gamma_v(2N + 1)$ [see Eqs. (15) and (16)]. The spin–rotation constant γ_v may be positive or negative.

$^2\Pi$ States

In a $^2\Pi$ state, $\Lambda = \pm 1$ and $S = \frac{1}{2}$ so that $\Sigma = \pm\frac{1}{2}$. Hence Ω can take on the values $-\frac{3}{2}$, $-\frac{1}{2}$, $\frac{1}{2}$, and $\frac{3}{2}$. Except for the $J = \frac{1}{2}$ level (which can be exclusively associated with $|\Omega| = \frac{1}{2}$), each J level has four rotational levels associated with it, two of each parity. We use as basis set wave functions

$$\left| n\,^2\Pi_{\frac{1}{2}}\, v\, JMp^\pm \right\rangle = \frac{1}{\sqrt{2}} \left[|n1\rangle \left| \tfrac{1}{2} -\tfrac{1}{2} \right\rangle \left| J\tfrac{1}{2}M \right\rangle \pm |n-1\rangle \left| \tfrac{1}{2}\tfrac{1}{2} \right\rangle \left| J -\tfrac{1}{2}M \right\rangle \right] |v\rangle$$

(24)

and

$$\left| n\,^2\Pi_{\frac{3}{2}}\, v\, JMp^\pm \right\rangle = \frac{1}{\sqrt{2}} \left[|n1\rangle \left| \tfrac{1}{2}\tfrac{1}{2} \right\rangle \left| J\tfrac{3}{2}M \right\rangle \pm |n-1\rangle \left| \tfrac{1}{2} -\tfrac{1}{2} \right\rangle \left| J -\tfrac{3}{2}M \right\rangle \right] |v\rangle$$

(25)

The contribution to the energy caused by nuclear rotation is expressed as

$$\mathcal{H}_{\text{rot}} = B(r)\mathbf{R}^2$$

$$= B(r)\left[(\mathbf{J} - \mathbf{S}) - \mathbf{L} \right]^2$$

$$= B(r)\left[\mathbf{J}^2 - 2J_zS_z + \mathbf{S}^2 - 2(J_z - S_z)L_z + L_z^2 \right]$$

$$\quad - B(r)\left[J_+S_- + J_-S_+ \right] - B(r)\left[(J_+ - S_+)L_- + (J_- - S_-)L_+ \right]$$

$$\quad + \frac{1}{2}B(r)\left[L_+L_- + L_-L_+ \right]$$

(26)

In the last line of Eq. (26) the first two terms contribute to the first-order energy, the third term connects states differing by one unit in Λ and the last term has no J dependence and once again contributes only an overall shift to the total energy. Hence we write

$$\mathcal{H}_{\text{rot}}^{(v)} = B_v\left[\mathbf{J}^2 - 2J_zS_z + \mathbf{S}^2 - 2(J_z - S_z)L_z + L_z^2 \right] - B_v\left[J_+S_- + J_-S_+ \right] \quad (27)$$

We also must introduce the spin–orbit interaction which we approximate[24] as

$$\mathcal{H}_{SO} = A(r)\mathbf{L}\cdot\mathbf{S}$$

$$= A(r)L_zS_z + \frac{1}{2}A(r)\left[L_+S_- + L_-S_+ \right]$$

(28)

Only the $A(r)L_zS_z$ term contributes to the energy in first order so that

$$\mathcal{H}_{SO}^{(v)} = A_v L_z S_z \tag{29}$$

C. The rotational energy levels of a $^2\Pi$ state are the same for each parity block, that is, $E(^2\Pi; vJp^+) = E(^2\Pi; vJp^-)$. Hence they occur in doubly degenerate pairs. Show that for $J > \frac{1}{2}$

$$E(^2\Pi; vJ) = B_v \left[\left(J - \tfrac{1}{2} \right) \left(J + \tfrac{3}{2} \right) \pm \tfrac{1}{2} X \right] \tag{30}$$

where

$$X = \left[4 \left(J - \tfrac{1}{2} \right) \left(J + \tfrac{3}{2} \right) + (Y - 2)^2 \right]^{\frac{1}{2}}$$

$$= \left[4 \left(J + \tfrac{1}{2} \right)^2 + Y(Y - 4) \right]^{\frac{1}{2}} \tag{31}$$

and

$$Y = \frac{A_v}{B_v} \tag{32}$$

Hint: Show that the 2×2 secular determinant is

$$\begin{vmatrix} B_v \left[J(J+1) - \tfrac{7}{4} \right] + \tfrac{1}{2} A_v - E & -B_v \left[J(J+1) - \tfrac{3}{4} \right]^{\frac{1}{2}} \\ -B_v \left[J(J+1) - \tfrac{3}{4} \right]^{\frac{1}{2}} & B_v \left[J(J+1) + \tfrac{1}{4} \right] - \tfrac{1}{2} A_v - E \end{vmatrix} = 0$$

$$\tag{33}$$

and carry out its evaluation, using the identity $J(J+1) - \frac{3}{4} = \left(J - \tfrac{1}{2} \right) \left(J + \tfrac{3}{2} \right)$.

D. The energy levels associated with the plus sign in Eq. (30) are called F_2; those with the minus sign F_1. Show that

$$|\psi(F_2)\rangle = a_J \left| ^2\Pi_{\frac{1}{2}} vJ \right\rangle + b_J \left| ^2\Pi_{\frac{3}{2}} vJ \right\rangle \tag{34}$$

$$|\psi(F_1)\rangle = -b_J \left| ^2\Pi_{\frac{1}{2}} vJ \right\rangle + a_J \left| ^2\Pi_{\frac{3}{2}} vJ \right\rangle \tag{35}$$

where

$$a_J = \left[\frac{X + (Y - 2)}{2X} \right]^{\frac{1}{2}} \tag{36}$$

$$b_J = \left[\frac{X - (Y - 2)}{2X} \right]^{\frac{1}{2}} \tag{37}$$

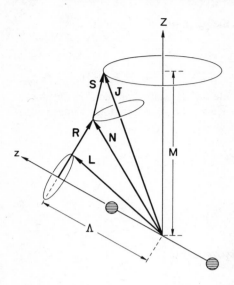

FIGURE 2   Vector diagram for Hund's case (b) coupling.

For large rotational values, $F_1(J)$ is the term series that behaves as $B_v N(N+1)$ with $J = N + \frac{1}{2}$ while $F_2(J)$ is the term series with $J = N - \frac{1}{2}$.

It is informative to compare the value of $(Y - 2)^2$ to $4\left(J - \frac{1}{2}\right)\left(J + \frac{3}{2}\right) = 4J(J + 1) - 3$. When the former dominates, $b_J \approx 0$ and we can assign the F_1 and F_2 levels as belonging to separate $^2\Pi_{\frac{3}{2}}$ and $^2\Pi_{\frac{1}{2}}$ states. It is for this situation that the molecular behavior follows closely the case (a) coupling scheme. However, with increasing rotation the $B_v(\mathbf{J} - \mathbf{L}) \cdot \mathbf{S} = B_v \mathbf{N} \cdot \mathbf{S}$ coupling term from the rotational part of the Hamiltonian grows in importance compared to the $A_v \mathbf{L} \cdot \mathbf{S}$ spin–orbit coupling term. This causes $\mathbf{S}$ to uncouple from the nuclear axis and recouple to $\mathbf{N}$ to give the resultant total angular momentum $\mathbf{J}$. This coupling scheme, called Hund's case (b) is shown in Figure 2.

In the case (b) limit Eqs. (36) and (37) show that $a_J = b_J = (2)^{-\frac{1}{2}}$; that is, the wave functions for the F_1 and F_2 energy levels become an equally weighted linear combination of the case (a) wave functions[25]. In general, the coupling in $^2\Pi$ states is intermediate between case (a) and case (b), going away from case (a) and toward case (b) with increasing rotation. If $A_v > 0$, the energy ordering is said to be *regular*, while for $A_v < 0$, the energy ordering is said to be *inverted*.

E. Construct a plot of $E(^2\Pi; vJ)$ versus Y for $B_v = 1$ cm^{-1}, $-20 \le Y \le 20$ and $J = \frac{3}{2}, \frac{5}{2}$, and $\frac{7}{2}$. Optional: Include the level $J = \frac{1}{2}$ in your plot.

Most diatomic molecules are well described by intermediate behavior between the limits of Hund's case (a) and Hund's case (b) coupling. However, there are two other

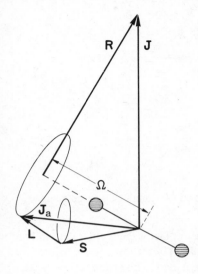

FIGURE 3   Vector diagram for Hund's case (c) coupling.

coupling schemes, introduced by Hund[2], which also are of importance. For heavy nuclei, the atomic spin–orbit interaction often becomes large and the interaction between $\mathbf{L}$ and $\mathbf{S}$ may exceed that between $\mathbf{L}$ and the internuclear axis. In this case, called Hund's case (c) coupling, $\mathbf{L}$ and $\mathbf{S}$ add vectorially to form the resultant $\mathbf{J}_a$ which then couples to the internuclear axis to make the projection Ω. The nuclear rotational angular momentum $\mathbf{R}$ then adds vectorially to Ω to form the total angular momentum $\mathbf{J}$ (see Figure 3). The $\Omega = 0$ states are nondegenerate and we speak of 0^+ and 0^- states in strict analogy to Σ^+ and Σ^- states. The $\Omega \neq 0$ states would be degenerate for a nonrotating molecule but split under rotation into two components. This is called Ω-type doubling in analogy with Λ-type doubling.

An opposite extreme is described by Hund's case (d) coupling (see Figure 4) in which the coupling between $\mathbf{L}$ and $\mathbf{R}$ is much larger than that between $\mathbf{L}$ and the internuclear axis. Hund's case (d) may be regarded as what happens to Hund's case (b) molecules in the limit of very large rotation (L-uncoupling). Hund's case (d) coupling also serves as an excellent starting point in describing the electronic states of many Rydberg molecules where the Rydberg electron is atomic-like and weakly interacts with the molecular core.

Parity Classification of Rotational Levels

Each wave function associated with an energy level may be classified as even or odd according to whether it remains unchanged or changes sign on inversion through the origin of the spatial coordinates of all particles. Since the molecular Hamiltonian is invariant under this symmetry operation, only states of the same parity have nonvanishing matrix elements connecting them—a fact we already employed to

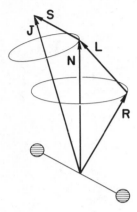

FIGURE 4   Vector diagram for Hund's case (d) coupling.

advantage when we used a case (a) parity basis set to factor the secular determinant to find the rotational energy levels of $^1\Sigma$, $^1\Pi$, $^2\Sigma$, and $^2\Pi$ electronic states. Moreover, because the electric dipole operator changes sign under inversion of the spatial coordinates of all particles, electric dipole transitions can only connect levels of opposite parity. Therefore, it is of great importance to establish the parity of different rotational levels. This is accomplished by considering how the inversion operator $\mathbf{i}_{sp}$ acts on the case (a) wave functions, a subtle topic that one might want to skim on first reading[26].

It can be shown from simple geometric arguments that the inversion of the spatial coordinates is equivalent to reflection of the molecule-fixed electronic coordinates followed by rotation of the molecular frame by $180°$ about an axis through the origin and perpendicular to the reflection plane. In the language of group theory we can write $\mathbf{i}_{sp} = C_2(y)\sigma_v(xz)$ or $\mathbf{i}_{sp} = C_2(x)\sigma_v(yz)$. We choose the former as it is more convenient[27] to carry out these operations with our choice of $\chi = 0$. It may be shown[11, 13, 16, 19] that

$$\sigma_v(xz)\,|n\Lambda\rangle = (-1)^{\Lambda+s}\,|n-\Lambda\rangle \tag{38}$$

and

$$\sigma_v(xz)\,|S\Sigma\rangle = (-1)^{S-\Sigma}\,|S-\Sigma\rangle \tag{39}$$

where $s = 1$ for Σ^- states, $s = 0$ for all other states. We outline how Eqs. (38) and (39) come about.

For $\Lambda \neq 0$ states, $|n\Lambda\rangle$ transforms as $Y_{L\Lambda}(\theta, \chi)$ (L unspecified) and under the reflection in the xz plane, $\chi \rightarrow -\chi$ so that

$$Y_{L\Lambda}(\theta, \chi) \rightarrow Y_{L\Lambda}(\theta, -\chi) = Y_{L\Lambda}^*(\theta, \chi) = (-1)^\Lambda Y_{L-\Lambda}(\theta, \chi) \tag{40}$$

For $\Lambda = 0$ states, we already denote whether the orbital part of the electronic wave function is unchanged or changes signs under reflection in a plane containing the internuclear axis by calling the former Σ^+ states and the latter Σ^- states. Hence $s = 0$ except for Σ^- states in which case $s = 1$.

Under the action of $\sigma_v(xz)$, $y \to -y$, which is the same as an inversion in the molecular frame $x \to -x$, $y \to -y$, $z \to -z$ followed by a rotation by π about the y axis, $C_2(y)$. Under inversion the spin-up wave function $\alpha = \left|\frac{1}{2}, \frac{1}{2}\right\rangle$ and the spin-down wave function $\beta = \left|\frac{1}{2}, -\frac{1}{2}\right\rangle$ are unchanged, whereas under $C_2(y)$,

$$\left|\tfrac{1}{2}, \tfrac{1}{2}\right\rangle \to D^{\frac{1}{2}}_{\frac{1}{2}\frac{1}{2}}(0, \pi, 0) \left|\tfrac{1}{2}, \tfrac{1}{2}\right\rangle + D^{\frac{1}{2}}_{-\frac{1}{2}\frac{1}{2}}(0, \pi, 0) \left|\tfrac{1}{2}, -\tfrac{1}{2}\right\rangle$$

$$= \left|\tfrac{1}{2}, -\tfrac{1}{2}\right\rangle \tag{41}$$

$$\left|\tfrac{1}{2}, -\tfrac{1}{2}\right\rangle \to D^{\frac{1}{2}}_{\frac{1}{2}-\frac{1}{2}}(0, \pi, 0) \left|\tfrac{1}{2}, \tfrac{1}{2}\right\rangle + D^{\frac{1}{2}}_{-\frac{1}{2}-\frac{1}{2}}(0, \pi, 0) \left|\tfrac{1}{2}, -\tfrac{1}{2}\right\rangle$$

$$= -\left|\tfrac{1}{2}, \tfrac{1}{2}\right\rangle \tag{42}$$

therefore:

$$\sigma_v(xy) \left|\tfrac{1}{2}, \sigma\right\rangle = (-1)^{\frac{1}{2}-\sigma} \left|\tfrac{1}{2}, -\sigma\right\rangle \tag{43}$$

The spin state $|S\Sigma\rangle$ is a determinantal wave function of n indistinguishable electrons. However, we are only interested in the symmetry properties, so we shall take some liberties. The $|S\Sigma\rangle$ state may be regarded as formed from a superposition of coupled states $|(S_{n-1} s_n)S\Sigma\rangle$, where $|S_{n-1}, \Sigma_{n-1}\rangle$ may also be regarded as a superposition of the coupled states $|(S_{n-2} s_{n-1})S_{n-1}\Sigma_{n-1}\rangle$, and so on. Hence the uncoupled component $|s_1\sigma_1\rangle |s_2\sigma_2\rangle \cdots |s_n\sigma_n\rangle$ occurs in $|S\Sigma\rangle$ with a coefficient whose dependence on the spin projections σ_i is contained in the product of Clebsch–Gordan coefficients, $\langle S_{n-1}, \Sigma_{n-1}, s_n\sigma_n | S\Sigma\rangle \langle S_{n-2}, \Sigma_{n-2}, s_{n-1}\sigma_{n-1} | S_{n-1}\Sigma_{n-1}\rangle \times \cdots \times \langle s_1\sigma_1, s_2\sigma_2 | S_2\Sigma_2\rangle$. On reflection we interchange α and β in the uncoupled state so that $\Sigma \to -\Sigma$, and we pick up a phase factor $(-1)^x$, where x equals the number of β spins in $|S\Sigma\rangle$. Since the number of α and β spins sum to n and the quantity, the number of α spins minus the number of β spins, multiplied by $\frac{1}{2}$ equals Σ, it follows that $x = \frac{1}{2}n - \Sigma$. From the product of Clebsch–Gordan coefficients we also pick up a second phase factor $(-1)^y$ where

$$y = (S - S_{n-1} - s_n) + (S_{n-1} - S_{n-2} - s_{n-1}) + \cdots + (S_2 - s_2 - s_1)$$

$$= S - \sum_i s_i = S - \frac{1}{2}n \tag{44}$$

because the magnetic quantum numbers of each Clebsch–Gordan coefficient have been reversed [see Eq. (2.26)]. Hence

$$\sigma_v(xz) |S\Sigma\rangle = (-1)^{x+y} |S -\Sigma\rangle = (-1)^{S-\Sigma} |S -\Sigma\rangle \tag{45}$$

Finally we turn to the behavior of $|J\Omega M\rangle$ under spatial inversion. The Euler angles (ϕ, θ, χ) relate the space-fixed frame $F = XYZ$ to the molecule-fixed frame $g = xyz$ according to

$$
\begin{bmatrix} X \\ Y \\ Z \end{bmatrix} = \begin{bmatrix} c\phi\,c\theta\,c\chi - s\phi\,s\chi & -c\phi\,c\theta\,s\chi - s\phi\,c\chi & c\phi\,s\theta \\ s\phi\,c\theta\,c\chi + c\phi\,s\chi & -c\phi\,c\theta\,s\chi + c\phi\,c\chi & s\phi\,s\theta \\ s\theta\,s\chi & s\theta\,s\chi & c\theta \end{bmatrix} \begin{bmatrix} x \\ y \\ z \end{bmatrix}
$$

(46)

where $\sin a$ and $\cos a$ have been abbreviated $s\,a$ and $c\,a$, respectively. Under spatial inversion $X \to -X$, $Y \to -Y$, and $Z \to -Z$. In order to preserve Eq. (46), it is readily seen that $x \to x$, $y \to -y$, and $z \to z$ as well as $\phi \to \pi + \phi$, $\theta \to \pi - \theta$, and $\chi \to \pi - \chi$. As Table 6.3 shows, this corresponds to the group operation $C_2(y)$ on the Euler angles. From Eq. (6.59) it follows that

$$
\mathbf{i}_{sp}|J\Omega M\rangle = (-1)^{J-\Omega}|J -\Omega M\rangle
$$

(47)

The vibrational wave function $|v\rangle$ depends solely on the internuclear distance r. Therefore, $|v\rangle$ is unchanged by spatial inversion.

Putting together all the bits, we have

$$
\mathbf{i}_{sp}|n\Lambda\rangle\,|v\rangle\,|S\Sigma\rangle\,|J\Omega M\rangle
$$

$$
= (-1)^{\Lambda+s}(-1)^{S-\Sigma}(-1)^{J-\Omega}|n -\Lambda\rangle\,|v\rangle\,|S -\Sigma\rangle\,|J -\Omega M\rangle
$$

$$
= (-1)^{J-S+s}|n -\Lambda\rangle\,|v\rangle\,|S -\Sigma\rangle\,|J -\Omega M\rangle
$$

(48)

Thus the levels of the $p^{\pm}$ parity block have the parity $\pm(-1)^{J-S+s}$ under spatial inversion. We follow the convention[28] of designating the parity levels as e or f according to the following. For integral J values, e levels have parity $(-1)^J$ and f levels $(-1)^{J+1}$; for half-integral J values, e levels have parity $(-1)^{J-\frac{1}{2}}$ and f levels $(-1)^{J+\frac{1}{2}}$.

Orientation of the Electronic Charge Cloud

In the case (a) limit the electronic charge distribution has strict cylindrical symmetry about the internuclear axis as both **L** and **S** precess uniformly about this axis. However, for $\Lambda \neq 0$ states, molecular rotation causes the wave function associated with the p^+ and p^- Λ components of the same J to be such that one is symmetric (A') while the other is antisymmetric (A'') with respect to reflection in the plane of rotation of the molecule[29]. One measure of this is the electron charge cloud asymmetry[30], given by

$$
\Delta = \left\langle \hat{x}^2 - \hat{y}^2 \right\rangle = \left\langle \cos^2 \chi - \sin^2 \chi \right\rangle = \left\langle \cos 2\chi \right\rangle
$$

(49)

where χ is the azimuthal angle of the electron charge distribution in the molecular frame measured from the line of nodes (which we have taken to coincide with the y axis). The dependence of the electronic orbital angular momentum part of the case (a) wave function on χ has the form $|n\Lambda\rangle = (2\pi)^{-\frac{1}{2}} \exp(i\Lambda\chi)$ as we have already noted.

F. Show that

$$\Delta = \pm a_J b_J = \pm \left[\frac{X^2 - (Y-2)^2}{4X^2} \right]^{\frac{1}{2}} \tag{50}$$

by explicit evaluation of Δ using the F_1 and F_2 wave functions. *Hint*: Present arguments that

$$\left\langle \psi\left(^2\Pi_{\frac{1}{2}}\right) \middle| \cos 2\chi \middle| \psi\left(^2\Pi_{\frac{1}{2}}\right) \right\rangle = \left\langle \psi\left(^2\Pi_{\frac{3}{2}}\right) \middle| \cos 2\chi \middle| \psi\left(^2\Pi_{\frac{3}{2}}\right) \right\rangle = 0$$

while

$$\left\langle \psi\left(^2\Pi_{\frac{1}{2}}\right) \middle| \cos 2\chi \middle| \psi\left(^2\Pi_{\frac{3}{2}}\right) \right\rangle = \left\langle \psi\left(^2\Pi_{\frac{3}{2}}\right) \middle| \cos 2\chi \middle| \psi\left(^2\Pi_{\frac{1}{2}}\right) \right\rangle = \frac{1}{2}$$

We conclude that the Δ value for each pair of Λ components of F_1 and F_2 is oppositely signed[31]. For $J = \frac{1}{2}$, $|\Delta| = 0$, and the electronic charge distribution is cylindrically symmetric. As J increases, $|\Delta| \to \frac{1}{2}$; the charge distribution points for one Λ component along $\hat{x}$ (along $\mathbf{J}$ in the high J limit[32]) the other along $\hat{y}$ (in the plane of rotation); both are perpendicular to the internuclear axis. The importance of this is that a preferential population of one of the two Λ components in a particle, surface, or photon collision process serves to tell us about the dynamics of this process in terms of the asymmetry produced in the electronic charge distribution of the product[33].

G. Find the values of $|\Delta|$ versus J for $\frac{1}{2} \leq J \leq \frac{9}{2}$ in the case of the two $^2\Pi$ ground state radicals, OH and NO, given that $Y = -7.4$ for OH $X^2\Pi v = 0$ and $Y = 72.9$ for NO $X^2\Pi v = 0$. This comparison explains why it is more likely to observe preferential population of the Λ components in OH $X^2\Pi$ than in NO $X^2\Pi$ for low to moderate J values in a "collision process."

In addition to characterizing the two Λ-doublet levels by the label e/f to denote the total parity, it is useful in the high J limit to label Λ-doublet levels by $\Lambda(A')$ and $\Lambda(A'')$ depending on whether the electronic wave function is symmetric or antisymmetric with respect to reflection in the plane of rotation of the molecule of the spatial coordinate of all electrons[29]. Even if the Λ doublets are nearly degenerate in energy, electric-dipole-allowed transitions access separately the two Λ-doublet levels of opposite parity, making the population differences, if present, readily observable. For example, irrespective of multiplicity, the main-branch Q lines of a $\Pi - \Sigma^+$ transition probe the $\Pi(A'')$ levels in the high J limit, while the main-branch P and R lines probe the $\Pi(A')$ levels in the high J limit. For a $\Pi - \Sigma^-$ transition this is reversed: Q lines probe $\Pi(A')$ levels while P and R lines probe $\Pi(A'')$ levels[29].

Λ-Type Doubling

In the foregoing we have not treated explicitly terms in the molecular Hamiltonian that couple different electronic states. We consider here interactions between Σ and Π states of the same multiplicity by means of the Van Vleck transformation. This causes additional adjustable parameters to appear in the block of levels having the same parity, J, S, and Λ. In particular, for $^1\Pi$ and $^2\Pi$ electronic states this removes the degeneracy of the Λ components and causes them to differ in energy.

H. For a $^1\Pi$ state the terms in the Hamiltonian responsible for causing interaction between the $^1\Pi$ state and other $^1\Sigma^+$ and $^1\Sigma^-$ states are

$$\mathcal{H}_1 = -B(r)\,[\,J_+L_- + J_-L_+\,] \tag{51}$$

[see Eq. (10)]. Use the Van Vleck transformation to show that for a v, J level of p^+ parity

$$\mathcal{H}^{(2)}_{\Omega=1,\Omega=1} = q_v\left(^1\Sigma^+\right) J\,(J+1) \tag{52}$$

and for a v, J level of p^- parity

$$\mathcal{H}^{(2)}_{\Omega=1,\Omega=1} = q_v\left(^1\Sigma^-\right) J\,(J+1) \tag{53}$$

where

$$q_v\left(^1\Sigma^\pm\right) = \sum_{n'v'} \frac{\left\langle n\,^1\Pi_1 v\right| B(r)L_+ \left| n'\,^1\Sigma_0^\pm v'\right\rangle \left\langle n'\,^1\Sigma_0^\pm v'\right| B(r)L_- \left| n\,^1\Pi_1 v\right\rangle}{E(n\,^1\Pi_1 v) - E(n'\,^1\Sigma_0^\pm v')} \tag{54}$$

Thus to a good first approximation the Λ-type splitting for a $^1\Pi$ state is given by

$$E\left(^1\Pi_1 vJe\right) - E\left(^1\Pi_1 vJf\right) = \left[q_v(^1\Sigma^+) - q_v(^1\Sigma^-)\right] J(J+1) \tag{55}$$

that is, increases quadratically with rotation. Please also note that $^1\Sigma^+$ states can perturb only the Λ components of e symmetry and $^1\Sigma^-$ states those of f symmetry. Which Λ component has the lower energy depends on the relative size and sign of $q_v\left(^1\Sigma^+\right)$ and $q_v\left(^1\Sigma^-\right)$, and not on some mechanical model involving the moment of inertia difference of the electron charge distributions associated with the two Λ components (a common misconception).

The treatment of Λ doubling in $^2\Pi$ states proceeds along similar lines, but the splitting does not have a simple analytical form. We introduce three Λ-doubling parameters[5]:

$$o_v = \sum_{n'v'} \frac{\left|\left\langle n\,^2\Pi\,v\left|\,\tfrac{1}{2}A(r)L_+\,\right|n'\,^2\Sigma^\pm v'\right\rangle\right|^2}{E_{nv} - E_{n'v'}}$$

$$p_v = 4\sum_{n'v'} \frac{\left\langle n\,^2\Pi\,v\left|\,\tfrac{1}{2}A(r)L_+\,\right|n'\,^2\Sigma^\pm v'\right\rangle\left\langle n\,^2\Pi\,v\left|\,B(r)L_+\,\right|n'\,^2\Sigma^\pm v'\right\rangle}{E_{nv} - E_{n'v'}}$$

$$q_v = 2\sum_{n'v'} \frac{\left|\left\langle n\,^2\Pi\,v\left|\,B(r)L_+\,\right|n'\,^2\Sigma^\pm v'\right\rangle\right|^2}{E_{nv} - E_{n'v'}} \tag{56}$$

and consider the correction Hamiltonian to be

$$\mathcal{H}_1 = -B(r)(J_+L_- + J_-L_+) + \left[B(r) + \tfrac{1}{2}A(r)\right](L_+S_- + L_-S_+) \tag{57}$$

The sum in Eq. (56) is over all vibrational levels of all $^2\Sigma^+$ and $^2\Sigma^-$ electronic states. The Van Vleck transformation shows that we can replace Eq. (57) by an effective Λ-doubling Hamiltonian for the $^2\Pi$ state in level v of the form[34]

$$\mathcal{H}_\Lambda^{(v)} = \tfrac{1}{2}q_v(J_+ + J_-)^2$$

$$-\left(\tfrac{1}{2}q_v + \tfrac{1}{4}p_v\right)\left[(J_+ + J_-)(S_+ + S_-) + (S_+ + S_-)(J_+ + J_-)\right]$$

$$+\left(\tfrac{1}{2}q_v + \tfrac{1}{2}p_v + o_v\right)(S_+ + S_-)^2 \tag{58}$$

However, the Λ-doubling parameter o_v cannot be determined independently from spectroscopic data (its variation is totally correlated with variation in the location of the vibrational level origin and in the value of the spin–orbit constant A_v for the $^2\Pi$ state). Consequently, we omit it from further consideration. For the 2×2 block of p^+ parity $\mathcal{H}_\Lambda^{(v)}$ contributes the additional off-diagonal terms to the secular determinant

$$\left\langle ^2\Pi_{\frac{3}{2}}p^+\left|\,\mathcal{H}_\Lambda^{(v)}\,\right|^2\Pi_{\frac{3}{2}}p^+\right\rangle = \tfrac{1}{2}q_v\left[\left(J + \tfrac{1}{2}\right)^2 - 1\right] \tag{59}$$

$$\left\langle ^2\Pi_{\frac{1}{2}}p^+\left|\,\mathcal{H}_\Lambda^{(v)}\,\right|^2\Pi_{\frac{1}{2}}p^+\right\rangle = \tfrac{1}{2}q_v\left[\left(J + \tfrac{1}{2}\right) + 1\right]^2 + \tfrac{1}{2}p_v\left[\left(J + \tfrac{1}{2}\right) + 1\right] \tag{60}$$

$$\left\langle {}^2\Pi_{\frac{1}{2}}p^+ \left| \mathcal{H}_\Lambda^{(v)} \right| {}^2\Pi_{\frac{3}{2}}p^+ \right\rangle = \left\langle {}^2\Pi_{\frac{3}{2}}p^+ \left| \mathcal{H}_\Lambda^{(v)} \right| {}^2\Pi_{\frac{1}{2}}p^+ \right\rangle$$

$$= \tfrac{1}{2}q_v \left[\left(J + \tfrac{1}{2}\right)^2 - 1 \right]^{\frac{1}{2}} \left[\left(J + \tfrac{1}{2}\right) + 1 \right]$$

$$+ \tfrac{1}{4}p_v \left[\left(J + \tfrac{1}{2}\right)^2 - 1 \right]^{\frac{1}{2}} \tag{61}$$

and for the 2×2 block of p^- parity

$$\left\langle {}^2\Pi_{\frac{3}{2}}p^- \left| \mathcal{H}_\Lambda^{(v)} \right| {}^2\Pi_{\frac{3}{2}}p^- \right\rangle = \tfrac{1}{2}q_v \left[\left(J + \tfrac{1}{2}\right)^2 - 1 \right] \tag{62}$$

$$\left\langle {}^2\Pi_{\frac{1}{2}}p^- \left| \mathcal{H}_\Lambda^{(v)} \right| {}^2\Pi_{\frac{1}{2}}p^- \right\rangle = \tfrac{1}{2}q_v \left[\left(J + \tfrac{1}{2}\right) \right]^2 - \tfrac{1}{2}p_v \left[\left(J + \tfrac{1}{2}\right) - 1 \right] \tag{63}$$

$$\left\langle {}^2\Pi_{\frac{1}{2}}p^- \left| \mathcal{H}_\Lambda^{(v)} \right| {}^2\Pi_{\frac{3}{2}}p^- \right\rangle = \left\langle {}^2\Pi_{\frac{3}{2}}p^- \left| \mathcal{H}_\Lambda^{(v)} \right| {}^2\Pi_{\frac{1}{2}}p^- \right\rangle$$

$$= -\tfrac{1}{2}q_v \left[\left(J + \tfrac{1}{2}\right)^2 - 1 \right]^{\frac{1}{2}} \left[\left(J + \tfrac{1}{2}\right) - 1 \right]$$

$$+ \tfrac{1}{4}p_v \left[\left(J + \tfrac{1}{2}\right)^2 - 1 \right]^{\frac{1}{2}} \tag{64}$$

For ${}^2\Pi$ states well approximated by case (a) coupling, that is, states for which $(Y - 2)^2 >> 4\left(J - \tfrac{1}{2}\right)\left(J + \tfrac{3}{2}\right)$, the Λ-type doubling for the ${}^2\Pi_{\frac{1}{2}}$ component varies linearly with J, whereas for the ${}^2\Pi_{\frac{3}{2}}$ component it varies with the third power of J and for low J values is very small compared to that of ${}^2\Pi_{\frac{1}{2}}$.

Rotational Line Strength

The wave function associated with each rotational level of a diatomic molecule can be expressed as a linear combination of case (a) wave functions. We write

$$\left| n\,{}^{2S+1}\Lambda_\Omega\, v\, JMp^\pm \right\rangle = \sum_\Omega a_{n\Omega}(p^\pm) \left| n\,{}^{2S+1}\Lambda_\Omega \right\rangle |v\rangle |J\Omega M\rangle \tag{65}$$

where $\left| n\,{}^{2S+1}\Lambda_\Omega \right\rangle$ is the electronic part, $|v\rangle$ is the vibrational part, and $|J\Omega M\rangle$ is the rotational part. In Eq. (65) the sum extends over all possible values of $\Omega = \Lambda + \Sigma$ from $-\Lambda - S$ to $\Lambda + S$ so that there are in general $(2 - \delta_{0\Lambda})(2S + 1)$ different $a_{n\Omega}(p^\pm)$ coefficients. The $a_{n\Omega}(p^\pm)$ are just the columns of the unitary matrix (eigenvectors) that diagonalizes the secular matrix of the molecular Hamiltonian for

TABLE 1   Rotational Line Strengths for a $^2\Sigma - {}^2\Sigma$ Transition.[a]

Branch	$S(J'; J)$
$P_1(J), \; P_2(J)$	$\dfrac{(2J-1)(2J+1)}{4J}$
$Q_{12}(J), \; Q_{21}(J)$	$\dfrac{2J+1}{4J(J+1)}$
$R_1(J), \; R_2(J)$	$\dfrac{(2J+1)(2J+3)}{4(J+1)}$

[a] These expressions were first derived by R. S. Mulliken, *Phys. Rev.*, **30**, 138 (1927).

parity block $p^{\pm}$. The rotational line strength factor for the $^{2S'+1}\Lambda'_{\Omega'} J' \leftrightarrow {}^{2S+1}\Lambda_{\Omega} J$ transition is calculated[35] in exact analogy to Eq. (6.123):

$$S(J'; J) = (2J' + 1)(2J + 1)$$

$$\times \left| \sum_{\Omega'} \sum_{\Omega} a_{n'\Omega'}(p^{\pm}) a_{n\Omega}(p^{\pm}) \begin{pmatrix} J & 1 & J' \\ \Omega & \Omega' - \Omega & -\Omega' \end{pmatrix} \right|^2 \quad (66)$$

where $S(J'; J)$ vanishes unless the upper and lower levels are of opposite parity.

For $^1\Lambda' J' \leftrightarrow {}^1\Lambda J$ transitions ($^1\Sigma^{\pm} - {}^1\Sigma^{\pm}$, $^1\Sigma^{\pm} - {}^1\Pi$, $^1\Pi - {}^1\Sigma^{\pm}$, etc.), Eq. (66) reduces to the Hönl–London factors, where Λ takes the place of K'' in Table 6.6. For electric-dipole-allowed transitions involving electronic states of higher multiplicity, it is convenient in general to evaluate Eq. (66) with a simple computer program[35, 36]. Analytical expressions can be derived in some special cases, but these expressions become increasingly complex as the multiplicity increases and the coupling in the $^{2S+1}\Lambda$ state deviates from the case (a) and case (b) limiting forms[9, 35, 36].

For reference purposes we list the rotational line strength of a $^2\Sigma - {}^2\Sigma$ transition in Table 1 and those for $^2\Sigma - {}^2\Pi$ and $^2\Pi - {}^2\Sigma$ transitions in Table 2. The branches are designated P, Q, or R according to whether $\Delta J = -1, 0$, or $+1$. In addition, subscripts 1 and 2 are added to denote the F_1 and F_2 levels, first the upper level, then the lower one. If both subscripts are the same, it is traditional to omit one (e.g., $P_{11} = P_1, P_{22} = P_2$, etc.). For a $^2\Sigma$ state the F_1 levels are those for which $J = N + \frac{1}{2}$ and the F_2 levels are those for which $J = N - \frac{1}{2}$. For a $^2\Pi$ state the F_1 and F_2 levels are as defined in Eqs. (34) and (35), that is, in the case (a) limit F_1 is associated with $^2\Pi_{\frac{1}{2}}$ and F_2 with $^2\Pi_{\frac{3}{2}}$ for $Y \gg 0$, the reverse for $Y \ll 0$. In applying the rotational line strengths to molecular problems, one must take care in many cases to distinguish whether the line transition refers to a specific pair of Λ components of a ΔJ transition or involves a sum over two (often unresolved) Λ component pairs of a ΔJ transition. The line strengths are chosen to satisfy the sum rule[37]

$$\Sigma S(J', J'') = (2 - \delta_{0,\Lambda'} \delta_{0,\Lambda''}) (2S + 1)(2J + 1) \quad (67)$$

TABLE 2   Rotational Line Strengths for $^2\Sigma - {}^2\Pi$ and $^2\Pi - {}^2\Sigma$ Transitions[a]

Branch		
$^2\Sigma - {}^2\Pi$	$^2\Pi - {}^2\Sigma$	$S(J'; J)$[b]
$R_1(J)$	$P_1(J+1)$	$\dfrac{(2J+1)^2 + (2J+1)X^{-1}(4J^2+4J-7+2Y)}{32(J+1)}$
$Q_1(J)$	$Q_1(J)$	
		$\dfrac{(2J+1)[(4J^2+4J-1)+X^{-1}(8J^3+12J^2-2J-7+2Y)]}{32J(J+1)}$
$P_1(J)$	$R_1(J-1)$	$\dfrac{(2J+1)^2 + (2J+1)X^{-1}(4J^2+4J+1-2Y)}{32J}$
$R_{12}(J)$	$P_{21}(J)$	$\dfrac{(2J+1)^2 - (2J+1)X^{-1}(4J^2+4J-7+2Y)}{32(J+1)}$
$Q_{12}(J)$	$Q_{21}(J)$	
		$\dfrac{(2J+1)[(4J^2+4J-1)-X^{-1}(8J^3+12J^2-2J-7+2Y)]}{32J(J+1)}$
$P_{12}(J)$	$R_{21}(J-1)$	$\dfrac{(2J+1)^2 - (2J+1)X^{-1}(4J^2+4J+1-2Y)}{32J}$
$R_{21}(J)$	$P_{12}(J+1)$	$\dfrac{(2J+1)^2 - (2J+1)X^{-1}(4J^2+4J+1-2Y)}{32(J+1)}$
$Q_{21}(J)$	$Q_{12}(J)$	
		$\dfrac{(2J+1)[(4J^2+4J-1)-X^{-1}(8J^3+12J^2-2J+1-2Y)]}{32J(J+1)}$
$P_2(J)$	$R_{12}(J-1)$	$\dfrac{(2J+1)^2 - (2J+1)X^{-1}(4J^2+4J-7+2Y)}{32J}$
$R_2(J)$	$P_2(J+1)$	$\dfrac{(2J+1)^2 + (2J+1)X^{-1}(4J^2+4J+1-2Y)}{32(J+1)}$
$Q_2(J)$	$Q_2(J)$	
		$\dfrac{(2J+1)[(4J^2+4J-1)+X^{-1}(8J^3+12J^2-2J+1-2Y)]}{32J(J+1)}$
$P_2(J)$	$R_2(J-1)$	$\dfrac{(2J+1)^2 + (2J+1)X^{-1}(4J^2+4J-7+2Y)}{32J}$

[a]These expressions were first derived by L. T. Earls, *Phys. Rev.*, **48**, 423 (1935).

[b]$X = \left[4\left(J - \tfrac{1}{2}\right)\left(J + \tfrac{3}{2}\right) + (Y-2)^2 \right]^{\tfrac{1}{2}}$.

where the summation is over all allowed transitions from or to the group of $(2 - \delta_{0,\Lambda})(2S + 1)$ J levels with the same value of J' or J''. The sum rule is symmetric in J' and J''; however, Eq. (67) is not valid for the first few rotational levels where the full spin multiplicity is not present.

I. Show that $S(J'; J) = (2J + 1)/[4J(J + 1)]$ for the $Q_{12}(J)$ and $Q_{21}(J)$ branch members of a $^2\Sigma - {}^2\Sigma$ transition (second entry in Table 1). *Hint*: Do not forget that only levels of opposite parity can be connected in an electric-dipole-allowed transition.

By associating Λ of the diatomic molecule with K of an asymmetric top molecule, we can classify electronic dipole transitions in diatomic molecules as parallel or perpendicular transitions. In a parallel transition, $\Delta\Lambda = 0$ (e.g., $\Sigma - \Sigma$, $\Pi - \Pi$, etc.) and the transition dipole moment lies along the internuclear axis (the z axis)[38]. In the high J limit, only P and R branches have intensity. In a perpendicular transition $\Delta\Lambda = \pm 1$ (e.g., $\Sigma - \Pi$, $\Pi - \Sigma$, etc.) and the transition dipole moment is perpendicular to the internuclear axis. In the high J limit, P and R branch members have intensity only when the transition dipole moment lies in the plane of rotation (perpendicular to the internuclear axis), and the Q branch members have intensity only when the transition dipole moment lies along **J**. These simple considerations are useful in determining how a diatomic molecule will interact with a beam of plane polarized (or unpolarized) light.

NOTES AND REFERENCES

1. G. Herzberg, *Spectra of Diatomic Molecules*, Van Nostrand, New York, 1950.

2. F. Hund, *Z. Phys.*, **42**, 93 (1927); **63**, 719 (1930); *Handb. d. Phys.*, **24**, I, 561 (1933).

3. R. de L. Kronig, *Z. Phys.*, **46**, 814 (1928); **50**, 347 (1928); *Band Spectra and Molecular Structure*, Cambridge University Press, Cambridge, England, 1930.

4. E. L. Hill and J. H. Van Vleck, *Phys. Rev.*, **32**, 250 (1928); J. H. Van Vleck, *Phys. Rev.*, **33**, 467 (1929); **40**, 544 (1932); *Rev. Mod. Phys.*, **23**, 213 (1951).

5. R. S. Mulliken and A. Christy, *Phys. Rev.*, **38**, 87 (1930); R. S. Mulliken, *Rev. Mod. Phys.*, **4**, 51, (1932).

6. Y.-N. Chiu, *J. Chem. Phys.*, **42**, 2671 (1965); **45**, 2969 (1966); **57**, 4056 (1972).

7. K. F. Freed, *J. Chem. Phys.*, **45**, 4214 (1966).

8. R. T. Pack and J. Hirschfelder, *J. Chem. Phys.*, **49**, 4009 (1968).

9. I. Kovács, *Rotational Structure in the Spectra of Diatomic Molecules*, American Elsevier, New York, 1969.

10. A. Carrington, D. H. Levy, and T. A. Miller, *Adv. Chem. Phys.*, **18**, 149 (1970).

11. J. T. Hougen, "The Calculation of Rotational Energy Levels and Rotational Intensities in Diatomic Molecules," N.B.S. Monograph 115, Washington, DC, 1970.

12. L. Veseth, *Theor. Chim. Acta*, **18**, 368 (1970); *J. Phys. B*, **3**, 1677 (1970); *J. Phys. B*, **4**, 20 (1971); *J. Molec. Spectrosc.*, **38**, 228 (1971).

13. I. Röeggen, *Theor. Chim. Acta*, **21**, 398 (1971).

14. R. N. Zare, A. L. Schmeltekopf, W. J. Harrop, and D. L. Albritton, *J. Molec. Spectrosc.*, **46**, 37 (1973).

15. M. Mizushima, *The Theory of Rotating Diatomic Molecules*, Wiley, New York, 1975.

16. B. R. Judd, *Angular Momentum Theory for Diatomic Molecules*, Academic Press, New York, 1975.

17. J. M. Brown, M. Kaise, C. M. L. Kerr, and D. J. Milton, *Molec. Phys.*, **36**, 553 (1978); J. M. Brown, E. A. Colbourn, J. K. G. Watson, and F. D. Wayne, *J. Molec. Spectrosc.*, **74**, 294 (1979).

18. P. R. Bunker, *Molecular Symmetry and Spectroscopy*, Academic Press, New York, 1979.

19. M. Larsson, *Physica Scripta*, **23**, 835 (1981).

20. H. Lefebvre-Brion and R. W. Field, *Perturbations in the Spectra of Diatomic Molecules*, Academic Press, Orlando, FL, 1986.

21. P. D. A. Mills, C. M. Western, and B. J. Howard, *J. Phys. Chem.*, **90**, 3331 (1986).

22. However, one must not forget about nuclear spin statistics, which can cause every other level to be missing in homonuclear diatomic molecules composed of zero spin nuclei. For a review of hyperfine structure in the electronic spectra of diatomic molecules, see T. M. Dunn, in *Molecular Spectroscopy, Modern Research*, K. N. Rao and C. W. Mathews, eds., Academic Press, New York, 1972, pp. 231–257.

23. S. Green and R. N. Zare, *J. Molec. Spectrosc.*, **64**, 217 (1977).

24. This is the phenomenological spin–orbit Hamiltonian, which can only connect electronic states of the same multiplicity.   To include singlet–triplet, doublet–quartet (etc.) interactions, we must replace $\mathcal{H}_{SO}$ by the microscopic spin–orbit Hamiltonian.   See P. R. Fontana, *Phys. Rev.*, **125**, 220 (1962); K. Kayama and J. C. Baird, *J. Chem. Phys.*, **46**, 2604 (1967); L. Veseth, *Theor. Chim. Acta*, **18**, 368 (1970).

25. The $^2\Sigma$ electronic state follows case (b) coupling, and its wave function is also an equally weighted linear combination of the two case (a) wave functions $|n0\rangle \left| S\frac{1}{2} \right\rangle \left| J\frac{1}{2}M \right\rangle$ and $|n0\rangle \left| S - \frac{1}{2} \right\rangle \left| J - \frac{1}{2}M \right\rangle$.

26. The arguments present many chances for error. After considering this problem at some length, Chiu[6] writes in footnote 25 of the 1966 paper: "Thus we see that nature, noble and true, retains her supreme symmetry in spite of our sometimes misleading, humble effort of description which has yet to strive for perfection." Indeed, this problem did not get fully sorted out until the work of Larsson[19] in 1981.

27. For the choice $\chi = \pi/2$, the operations $\mathbf{i}_{sp} = C_2(x)\sigma_v(yz)$ are the more convenient. However, in either case the same result [Eq. (48)] is obtained.

28. J. M. Brown, J. T. Hougen, K.-P. Huber, J. W. C. Johns, I. Kopp, H. Lefebvre-Brion, A. J. Merer, D. A. Ramsay, J. Rostas, and R. N. Zare, *J. Molec. Spectrosc.*, **55**, 500 (1975).

29. M. H. Alexander and P. J. Dagdigian, *J. Chem. Phys.*, **80**, 4325 (1984). For $^1\Pi$ states the e levels are symmetric with respect to reflection in the plane of rotation and the f levels are antisymmetric. For $^2\Pi$ states the F_1 e and F_2 f levels are symmetric and the F_1 f and F_2 e levels are antisymmetric. The $\Lambda(A')$ and $\Lambda(A'')$ nomenclature to describe the Λ-doublet levels in rotating linear molecules is proposed and discussed by M. H. Alexander et al., *J. Chem. Phys.* (in press).

30. W. D. Gwinn, B. E. Turner, W. M. Goss, and G. L. Blackman, *Astrophys. J.*, **179**, 789 (1973).

31. As Alexander and Dagdigian[29] show, the electron distribution in the e Λ-doublet level is oriented preferentially in the plane of rotation in the case of a singly filled π orbital but along **J** in the case of a π^3 configuration.

32. S. Green and R. N. Zare, *Chem. Phys.*, **7**, 62 (1975).

33. For example, see R. Vasudev, R. N. Zare, and R. N. Dixon, *J. Chem. Phys.*, **80**, 4863 (1984), and P. Andersen and E. W. Rothe, *J. Chem. Phys.*, **82**, 3634 (1985).

34. J. M. Brown, D. J. Milton, J. K. G. Watson, R. N. Zare, D. L. Albritton, M. Horani, and J. Rostas, *J. Molec. Spectrosc.*, **90**, 139 (1981).

35. R. N. Zare, in *Molecular Spectroscopy: Modern Research*, K. N. Rao and C. W. Mathews, eds., Academic Press, New York, 1972, pp. 207–221.

36. E. E. Whiting, J. A. Paterson, I. Kovaćs, and R. W. Nicholls, *J. Molec. Spectrosc.*, **47**, 84 (1973).

37. E. E. Whiting, A. Schadee, J. B. Tatum, J. T. Hougen, and R. W. Nicholls, *J. Molec. Spectrosc.*, **80**, 249 (1980); E. E. Whiting and R. W. Nicholls, *Astrophys. J.* (Suppl. 235) **27**, 1 (1974).

38. It might be thought that the parallel transition classification also applies to $\Delta\Omega = 0$ transitions. However, for $\Omega = \frac{1}{2} \leftrightarrow \Omega = \frac{1}{2}$ transitions, this need not be the case; see L. Brus, *J. Molec. Spectrosc.*, **64**, 376 (1977).

APPLICATION 16
MOLECULAR REORIENTATION IN LIQUIDS

We consider here the nature of molecular rotation in liquids. In the limit that the motion of the molecule can be described as a random walk over small angular orientations, the Debye small-angle rotational diffusion model[1, 2] is an appropriate description. We follow closely Favro's treatment[3] (see also Freed[4] and Huntress[5]). Let $P(\Omega, t)\, d\Omega$ denote the probability of finding the molecule pointing into the solid angle element $d\Omega = d\phi \sin \theta d\theta d\chi$ at time t. The orientational probability at some later time $t + \Delta t$ can be found from $P(\Omega, t)\, d\Omega$ by including the contributions due to all possible rotations $\Delta\Omega$ that occur in the time increment Δt:

$$P(\Omega, t + \Delta t)\, d\Omega = \int p(\Delta\Omega, \Delta t)\, d(\Delta\Omega)\, P(\Omega_0, t)\, d\Omega_0 \tag{1}$$

where Ω_0 is the orientation that when rotated by $\Delta\Omega$ gives Ω, and $p(\Delta\Omega, \Delta t)$ is the probability density of the rotation by $\Delta\Omega$ taking place during Δt. The integration in Eq. (1) can be taken over Ω_0 or $\Delta\Omega$, since these two variables are not independent but are related by $\Omega = \Delta\Omega + \Omega_0$. It follows that the rotation $R(\Delta\Omega)$ carries $P(\Omega_0, t)\, d\Omega_0$ into $P(\Omega, t)\, d\Omega$, that is,

$$P(\Omega, t)\, d\Omega = R(\Delta\Omega)\, P(\Omega_0, t)\, d\Omega_0$$

$$= \exp(-i\theta \cdot \mathbf{L})\, P(\Omega_0, t)\, d\Omega_0 \tag{2}$$

or

$$P(\Omega_0, t)\, d\Omega_0 = \exp(i\theta \cdot \mathbf{L})\, P(\Omega, t)\, d\Omega \tag{3}$$

where θ is the angle conjugate to $\mathbf{L}$ which brings Ω_0 to Ω.

A. Show that

$$\frac{\partial P(\Omega, t)}{\partial t} = -\mathbf{L} \cdot \underset{\sim}{\mathbf{D}} \cdot \mathbf{L}\, P(\Omega, t) \tag{4}$$

where

$$\underset{\sim}{\mathbf{D}} = \frac{1}{2\Delta t} \int \theta\theta p(\theta, \Delta t)\, d^3\theta \tag{5}$$

is the second moment, $\frac{1}{2}\langle\theta\theta\rangle/\Delta t$, or so-called diffusion tensor. Here $\langle\cdots\rangle$ denotes an ensemble average. *Hint*: Substitute Eqs. (2) and (3) into Eq. (1), expand the exponential in the integrand, develop $P(\Omega, t + \Delta t)$ in a Taylor series in Δt, and retain only terms first order in Δt and second order in θ. Use the fact that the first moment

$$\langle \theta \rangle / \Delta t = \frac{1}{\Delta t} \int \theta p(\theta, t) \, d^3\theta \tag{6}$$

vanishes for a random walk.

In terms of its Cartesian components, Eq. (4) becomes

$$\frac{\partial P(\Omega, t)}{\partial t} = -H P(\Omega, t) \tag{7}$$

where

$$H = \sum_{i,j} L_i D_{ij} L_j \tag{8}$$

Except for a missing factor of $i/\hbar$ on the right-hand side, Eq. (7) is the same as the Schrödinger equation for a rigid body! Thus all the results already developed for the general motion of an asymmetric top can be applied to the anisotropic rotation of molecules in liquids (in the diffusion model limit).

In particular, let us work in the coordinate frame that diagonalizes the diffusion tensor. Then

$$H = \sum_{i} D_i L_i^2 \tag{9}$$

and the principal diffusion constants D_i play the same role as $1/(2 I_i)$, where I_i is the ith principal moment of inertia. For a completely anisotropic diffusor $D_1 \neq D_2 \neq D_3$; for a symmetric diffusor $D_1 = D_2 = D_\perp$ and $D_3 = D_\parallel$ $(D_\perp \neq D_\parallel)$; and for a spherical diffusor $D_1 = D_2 = D_3 = D$.

The solution to Eq. (7) may be written as

$$P(\Omega, t) = \int P(\Omega_0) G(\Omega_0 | \Omega, t) \, d\Omega_0 \tag{10}$$

where $P(\Omega_0)$ is the probability density that at some initial time $t = 0$ the diffusor was at some initial orientation Ω_0 and $G(\Omega_0 | \Omega, t)$ is the conditional probability density that if the diffusor was initially at Ω_0 then at time t it will be found at Ω. This latter function is a Green's function, which describes the evolution of the motion of the diffusor from its original orientation; it is subject to the initial condition that

$$G(\Omega_0 | \Omega, 0) = \delta^3(\Omega - \Omega_0) \tag{11}$$

Because $P(\Omega, t)$ satisfies the diffusion equation, so must $G(\Omega_0 | \Omega, t)$. Let $\psi_n(\Omega)$ be the eigenfunctions of H with eigenvalues E_n. Then $G(\Omega_0 | \Omega, t)$ can be expanded in the $\psi_n(\Omega)$ since the latter constitute a complete set.

B. Show that

$$G(\Omega_0|\Omega,t) = \sum_n \psi_n^*(\Omega_0)\psi_n(\Omega)\exp(-E_n t) \tag{12}$$

Use this result to express the Green's function for a symmetric diffusor as

$$G(\Omega_0|\Omega,t) = \sum_{L=0}^{\infty} \sum_{M,K=-L}^{L} \left(\frac{2L+1}{8\pi^2}\right)$$

$$\times D_{MK}^L(\phi_0,\theta_0,\chi_0)\, D_{MK}^{L*}(\phi,\theta,\chi)$$

$$\times \exp\{-[D_\perp L(L+1) + (D_\parallel - D_\perp)K^2]t\} \tag{13}$$

As $t \to 0$, $G(\Omega_0|\Omega,t)$ becomes $\sum_{L,M,K} D_{MK}^L(\phi_0,\theta_0,\chi_0)\, D_{MK}^{L*}(\phi,\theta,\chi) = \delta^3(\Omega - \Omega_0)$ and Eq. (11) is satisfied. The Green's function for an asymmetric diffusor has a more complex form and will not be pursued here[3–5].

The description of many physical properties of molecules in liquids, from light scattering[6] to fluorescence depolarization[7] to electron spin[4] and nuclear magnetic relaxation[5], is most conveniently formulated in terms of time-correlation functions $\langle A^*[\Omega(t)]A[\Omega(0)]\rangle$ where A is some dynamical variable that depends on the molecular orientation and the angular brackets $\langle \cdots \rangle$ denote an ensemble average (see Application 10). Explicitly

$$\langle A^*(t+\tau)A(t)\rangle = \lim_{t \to \infty} \frac{1}{T}\int_0^T A^*(t+\tau)A(t)\,dt \tag{14}$$

It is always possible to expand A in terms of the symmetric top wave functions $\psi_{LKM}(\Omega)$. Thus we are interested in general in the time-correlation functions of the form

$$F_{MK}^L(t) = \langle \psi_{LKM}^*[\Omega(t)]\psi_{LKM}[\Omega(0)]\rangle \tag{15}$$

where $\Omega(t)$ specifies the orientation of the molecule at time t. These correlation functions can be evaluated from $G(\Omega_0|\Omega,t)$ according to

$$F_{MK}^L(t) = \int d\Omega_0 \int d\Omega\, [G(\Omega_0|\Omega,t)P(\Omega_0)\psi_{LKM}(\Omega_0)]^*\psi_{LKM}(\Omega) \tag{16}$$

C. Show that for a symmetric diffusor having an isotropic initial distribution

$$F_{MK}^L(t) = \frac{1}{8\pi^2}\exp\{-[L(L+1)D_\perp + K^2(D_\parallel - D_\perp)]t\} \tag{17}$$

As $t \rightarrow \infty$, the $F_{MK}^L(t)$ approach zero except for $F_{00}^0(t)$, which approaches $(1/8\pi^2)$; that is, the symmetric diffusors become uniformly distributed. The symmetric diffusor correlation functions $F_{MK}^L(t)$ for $L \neq 0$ and $K \neq 0$ are characterized by two lifetimes (relaxation times), one equal to $[L(L+1)D_\perp]^{-1}$ and the other to $[K^2(D_\parallel - D_\perp)]^{-1}$. In the most general case of an asymmetric diffusor, the rotational relaxation is more complex and the fluorescence depolarization and light scattering correlation function are both characterized by a superposition of five exponentials[6, 7].

NOTES AND REFERENCES

1. P. Debye, *Polar Molecules*, Reinhold, New York, 1929.

2. P. S. Hubbard, *Phys. Rev.*, **131**, 1155 (1963).

3. L. D. Favro, *Fluctuation Phenomena in Solids*, R. E. Burgess, ed., Academic Press, New York, 1965, pp. 79–101; *Phys. Rev.*, **119**, 53 (1960).

4. J. H. Freed, *J. Chem. Phys.*, **41**, 2077 (1964).

5. W. T. Huntress, Jr., *Adv. Magn. Res.*, **4**, 1 (1970); *J. Chem. Phys.*, **48**, 3524 (1968).

6. B. J. Berne and R. Pecora, *Dynamic Light Scattering*, Wiley, New York, 1976.

7. G. G. Belford, R. L. Belford and G. Weber, *Proc. Natl. Acad. Sci. USA*, **69**, 1392 (1972); T. J. Chuang and K. B. Eisenthal, *J. Chem. Phys.*, **57**, 5094 (1972); M. Ehrenberg and R. Rigler, *Chem. Phys. Lett.*, **14**, 539 (1972); R. Rigler and M. Ehrenberg, *Quart. Rev. Biophys.*, **6**, 139 (1973). See also K. Razi Naqvi, *Chem. Phys. Lett.*, **136**, 407 (1987).

COMPUTER PROGRAMS FOR
3-*j*, 6-*j*, AND 9-*j* SYMBOLS

Simple (inelegant and unoptimized) FORTRAN function subprogram listings are presented for calculating 3-j, 6-j, and 9-j coefficients[1] using (1) twice the arguments as integers, (2) real arguments, and (3) integer arguments. The arguments for the 6-j and 9-j coefficients are referred to by matrix element notation. In particular:

XNINEJ (JJ11, JJ12, JJ13, JJ21, JJ22, JJ23, JJ31, JJ32, JJ33)

$$= \left\{ \begin{array}{ccc} \dfrac{JJ11}{2} & \dfrac{JJ12}{2} & \dfrac{JJ13}{2} \\[2ex] \dfrac{JJ21}{2} & \dfrac{JJ22}{2} & \dfrac{JJ23}{2} \\[2ex] \dfrac{JJ31}{2} & \dfrac{JJ32}{2} & \dfrac{JJ33}{2} \end{array} \right\}$$

REAL9J (RJ11, RJ12, RJ13, RJ21, RJ22, RJ23, RJ31, RJ32, RJ33)

$$= \left\{ \begin{array}{ccc} RJ11 & RJ12 & RJ13 \\ RJ21 & RJ22 & RJ23 \\ RJ31 & RJ32 & RJ33 \end{array} \right\}$$

VINT9J (IJ11, IJ12, IJ13, IJ21, IJ22, IJ23, IJ31, IJ32, IJ33)

$$
= \begin{Bmatrix}
IJ11 & IJ12 & IJ13 \\
IJ21 & IJ22 & IJ23 \\
IJ31 & IJ32 & IJ33
\end{Bmatrix}
$$

SIXJ (JJ11, JJ12, JJ13, JJ21, JJ22, JJ23)

$$
= \begin{Bmatrix}
\dfrac{JJ11}{2} & \dfrac{JJ12}{2} & \dfrac{JJ13}{2} \\
\dfrac{JJ21}{2} & \dfrac{JJ22}{2} & \dfrac{JJ23}{2}
\end{Bmatrix}
$$

REAL6J (RJ11, RJ12, RJ13, RJ21, RJ22, RJ23)

$$
= \begin{Bmatrix}
RJ11 & RJ12 & RJ13 \\
RJ21 & RJ22 & RJ23
\end{Bmatrix}
$$

VINT6J (IJ11, IJ12, IJ13, IJ21, IJ22, IJ23)

$$
= \begin{Bmatrix}
IJ11 & IJ12 & IJ13 \\
IJ21 & IJ22 & IJ23
\end{Bmatrix}
$$

In a similar manner we also have

THRJ (JJ1, JJ2, JJ3, MM1, MM2, MM3)

$$
= \begin{pmatrix}
\dfrac{JJ1}{2} & \dfrac{JJ2}{2} & \dfrac{JJ3}{2} \\
\dfrac{MM1}{2} & \dfrac{MM2}{2} & \dfrac{MM3}{2}
\end{pmatrix}
$$

REAL3J (RJ1, RJ2, RJ3, RM1, RM2, RM3)

$$
= \begin{pmatrix}
RJ1 & RJ2 & RJ3 \\
RM1 & RM2 & RM3
\end{pmatrix}
$$

and

$$\text{VINT3J (IJ1, IJ2, IJ3, IM1, IM2, IM3)}$$

$$= \begin{pmatrix} \text{IJ1} & \text{IJ2} & \text{IJ3} \\ \text{IM1} & \text{IM2} & \text{IM3} \end{pmatrix}$$

Subroutine SETUP must be called once prior to the first usage of any of these functions. SETUP calculates the factorials and stores them in a common block used by the other function subprograms. The size of the arguments for the 3-j, 6-j, and 9-j coefficients is limited by the size of the array in the common block and by round off error. Arguments less than approximately 75 are satisfactory.

The Wigner 3-j symbol is evaluated using the expression[2]

$$\begin{pmatrix} j_1 & j_2 & j_3 \\ m_1 & m_2 & m_3 \end{pmatrix} = (-1)^{j_1 - j_2 - m_3} \Delta(j_1, j_2, j_3) \, \text{w3j}(j_1, j_2, j_3, m_1, m_2, m_3)$$

$$(A\text{-}1)$$

where

$$\Delta(a, b, c) = \left[\frac{(a+b-c)!(a-b+c)!(-a+b+c)!}{(a+b+c+1)!} \right]^{\frac{1}{2}} \qquad (A\text{-}2)$$

$$\text{w3j}(j_1, j_2, j_3, m_1, m_2, m_3)$$

$$= [(j_1 + m_1)!(j_1 - m_1)!(j_2 + m_2)!(j_2 - m_2)!(j_3 + m_3)!(j_3 - m_3)!]^{\frac{1}{2}}$$

$$\times \sum_{\nu} \frac{(-1)^{\nu}}{\nu!(j_1 + j_2 - j_3 - \nu)!(j_1 - m_1 - \nu)!(j_2 + m_2 - \nu)!}$$

$$\times \frac{1}{(j_3 - j_2 + m_1 - \nu)!(j_3 - j_1 - m_2 - \nu)!} \qquad (A\text{-}3)$$

and the summation is over all positive integer values of ν such that no factorial in Eq. (A-3) is negative. The Wigner 6-j symbol is calculated by evaluating the expression[2]

$$\begin{Bmatrix} j_1 & j_2 & j_3 \\ \ell_1 & \ell_2 & \ell_3 \end{Bmatrix}$$

$$= \Delta(j_1, j_2, j_3) \Delta(j_1, \ell_2, \ell_3) \Delta(\ell_1, j_2, \ell_3) \Delta(\ell_1, \ell_2, j_3)$$

$$\times \text{w6j}(j_1, j_2, j_3, \ell_1, \ell_2, \ell_3) \qquad (A\text{-}4)$$

where the Δ functions are defined in Eq. (A-2),

$$w6j(j_1, j_2, j_3, \ell_1, \ell_2, \ell_3)$$

$$= \sum_\nu \frac{(-1)^\nu (\nu + 1)!}{(\nu - j_1 - j_2 - j_3)!(\nu - j_1 - \ell_2 - \ell_3)!(\nu - \ell_1 - j_2 - \ell_3)!}$$

$$\times \frac{1}{(\nu - \ell_1 - \ell_2 - j_3)!(j_1 + j_2 + \ell_1 + \ell_2 - \nu)!}$$

$$\times \frac{1}{(j_2 + j_3 + \ell_2 + \ell_3 - \nu)!(j_3 + j_1 + \ell_3 + \ell_1 - \nu)!} \tag{A-5}$$

and again the summation in Eq. (A-5) is over all positive integer values of ν such that none of the factorials have negative arguments. The Wigner 9-j symbol is calculated by evaluating the sum over the triple product of Wigner 6-j symbols using Eq. (4.24). An alternative algorithm for the calculation of the Wigner 9-j symbol is the numerical evaluation of its algebraic expression[3]:

$$\left\{ \begin{matrix} j_{11} & j_{12} & j_{13} \\ j_{21} & j_{22} & j_{23} \\ j_{31} & j_{32} & j_{33} \end{matrix} \right\}$$

$$= (-1)^{j_{13} + j_{23} - j_{33}} \frac{d(j_{21}, j_{11}, j_{31}) d(j_{12}, j_{22}, j_{32}) d(j_{33}, j_{31}, j_{32})}{d(j_{21}, j_{22}, j_{23}) d(j_{12}, j_{11}, j_{13}) d(j_{33}, j_{13}, j_{23})}$$

$$\times w9j(j_{11}, j_{12}, j_{13}, j_{21}, j_{22}, j_{23}, j_{31}, j_{32}, j_{33}) \tag{A-6}$$

where

$$d(a, b, c) = \left[\frac{(a - b + c)!(a + b - c)!(a + b + c + 1)!}{(b + c - a)!} \right]^{\frac{1}{2}} \tag{A-7}$$

$$w9j(j_{11}, j_{12}, j_{13}, j_{21}, j_{22}, j_{23}, j_{31}, j_{32}, j_{33})$$

$$= \sum_{\tau, \mu, \nu} \frac{(-1)^{\tau + \mu + \nu}}{\tau! \mu! \nu!} \times \frac{(2 j_{23} - \tau)!(2 j_{11} - \nu)!}{(2 j_{32} + 1 + \mu)!(j_{11} + j_{21} + j_{31} + 1 - \nu)!}$$

$$\times \frac{(j_{21} + j_{22} - j_{23} + \tau)!(j_{13} + j_{33} - j_{23} + \tau)!}{(j_{22} + j_{23} - j_{21} - \tau)!(j_{13} + j_{23} - j_{33} - \tau)!}$$

$$\times \frac{(j_{22} + j_{32} - j_{12} + \mu)!(j_{31} + j_{32} - j_{33} + \mu)!}{(j_{12} + j_{22} - j_{32} - \mu)!(j_{31} + j_{33} - j_{32} - \mu)!}$$

$$\times \frac{(j_{12} + j_{13} - j_{11} + \nu)!}{(j_{11} + j_{21} - j_{31} - \nu)!(j_{11} + j_{13} - j_{12} - \nu)!}$$

$$\times \frac{(j_{11} + j_{21} + j_{33} - j_{32} - \mu - \nu)!}{(j_{21} + j_{32} - j_{12} - j_{23} + \tau + \mu)!(j_{12} - j_{23} - j_{11} + j_{33} + \tau + \nu)!}$$

$$(A\text{-}8)$$

and the summation in Eq. (A-8) is restricted to positive integer values of τ, μ, ν for which no factorial has a negative argument. However, there appears to be no significant gain in computation time using Eq. (A-8)[4].

These function subprograms have been extensively tested using orthonormality relations, and they have brought us many hours of pleasure.

NOTES AND REFERENCES

1. Attention is called to the existence of algorithms for evaluating $3\text{-}j$ and $6\text{-}j$ coefficients [K. Schulten and R. G. Gordon, *J. Math. Phys.*, **16**, 1961 (1975)] as well as $9\text{-}j$ coefficients [K. Schulten and R. G. Gordon, *J. Chem. Phys.*, **64**, 2918 (1976)] based on the exact solution of recursion relations in a particular order designed to ensure numerical stability even for large angular momentum arguments. These algorithms are particularly useful in the frequently occurring case, such as in molecular scattering calculations, where the evaluation of a whole set of related coefficients is needed. A computer program for carrying out this task is available elsewhere [K. Schulten and R. G. Gordon, *Computer Phys. Comm.*, **11**, 269 (1976)].

2. G. Racah, *Phys. Rev.*, **62**, 438 (1942).

3. A. P. Jucys and A. A. Bandzaitis, *Theory of Angular Momentum in Quantum Mechanics*, (Mokslas, Vilnius, 1977).

4. Daqing Zhao, unpublished results.

```
C       LIBRARY FUNCTIONS FOR 3-J,6-J,9-J SYMBOLS USED BY THE ZARELAB
C       FROM CHRIS GREENE(1/5/83 VERSION), JOYCE GUEST(9/27/85 VERSION),
C       AND ANDREW KUMMEL(8/9/86 VERSION).
C       _____
C       SUBROUTINE SETUP MUST BE CALLED PRIOR TO THE FIRST USAGE OF ANY
C       OF THESE FUNCTIONS.  SETUP CALCULATES THE FACTORIALS AND STORES
C       THEM IN A COMMON BLOCK USED BY THE OTHER FUNCTION SUBPROGRAMS.
C       _____
        FUNCTION XNINEJ(JJ11,JJ12,JJ13,JJ21,JJ22,JJ23,JJ31,JJ32,JJ33)
C       *** CALCULATES 9-J VALUE, EACH JII IS 2X ACTUAL J
        IMPLICIT REAL*8 (A-H,O-Z)
        REAL*4 RJ11,RJ12,RJ13,RJ21,RJ22,RJ23,RJ31,RJ32,RJ33
        J11 = JJ11
        J12 = JJ12
        J13 = JJ13
        J21 = JJ21
        J22 = JJ22
        J23 = JJ23
        J31 = JJ31
        J32 = JJ32
        J33 = JJ33
        GOTO 50
C
        ENTRY REAL9J(RJ11,RJ12,RJ13,RJ21,RJ22,RJ23,RJ31,RJ32,RJ33)
C       *** CALCULATES 9-J VALUE, EACH JII IS REAL
        J11 = INT(2.*RJ11)
        J12 = INT(2.*RJ12)
        J13 = INT(2.*RJ13)
        J21 = INT(2.*RJ21)
        J22 = INT(2.*RJ22)
        J23 = INT(2.*RJ23)
        J31 = INT(2.*RJ31)
        J32 = INT(2.*RJ32)
        J33 = INT(2.*RJ33)
        GOTO 50
C
        ENTRY VINT9J(IJ11,IJ12,IJ13,IJ21,IJ22,IJ23,IJ31,IJ32,IJ33)
C       *** CALCULATES 9-J VALUE, EACH JII IS AN INTEGER
        J11 = 2*IJ11
        J12 = 2*IJ12
        J13 = 2*IJ13
        J21 = 2*IJ21
        J22 = 2*IJ22
        J23 = 2*IJ23
        J31 = 2*IJ31
        J32 = 2*IJ32
        J33 = 2*IJ33
C
50      KMIN1 = IABS(J11-J33)
        KMIN2 = IABS(J32-J21)
        KMIN3 = IABS(J23-J12)
        KMAX1 = J11 + J33
        KMAX2 = J32 + J21
        KMAX3 = J23 + J12
C
```

```
      IF(KMIN2.GT.KMIN1) KMIN1 = KMIN2
      IF(KMIN3.GT.KMIN1) KMIN1 = KMIN3
      IF(KMAX2.LT.KMAX1) KMAX1 = KMAX2
      IF(KMAX3.LT.KMAX1) KMAX1 = KMAX3
C
      KMIN1 = KMIN1 + 1
      KMAX1 = KMAX1 + 1
      XNINEJ = 0.D0
      IF(KMIN1.GT.KMAX1) GOTO 1000
      DO 100 K1 = KMIN1,KMAX1,2
            K = K1 - 1
            S1 = SIXJ(J11,J21,J31,J32,J33,K)
            S2 = SIXJ(J12,J22,J32,J21,K,J23)
            S3 = SIXJ(J13,J23,J33,K,J11,J12)
            P = DFLOAT((K+1)*((-1)**K))
            XNINEJ = XNINEJ + P*S1*S2*S3
100   CONTINUE
1000  CONTINUE
      VINT9J = XNINEJ
      REAL9J = XNINEJ
      RETURN
      END
C     _____
      FUNCTION SIXJ(JJ11,JJ12,JJ13,JJ21,JJ22,JJ23)
C     *** CALCULATES 6-J VALUE, EACH JII IS 2X ACTUAL J
      IMPLICIT REAL*8 (A-H,O-Z)
      REAL*4 RJ11,RJ12,RJ13,RJ21,RJ22,RJ23
      DIMENSION FL(400)
      COMMON/FACTOR/FL
      J1D = JJ11
      J2D = JJ12
      J3D = JJ13
      J4D = JJ21
      J5D = JJ22
      J6D = JJ23
      GOTO 50
C
      ENTRY REAL6J(RJ11,RJ12,RJ13,RJ21,RJ22,RJ23)
C     *** CALCULATES 6-J VALUE, EACH JII IS REAL
      J1D = INT(2.*RJ11)
      J2D = INT(2.*RJ12)
      J3D = INT(2.*RJ13)
      J4D = INT(2.*RJ21)
      J5D = INT(2.*RJ22)
      J6D = INT(2.*RJ23)
      GOTO 50
C
      ENTRY VINT6J(IJ11,IJ12,IJ13,IJ21,IJ22,IJ23)
C     *** CALCULATES 6-J VALUE, EACH JII IS AN INTEGER
      J1D = 2*IJ11
      J2D = 2*IJ12
      J3D = 2*IJ13
      J4D = 2*IJ21
      J5D = 2*IJ22
      J6D = 2*IJ23
```

```
C
50      JM1 = (J1D+J2D+J3D)/2
        JM2 = (J1D+J5D+J6D)/2
        JM3 = (J4D+J2D+J6D)/2
        JM4 = (J4D+J5D+J3D)/2
        JX1 = (J1D+J2D+J4D+J5D)/2
        JX2 = (J2D+J3D+J5D+J6D)/2
        JX3 = (J3D+J1D+J6D+J4D)/2
C
        JM = JM1
        IF(JM2.GT.JM) JM = JM2
        IF(JM3.GT.JM) JM = JM3
        IF(JM4.GT.JM) JM = JM4
        JX = JX1
        IF(JX2.LT.JX) JX = JX2
        IF(JX3.LT.JX) JX = JX3
        KM = JM + 1
        KX = JX + 1
        IF(KM.GT.KX) GOTO 9998
        TERM1 = FRTSJL(J1D,J2D,J3D,J4D,J5D,J6D)
        SIXJ = 0.D0
        DO 10 I1 = KM,KX
             I = I1 - 1
             TERM2 = FL(I+2)-FL(I+1-JM1)-FL(I+1-JM2)-FL(I+1-JM3)
             TERM2 = TERM2-FL(I+1-JM4)-FL(JX1-I+1)-FL(JX2-I+1)
             TERM2 = TERM2-FL(JX3-I+1)
             TERM = 0.D0
                  TERM = DEXP(TERM1+TERM2)*DFLOAT((-1)**I)
             SIXJ = SIXJ+TERM
10      CONTINUE
C
        GOTO 9999
9998    CONTINUE
        SIXJ = 0.D0
9999    CONTINUE
        VINT6J = SIXJ
        REAL6J = SIXJ
        RETURN
        END
C       _____
        FUNCTION THRJ(JJ1,JJ2,JJ3,MM1,MM2,MM3)
C       *** CALCULATES 3-J VALUE, EACH JI/MI IS 2X ACTUAL JI/MI
        IMPLICIT REAL*8 (A-H,O-Z)
        REAL*4 RJ1,RJ2,RJ3,RM1,RM2,RM3
        DIMENSION FL(400)
        COMMON/FACTOR/FL
        J1D = JJ1
        J2D = JJ2
        J3D = JJ3
        M1D = MM1
        M2D = MM2
        M3D = MM3
        GOTO 50
C
        ENTRY REAL3J(RJ1,RJ2,RJ3,RM1,RM2,RM3)
```

```
C       *** CALCULATES 3-J VALUE, EACH JI/MI IS REAL
        J1D = INT(2.*RJ1)
        J2D = INT(2.*RJ2)
        J3D = INT(2.*RJ3)
        M1D = INT(2.*RM1)
        M2D = INT(2.*RM2)
        M3D = INT(2.*RM3)
        GOTO 50
C
        ENTRY VINT3J(IJ1,IJ2,IJ3,IM1,IM2,IM3)
C       *** CALCULATES 3-J VALUE, EACH JI/MI IS AN INTEGER
        J1D = 2*IJ1
        J2D = 2*IJ2
        J3D = 2*IJ3
        M1D = 2*IM1
        M2D = 2*IM2
        M3D = 2*IM3
C
50      X1 = DFLOAT(J1D)/2.D0
        X2 = DFLOAT(J2D)/2.D0
        X3 = DFLOAT(J3D)/2.D0
        Y1 = DFLOAT(M1D)/2.D0
        Y2 = DFLOAT(M2D)/2.D0
        Y3 = DFLOAT(M3D)/2.D0
C
        IF(IABS(M1D).GT.J1D) GOTO 9998
        IF(IABS(M2D).GT.J2D) GOTO 9998
        IF(IABS(M3D).GT.J3D) GOTO 9998

        IF(J1D+J2D-J3D.LT.0) GOTO 9998
        IF(J2D+J3D-J1D.LT.0) GOTO 9998
        IF(J3D+J1D-J2D.LT.0) GOTO 9998
        IF(M1D+M2D+M3D.NE.0) GOTO 9998
C
        KMIN = (J3D-J1D-M2D)/2
        KMIN1 = KMIN
        KMIN2 = (J3D-J2D+M1D)/2
        IF(KMIN2.LT.KMIN) KMIN = KMIN2
        KMIN = (-1)*KMIN
        KMAX = IFIX(SNGL(X1 + X2 - X3 + 0.1))
        KMAX1 = KMAX
        KMAX2 = IFIX(SNGL(X1 - Y1 + 0.1))
        KMAX3 = IFIX(SNGL(X2 + Y2 + 0.1))
        IF(KMAX2.LT.KMAX) KMAX = KMAX2
        IF(KMAX3.LT.KMAX) KMAX = KMAX3
        IF(KMIN.LT.0) KMIN = 0
        IF(KMIN.GT.KMAX) GOTO 9998
C
        JMIN = KMIN + 1
        JMAX = KMAX + 1
        TERM1 = FRONTL(X1,X2,X3,Y1,Y2,Y3)
        MSIGN = (-1)**((J1D-J2D-M3D)/2)
        SUM = 0.0D0
C
        DO 10 I1 = JMIN,JMAX
```

```
            I = I1 - 1
            TERM2 = FL(I1) + FL(KMIN1+I1) + FL(KMIN2 + I1)
            TERM2 = TERM2+FL(KMAX1-I+1)+FL(KMAX2-I+1)+FL(KMAX3-I+1)
            TERM = DEXP(TERM1-TERM2)
            TERM = TERM*MSIGN*((-1)**I)
            SUM = SUM + TERM
10     CONTINUE
C
       THRJ = SUM
       GOTO 9999
9998   THRJ = 0.0
9999   CONTINUE
       VINT3J = THRJ
       REAL3J = THRJ
       RETURN
       END
C      _____
C
       FUNCTION FRONTL(X1,X2,X3,Y1,Y2,Y3)
C      ***USED IN 3J CALCULATION
       IMPLICIT REAL*8 (A-H,O-Z)
       DIMENSION FL(400)
       COMMON/FACTOR/FL
C
       L1 = IFIX(SNGL(X1+X2-X3+1.1))
       L2 = IFIX(SNGL(X2+X3-X1+1.1))
       L3 = IFIX(SNGL(X3+X1-X2+1.1))
       L4 = IFIX(SNGL(X1+X2+X3+2.1))
       L5 = IFIX(SNGL(X1+Y1+1.1))
       L6 = IFIX(SNGL(X1-Y1+1.1))
       L7 = IFIX(SNGL(X2+Y2+1.1))
       L8 = IFIX(SNGL(X2-Y2+1.1))
       L9 = IFIX(SNGL(X3+Y3+1.1))
       L10= IFIX(SNGL(X3-Y3+1.1))
       FRONTL = FL(L1) + FL(L2) + FL(L3) - FL(L4) + FL(L5) + FL(L6)
       FRONTL = FRONTL + FL(L7) + FL(L8) + FL(L9) + FL(L10)
       FRONTL = FRONTL/2.D0
       RETURN
       END
C      _____
       FUNCTION FRTSJL(J1D,J2D,J3D,J4D,J5D,J6D)
       IMPLICIT REAL*8 (A-H,O-Z)
C      ***USED IN 6J CALCULATION
C
       FRTSJL = DL(J1D,J2D,J3D) + DL(J1D,J5D,J6D)
       FRTSJL = FRTSJL + DL(J4D,J2D,J6D) + DL(J4D,J5D,J3D)
       RETURN
       END
C      _____
C
       FUNCTION DL(J1D,J2D,J3D)
C      ***USED IN FRTSJL/6J CALCULATION
       IMPLICIT REAL*8 (A-H,O-Z)
       DIMENSION FL(400)
       COMMON/FACTOR/FL
```

```
C
      L1 = (J1D+J2D-J3D)/2
      L2 = (J2D+J3D-J1D)/2
      L3 = (J3D+J1D-J2D)/2
      L4 = (J1D+J2D+J3D)/2+1
      DL = FL(L1+1)+FL(L2+1)+FL(L3+1)-FL(L4+1)
      DL = DL/2.D0
      RETURN
      END
C    _____
      SUBROUTINE SETUP
C     ***CALCULATES AND STORES FACTORIALS FOR 3NJ PROGRAMS
C
      IMPLICIT REAL*8 (A-H,O-Z)
      DIMENSION FACL(400)
      COMMON/FACTOR/FACL
C     ***N CAN BE INCREASED OR DECREASED AS DESIRED
C
      N = 400
      FACL(1) = 0.D0
      DO 100 I=2,N,1
            I1 = I-1
            FACL(I) = FACL(I-1) + DLOG(DFLOAT(I1))
100   CONTINUE
      RETURN
      END
```

INDEX

Ground Level

One Quantum

Large J

Transition Moment

Forbidden Transition

Excited State

With permission from Professor Henry A. Bent, Department of Chemistry, North Carolina State University.